Advanced Inorganic Semiconductor Materials

Advanced Inorganic Semiconductor Materials

Editors

Sake Wang
Minglei Sun
Nguyen Tuan Hung

 Basel • Beijing • Wuhan • Barcelona • Belgrade • Novi Sad • Cluj • Manchester

Editors

Sake Wang
College of Science, Jinling Institute of Technology
Nanjing
China

Minglei Sun
Department of Physics and NANOlab Center of Excellence, University of Antwerp
Antwerp
Belgium

Nguyen Tuan Hung
Frontier Research Institute for Interdisciplinary Sciences, Tohoku University
Sendai
Japan

Editorial Office
MDPI
St. Alban-Anlage 66
4052 Basel, Switzerland

This is a reprint of articles from the Special Issue published online in the open access journal *Inorganics* (ISSN 2304-6740) (available at: https://www.mdpi.com/journal/inorganics/special_issues/T6UC181H88).

For citation purposes, cite each article independently as indicated on the article page online and as indicated below:

Lastname, A.A.; Lastname, B.B. Article Title. *Journal Name* **Year**, *Volume Number*, Page Range.

ISBN 978-3-7258-0723-9 (Hbk)
ISBN 978-3-7258-0724-6 (PDF)
doi.org/10.3390/books978-3-7258-0724-6

© 2024 by the authors. Articles in this book are Open Access and distributed under the Creative Commons Attribution (CC BY) license. The book as a whole is distributed by MDPI under the terms and conditions of the Creative Commons Attribution-NonCommercial-NoDerivs (CC BY-NC-ND) license.

Contents

About the Editors vii

Preface ix

Sake Wang, Minglei Sun and Nguyen Tuan Hung
Advanced Inorganic Semiconductor Materials
Reprinted from: *Inorganics* **2024**, *12*, 81, doi:10.3390/inorganics12030081 1

Xinhua Tian, Hao Chang, Hongxing Dong, Chi Zhang and Long Zhang
Fluorescence Resonance Energy Transfer Properties and Auger Recombination Suppression in Supraparticles Self-Assembled from Colloidal Quantum Dots
Reprinted from: *Inorganics* **2023**, *11*, 218, doi:10.3390/inorganics11050218 6

Xinhua Tian, Hao Chang, Hongxing Dong, Chi Zhang and Long Zhang
Correction: Tian et al. Fluorescence Resonance Energy Transfer Properties and Auger Recombination Suppression in Supraparticles Self-Assembled from Colloidal Quantum Dots. *Inorganics* 2023, *11*, 218
Reprinted from: *Inorganics* **2023**, *11*, 340, doi:10.3390/inorganics11080340 17

Lei Jiang, Yanbo Dong and Zhen Cui
Adsorption of Metal Atoms on SiC Monolayer
Reprinted from: *Inorganics* **2023**, *11*, 240, doi:10.3390/inorganics11060240 19

Junlei Zhou, Yuzhou Gu, Yue-E Xie, Fen Qiao, Jiaren Yuan, Jingjing He, Sake Wang, et al.
Strain Modulation of Electronic Properties in Monolayer SnP_2S_6 and GeP_2S_6
Reprinted from: *Inorganics* **2023**, *11*, 301, doi:10.3390/inorganics11070301 28

Liping Qiao, Zhongqi Ma, Fulong Yan, Sake Wang and Qingyang Fan
A First-Principle Study of Two-Dimensional Boron Nitride Polymorph with Tunable Magnetism
Reprinted from: *Inorganics* **2024**, *12*, 59, doi:10.3390/inorganics12020059 39

Hazem Deeb, Kristina Khomyakova, Andrey Kokhanenko, Rahaf Douhan and Kirill Lozovoy
Dependence of Ge/Si Avalanche Photodiode Performance on the Thickness and Doping Concentration of the Multiplication and Absorption Layers
Reprinted from: *Inorganics* **2023**, *11*, 303, doi:10.3390/inorganics11070303 51

Hassen Dakhlaoui, Walid Belhadj, Haykel Elabidi, Fatih Ungan and Bryan M. Wong
GaAs Quantum Dot Confined with a Woods–Saxon Potential: Role of Structural Parameters on Binding Energy and Optical Absorption
Reprinted from: *Inorganics* **2023**, *11*, 401, doi:10.3390/inorganics11100401 66

Syed Bilal Junaid, Furqanul Hassan Naqvi and Jae-Hyeon Ko
The Effect of Cation Incorporation on the Elastic and Vibrational Properties of Mixed Lead Chloride Perovskite Single Crystals
Reprinted from: *Inorganics* **2023**, *11*, 416, doi:10.3390/inorganics11100416 80

Zan Wang, Yunjiao Gu, Daniil Aleksandrov, Fenghua Liu, Hongbo He and Weiping Wu
Engineering Band Gap of Ternary $Ag_2Te_xS_{1-x}$ Quantum Dots for Solution-Processed Near-Infrared Photodetectors
Reprinted from: *Inorganics* **2024**, *12*, 1, doi:10.3390/inorganics12010001 94

Jie Lu, Zeyang Xiang, Kexiang Wang, Mengrui Shi, Liuxuan Wu, Fuyu Yan, Ranping Li, et al.
Bipolar Plasticity in Synaptic Transistors: Utilizing HfSe$_2$ Channel with Direct-Contact HfO$_2$ Gate Dielectrics
Reprinted from: *Inorganics* **2024**, *12*, 60, doi:10.3390/inorganics12020060 . . . . **106**

Hongming Xiang, Shu Yang, Emon Talukder, Chenyan Huang and Kaikai Chen
Research and Application Progress of Inverse Opal Photonic Crystals in Photocatalysis
Reprinted from: *Inorganics* **2023**, *11*, 337, doi:10.3390/inorganics11080337 . . . . **117**

Lin-Qing Zhang, Wan-Qing Miao, Xiao-Li Wu, Jing-Yi Ding, Shao-Yong Qin, Jia-Jia Liu, Ya-Ting Tian, et al.
Recent Progress in Source/Drain Ohmic Contact with β-Ga$_2$O$_3$
Reprinted from: *Inorganics* **2023**, *11*, 397, doi:10.3390/inorganics11100397 . . . . **135**

About the Editors

Sake Wang

Sake Wang earned his PhD in physics at Southeast University, China, in 2016. During this period, he was awarded the national scholarship for doctoral students. Since 2021, he has been an Associate Professor at Jinling Institute of Technology, China. He was a Visiting Scientist at Tohoku University, Japan, from 2019 to 2021. His current interests focus on theoretical studies of spin and valley transport, as well as valley-optoelectronic devices in two-dimensional materials. He is a PI of the National Science Foundation for Young Scientists of China and the Natural Science Foundation of Jiangsu Province. He has published 68 papers with more than 3,300 citations, and he published 27 of these papers as the first author or corresponding author. Four of his first-authored and corresponding-authored papers are ESI - Top 1% highly cited papers. He was ranked in the World's Top 2% most-cited scientists 2023 by Stanford University. In addition, he has served as an Associate Editor of the *Journal of Superconductivity and Novel Magnetism* (Springer Publishing) since 2020 and as a guest editor of the *Journal of Physics D: Applied Physics* (IOP Publishing) since 2023, as well as being an outstanding reviewer of three SCI journals.

Minglei Sun

Minglei Sun is a Senior Postdoctoral Fellow at the University of Antwerp, funded by the Fonds Wetenschappelijk Onderzoek (FWO). He obtained his Ph.D. from Southeast University in 2018. Following four years working as a Postdoctoral Research Associate at the King Abdullah University of Science and Technology, Dr. Sun joined the University of Antwerp in November 2022. His research focuses on the prediction of novel two-dimensional semiconducting materials, investigating their electronic, optical, and excitonic properties with a specific focus on their potential applications in energy conversion and storage. Dr. Sun's work has resulted in several highly cited papers featured in prestigious journals like *Chemistry of Materials*.

Nguyen Tuan Hung

Nguyen Tuan Hung received a Ph.D. in Physics from the Interdepartmental Doctoral Degree Program of Tohoku University in March 2019. Since April 2019, he has been an assistant professor at the Frontier Research Institute for Interdisciplinary Sciences (FRIS), Tohoku University. He has been a visiting scholar at the Chinese Academy of Sciences (2017) and the Massachusetts Institute of Technology (2023-2024). In addition, he received the Aoba Society Prize for the Promotion of Science from Tohoku University (2017) and Research Fellowships for Young Scientists from the Japan Society for the Promotion of Science (2018); he is a Prominent Research Fellow of Tohoku University (2021-2025).

Preface

The information technology revolution has been decisively based on the development and application of inorganic semiconductors. For conventional devices, silicon forms the basis of most electronic devices, whilst compound semiconductors such as GaAs are used for many optoelectronic applications. In 2010, the discovery of monolayer graphene led to the receipt of a Nobel Prize in Physics. As a result, more and more atomically thin two-dimensional (2D) inorganic materials have gained significant interest. In addition to their promising applications in various ultrathin, transparent, and flexible nanodevices, 2D materials could also serve as one of the ideal models for establishing clear structure–property relationships in the field of solid-state physics and nanochemistry.

This Book was originally published in *Inorganics* as a Special Issue titled "Advanced Inorganic Semiconductor Materials". The Book consists of nine articles and two topical reviews written by authors around the globe, e.g., China, the Republic of Korea, Russia, Saudi Arabia, Turkey, and the USA. It highlights the most current research and ideas in inorganic semiconductors, from traditional to novel two-dimensional semiconductor materials to one-dimensional quantum dots. It captures the diversity of studies in the literature, including those on experimental fabrication and characterization as well as the electronic, electrical, magnetic, optoelectronic, and thermal properties of inorganic semiconductors. These contributions will help readers increase their knowledge in the field of inorganic semiconductor materials and be a new source of inspiration for novel, focused investigations, which we sincerely hope will contribute to the next edition of this SI, "Advanced Inorganic Semiconductor Materials: 2nd Edition".

Sake Wang, Minglei Sun, and Nguyen Tuan Hung
Editors

Editorial

Advanced Inorganic Semiconductor Materials

Sake Wang [1,*], Minglei Sun [2] and Nguyen Tuan Hung [3]

1. College of Science, Jinling Institute of Technology, 99 Hongjing Avenue, Nanjing 211169, China
2. NANOlab Center of Excellence, Department of Physics, University of Antwerp, Groenenborgerlaan 171, 2020 Antwerp, Belgium; minglei.sun@uantwerpen.be
3. Frontier Research Institute for Interdisciplinary Sciences, Tohoku University, Sendai 980-8578, Japan; nguyen.tuan.hung.e4@tohoku.ac.jp
* Correspondence: isaacwang@jit.edu.cn

1. Introduction

The information technology revolution has been based decisively on the development and application of inorganic semiconductors. Conventional devices utilize bulk semiconductors in which charge carriers are free to move in all three spatial directions. For example, silicon forms the basis of the vast majority of electronic devices, whilst compound semiconductors such as GaAs are used for many optoelectronic applications [1]. Recently, with the global boom in graphene research, more and more atomically thin 2D inorganic materials have gained significant interest [2–4]. Besides their promising applications in various ultrathin, transparent, and flexible nanodevices, 2D materials could also serve as one of the ideal models for establishing clear structure–property relationships in the field of solid-state physics and nanochemistry.

Despite the significant advances in the previous decade, both opportunities and challenges remain in this field. This SI consists of nine articles and two topical reviews, which highlight the most current research and ideas in inorganic semiconductors, from traditional to novel 2D semiconductor materials to 1D quantum dots. It captures the diversity of studies including experimental fabrication and characterization, as well as the electronic, electrical, magnetic, optoelectronic, and thermal properties of inorganic semiconductors.

Citation: Wang, S.; Sun, M.; Hung, N.T. Advanced Inorganic Semiconductor Materials. *Inorganics* **2024**, *12*, 81. https://doi.org/10.3390/inorganics12030081

Received: 28 February 2024
Revised: 29 February 2024
Accepted: 4 March 2024
Published: 6 March 2024

Copyright: © 2024 by the authors. Licensee MDPI, Basel, Switzerland. This article is an open access article distributed under the terms and conditions of the Creative Commons Attribution (CC BY) license (https://creativecommons.org/licenses/by/4.0/).

2. An Overview of Published Articles

This section provides a brief overview of the 11 contributions, organizing them into discreet subsections that include CQDs, 2D materials, germanium (Ge) on silicon (Si) avalanche photodiode, etc.

2.1. CQDs

In the article by Tian et al. (Contribution 1), the authors combine the microemulsion method with the microfluidic chip to prepare spherical QDSPs with regular shapes and high packing density. The small inter-dot distance enables QDSPs to have the following unique optical properties:

1. This self-assembled QDSP structure enables energy transfer between CQDs through fluorescence resonance energy transfer, resulting in a red shift in the steady-state fluorescence spectra of the SPs.
2. The dynamics of the energy transfer process of individual SPs are investigated by time-resolved fluorescence spectroscopy. The fast FRET process promotes the rapid energy transfer between excitons, resulting in the decay rate of PL intensity gradually increasing with the increase in energy, and the PL spectrum red shifts with time.
3. The non-radiative Auger recombination of CQDs is suppressed as FRET rates increase and potentially improve stability at high temperatures. Through short-chain ligand

exchange, higher packed SPs with better temperature-dependent optical stability are achieved, which can be attributed to the increased FRET rate and suppressed Auger recombination in the SPs with smaller dot spacing.

An CQD's self-assembled superparticle structure is an ideal platform for the research of multiparticle systems, and its novel FRET effect and temperature-insensitive fluorescence emission characteristics promote the development of new-type optoelectronic devices, such as photonic materials, solar cells, and optical sensors.

2.2. Two-dimensional Materials

In separate first-principles studies, the three articles by Jiang et al. (Contribution 2), Zhou et al. (Contribution 3), and Qiao et al. (Contribution 4) discuss the mechanical, electronic, magnetic, and optical properties of three typical 2D materials, i.e., adsorption of metal atoms on monolayer SiC, monolayer SnP_2S_6 and GeP_2S_6 under strain, as well as two doped 2D boron nitride polymorphs B_5N_6Al and B_5N_6C sheets. The results are summarized as follows.

1. Metal–atom-adsorbed SiC systems have potential applications in spintronic devices and solar energy conversion photovoltaic devices.
2. The strain is an effective band engineering scheme crucial for designing and developing next-generation nanoelectronic and optoelectronic devices.
3. Doping different atoms induces tunable electronic and magnetic properties in the 2D boron nitride sheets.

2.3. Ge on Si Avalanche Photodiode

In the article by Deeb et al. (Contribution 5), the planar structure of Ge on Si avalanche photodiode is designed. The dependences of the breakdown voltage, gain, bandwidth, responsivity, and quantum efficiency on the reverse bias voltage for different doping concentrations and thicknesses of the absorption and multiplication layers of the germanium on silicon avalanche photodiode are presented. The article presents simulation results, discussions, and analysis of design considerations. The dependence of the photodetectors' operating characteristics on the doping concentration for multiplication and absorption layers is revealed for the first time. Based on the analysis and simulation results, the optimal design for separate-absorption-charge-multiplication Ge on Si avalanche photodiode is proposed.

2.4. GaAs Quantum Dot

In the article by Dakhlaoui et al. (Contribution 6), the authors present the first detailed study of OACs produced by a Woods–Saxon-like spherical quantum dot containing a hydrogenic impurity at its center. They use a finite difference method to solve the Schrödinger equation within the framework of the effective mass approximation. First, they compute energy levels and probability densities for different parameters governing the confining potential. Then they calculate dipole matrix elements and energy differences and discuss their role in OACs. The findings demonstrate the important role of these parameters in tuning the OAC to enable blue or red shifts and alter its amplitude. The simulations provide a guided path to fabricate new optoelectronic devices by adjusting the confining potential shape.

2.5. Lead Halide Perovskites

The article by Junaid et al. (Contribution 7) reports the vibrational, structural, and elastic properties of mixed halide single crystals of $MA_xFA_{1-x}PbCl_3$ at room temperature by introducing the FA cation at the A-site of the perovskite crystal structure. Powder X-ray diffraction analysis confirms that its cubic crystal symmetry is similar to that of $MAPbCl_3$ and $FAPbCl_3$ with no secondary phases, indicating a successful synthesis of the $MA_xFA_{1-x}PbCl_3$ mixed halide single crystals. Structural analysis confirms that the FA substitution increases the lattice constant with increasing FA concentration. Raman spectroscopy provides insight into the vibrational modes, revealing the successful incorporation of the FA cation into the system. Brillouin spectroscopy is used to investigate the changes in the elastic properties induced via the FA substitution. A monotonic decrease in the sound velocity and the elastic constant suggests that the incorporation of large FA cations causes distortion within the inorganic framework, altering bond lengths and angles and ultimately resulting in decreased elastic constants. An analysis of the absorption coefficient reveals lower attenuation coefficients as the FA content increases, indicating reduced damping effects and internal friction. The current findings can facilitate the fundamental understanding of mixed lead chloride perovskite materials and pave the way for future investigations to exploit the unique properties of mixed halide perovskites for advanced optoelectronic applications.

2.6. Silver-Based Chalcogenide Semiconductors

The article by Wang et al. (Contribution 8) reports a facile mixture precursor hot-injection colloidal route to prepare $Ag_2Te_xS_{1-x}$ ternary QDs with tunable PL emissions from 950 nm to 1600 nm via alloying band gap engineering. As a proof-of-concept application, the $Ag_2Te_xS_{1-x}$ QDs-based near-infrared PD is fabricated via solution processes to explore their photoelectric properties. The ICP-OES results reveal the relationship between the compositions of the precursor and the samples, which is consistent with Vegard's equation. Alloying broadens the absorption spectrum and narrows the band gap of Ag_2S QDs. The UPS results demonstrate the energy band alignment of $Ag_2Te_{0.53}S_{0.47}$ QDs. The solution-processed $Ag_2Te_xS_{1-x}$ QD-based PD exhibits a photoresponse to 1350 nm illumination. With an applied voltage of 0.5 V, the specific detective is 0.91×10^{10} Jones and the responsivity is 0.48 mA/W. The PD maintains a stable response under multiple optical switching cycles, with a rise time of 2.11 s and a fall time of 1.04 s, which indicates excellent optoelectronic performance.

2.7. Thin-Film Transistors Featuring Ferroelectric $HfO_2/HfSe_2$ Stack

The article by Lu et al. (Contribution 9) investigates dual-mode synaptic plasticity in TFTs featuring an $HfSe_2$ channel, coupled with an OD-HfO_2 layer structure. In these transistors, the application of negative gate pulses results in a notable increase in the post-synaptic current, while positive pulses lead to a decrease. This distinctive response can be attributed to the dynamic interplay of charge interactions, significantly influenced by the ferroelectric characteristics of the OD-HfO_2 layer. The findings from this study highlight the capability of this particular TFT configuration to closely mirror the intricate functionalities of biological neurons, paving the way for advancements in bio-inspired computing technologies.

2.8. Inverse Opal Photonic Crystals

The review by Xiang et al. (Contribution 10) introduces the preparation methods of three-dimensional inverse opal photonic crystals, summarizes the principle of photocatalysis and the advantages of inverse opal photonic crystals in the field of photocatalysis, as well as the modification methods to further improve the efficiency of photocatalysis. Finally, the application progress in the fields of sewage purification, hydrogen production, and carbon dioxide decomposition is introduced.

2.9. Ohmic Contact Based on β-Ga_2O_3

The review by Zhang et al. (Contribution 11) summarizes the ohmic contact techniques developed in past years. First, the basic theory of metal–semiconductor contact is introduced. After that, the representative literature related to ohmic contact on β-Ga_2O_3 is summarized and analyzed, including the electrical property, the interface microstructure, the ohmic contact formation mechanism, and contact reliability. In addition, promising alternative schemes, including novel annealing techniques, and Au-free contact materials which are compatible with the CMOS process are expected and discussed. This review provides a theoretical basis understanding of ohmic contact on β-Ga$_2$O$_3$ devices, as well as the development trends of ohmic contact schemes.

3. Summary

In summary, this SI "Advanced Inorganic Semiconductor Materials" combines the latest achievements in the field of inorganic semiconductor materials, along with two reviews of previously obtained results. Inorganic semiconductors exhibit a wide range of new and unusual properties, which can be employed to fabricate improved and novel electronic and electro-optical devices.

With this SI published in the "Inorganics Materials" section, and further published as a book, the editors hope that the completeness and high quality of the contributions collected here receive the visibility and attention they deserve. These contributions would help readers to increase their knowledge in the field of inorganic semiconductor materials and be a new source of inspiration for novel, focused investigations, which we sincerely hope will contribute to the next edition of this SI "Advanced Inorganic Semiconductor Materials: 2nd Edition" [5].

Funding: S. W. was funded by the China Scholarship Council (No. 201908320001), the Natural Science Foundation of Jiangsu Province (No. BK20211002), and Qinglan Project of Jiangsu Province of China. N.T.H. was funded by financial support from the Frontier Research Institute for Interdisciplinary Sciences, Tohoku University, Japan. M. S. was supported by funding from Research Foundation-Flanders (FWO; no. 12A9923N).

Acknowledgments: The authors would like to thank all the staff in MDPI Publishing and editors of *Inorganics* for the establishment and running of this SI, as well as reviewers around the globe who spent their valuable time thoroughly reviewing and improving the articles published in this SI. We also feel grateful to all the authors from China, the Republic of Korea, Russia, Saudi Arabia, Turkey, and the USA, for choosing this SI to publish their excellent science.

Conflicts of Interest: The authors declare no conflicts of interest.

Abbreviations

The following abbreviations are used in this manuscript:

GaAs	gallium arsenide
2D	two-dimensional
SI	Special Issue
QDSP	quantum dot supraparticle
CQD	colloid quantum dot
SP	supraparticles
FRET	fluorescence resonance energy transfer
OAC	optical absorption coefficient
MA	$CH_3NH_3^+$, methylammonium
FA	$CH(NH_2)_2^+$, formamidinium
QD	quantum dot
PL	photoluminescence
PD	photodetector
ICP	inductively coupled plasma
OES	optical emission spectrometer
UPS	ultraviolet photoelectron spectroscopy
TFT	thin-film transistors
OD	oxygen-deficient

List of Contributions

1. Tian, X.; Chang, H.; Dong, H.; Zhang, C.; Zhang, L. Fluorescence Resonance Energy Transfer Properties and Auger Recombination Suppression in Supraparticles Self-Assembled from Colloidal Quantum Dots. *Inorganics* **2023**, *11*, 218. https://doi.org/10.3390/inorganics11060240.

2. Jiang, L.; Dong, Y.; Cui, Z. Adsorption of Metal Atoms on SiC Monolayer. *Inorganics* **2023**, *11*, 240. https://doi.org/10.3390/inorganics11060240.
3. Zhou, J.; Gu, Y.; Xie, Y.-E.; Qiao, F.; Yuan, J.; He, J.; Wang, S.; Li, Y.; Zhou, Y. Strain Modulation of Electronic Properties in Monolayer SnP_2S_6 and GeP_2S_6. *Inorganics* **2023**, *11*, 301. https://doi.org/10.3390/inorganics11070301.
4. Qiao, L.; Ma, Z.; Yan, F.; Wang, S.; Fan, Q. A First-Principle Study of Two-Dimensional Boron Nitride Polymorph with Tunable Magnetism. *Inorganics* **2024**, *12*, 59. https://doi.org/10.3390/inorganics12020059.
5. Deeb, H.; Khomyakova, K.; Kokhanenko, A.; Douhan, R.; Lozovoy, K. Dependence of Ge/Si Avalanche Photodiode Performance on the Thickness and Doping Concentration of the Multiplication and Absorption Layers. *Inorganics* **2023**, *11*, 303. https://doi.org/10.3390/inorganics12020059.
6. Dakhlaoui, H.; Belhadj, W.; Elabidi, H.; Ungan, F.; Wong, B.M. GaAs Quantum Dot Confined with a Woods–Saxon Potential: Role of Structural Parameters on Binding Energy and Optical Absorption. *Inorganics* **2023**, *11*, 401. https://doi.org/10.3390/inorganics11100401.
7. Junaid, S.B.; Naqvi, F.H.; Ko, J.-H. The Effect of Cation Incorporation on the Elastic and Vibrational Properties of Mixed Lead Chloride Perovskite Single Crystals. *Inorganics* **2023**, *11*, 416. https://doi.org/10.3390/inorganics11100416.
8. Wang, Z.; Gu, Y.; Aleksandrov, D.; Liu, F.; He, H.; Wu, W. Engineering Band Gap of Ternary $Ag_2Te_xS_{1-x}$ Quantum Dots for Solution-Processed Near-Infrared Photodetectors. *Inorganics* **2024**, *12*, 1. https://doi.org/10.3390/inorganics12010001.
9. Lu, J.; Xiang, Z.; Wang, K.; Shi, M.; Wu, L.; Yan, F.; Li, R.; Wang, Z.; Jin, H.; Jiang, R. Bipolar Plasticity in Synaptic Transistors: Utilizing $HfSe_2$ Channel with Direct-Contact HfO_2 Gate Dielectrics. *Inorganics* **2024**, *12*, 60. https://doi.org/10.3390/inorganics12010060.
10. Xiang, H.; Yang, S.; Talukder, E.; Huang, C.; Chen, K. Research and Application Progress of Inverse Opal Photonic Crystals in Photocatalysis. *Inorganics* **2023**, *11*, 337. https://doi.org/10.3390/inorganics11080337.
11. Zhang, L.-Q.; Miao, W.-Q.; Wu, X.-L.; Ding, J.-Y.; Qin, S.-Y.; Liu, J.-J.; Tian, Y.-T.; Wu, Z.-Y.; Zhang, Y.; Xing, Q.; et al. Recent Progress in Source/Drain Ohmic Contact with $β$-Ga_2O_3. *Inorganics* **2023**, *11*, 337. https://doi.org/10.3390/inorganics11080337.

References

1. Kelsall, R.W.; Hamley, I.W.; Geoghegan, M. (Eds.) *Nanoscale Science and Technology*; John Wiley & Sons Ltd.: Chichester, UK, 2005.
2. Rao, C.N.R.; Waghmare, U.V. *2D Inorganic Materials beyond Graphene*; World Scientifc Publishing Europe Ltd.: London, UK, 2018.
3. Tian, H.; Ren, C.; Wang, S. Valleytronics in two-dimensional materials with line defect. *Nanotechnology* **2022**, *33*, 212001. [CrossRef] [PubMed]
4. Zhang, C.; Ren, K.; Wang, S.; Luo, Y.; Tang, W.; Sun, M. Recent progress on two-dimensional van der Waals heterostructures for photocatalytic water splitting: A selective review. *J. Phys. D Appl. Phys.* **2023**, *56*, 483001. [CrossRef]
5. Inorganics | Special Issue : Advanced Inorganic Semiconductor Materials: 2nd Edition. Available online: https://www.mdpi.com/journal/inorganics/special_issues/4L19I7955Z (accessed on 27 December 2023).

Disclaimer/Publisher's Note: The statements, opinions and data contained in all publications are solely those of the individual author(s) and contributor(s) and not of MDPI and/or the editor(s). MDPI and/or the editor(s) disclaim responsibility for any injury to people or property resulting from any ideas, methods, instructions or products referred to in the content.

Article

Fluorescence Resonance Energy Transfer Properties and Auger Recombination Suppression in Supraparticles Self-Assembled from Colloidal Quantum Dots

Xinhua Tian [1,2,†], Hao Chang [1,†], Hongxing Dong [1,3,4,*], Chi Zhang [3,5,*] and Long Zhang [1,2,3,4,*]

1. Key Laboratory of Materials for High-Power Laser, Shanghai Institute of Optics and Fine Mechanics, Chinese Academy of Sciences, Shanghai 201800, China
2. School of Physical and Technology, ShanghaiTech University, Shanghai 201210, China
3. School of Physics and Optoelectronic Engineering, Hangzhou Institute for Advanced Study, University of Chinese Academy of Sciences, Hangzhou 310024, China
4. CAS Center for Excellence in Ultra-Intense Laser Science, Shanghai 201800, China
5. Department of Mechanical and Aerospace Engineering, University of Missouri, Columbia, MO 65211, USA
* Correspondence: hongxingd@siom.ac.cn (H.D.); chizhang@ucas.ac.cn (C.Z.); lzhang@siom.ac.cn (L.Z.)
† These authors contributed equally to this work.

Citation: Tian, X.; Chang, H.; Dong, H.; Zhang, C.; Zhang, L. Fluorescence Resonance Energy Transfer Properties and Auger Recombination Suppression in Supraparticles Self-Assembled from Colloidal Quantum Dots. *Inorganics* 2023, 11, 218. https://doi.org/10.3390/inorganics11050218

Academic Editors: Sake Wang, Minglei Sun and Nguyen Tuan Hung

Received: 11 April 2023
Revised: 9 May 2023
Accepted: 15 May 2023
Published: 18 May 2023
Corrected: 18 August 2023

Copyright: © 2023 by the authors. Licensee MDPI, Basel, Switzerland. This article is an open access article distributed under the terms and conditions of the Creative Commons Attribution (CC BY) license (https://creativecommons.org/licenses/by/4.0/).

Abstract: Colloid quantum dots (CQDs) are recognized as an ideal material for applications in next-generation optoelectronic devices, owing to their unique structures, outstanding optical properties, and low-cost preparation processes. However, monodisperse CQDs cannot meet the requirements of stability and collective properties for device applications. Therefore, it is urgent to build stable 3D multiparticle systems with collective physical and optical properties, which is still a great challenge for nanoscience. Herein, we developed a modified microemulsion template method to synthesize quantum dot supraparticles (QD-SPs) with regular shapes and a high packing density, which is an excellent research platform for ultrafast optical properties of composite systems. The redshift of the steady-state fluorescence spectra of QD-SPs compared to CQD solutions indicates that fluorescence resonance energy transfer (FRET) occurred between the CQDs. Moreover, we investigated the dynamic processes of energy transfer in QD-SPs by time-resolved ultrafast fluorescence spectroscopy. The dynamic redshift and lifetime changes of the spectra further verified the existence of rapid energy transfer between CQDs with different exciton energies. In addition, compared with CQD solutions, the steady-state fluorescence lifetime of SPs increased and the fluorescence intensity decreased slowly with increasing temperature, which indicates that the SP structure suppressed the Auger recombination of CQDs. Our results provide a practical approach to enhance the coupling and luminescence stability of CQDs, which may enable new physical phenomena and improve the performance of optoelectronic devices.

Keywords: colloidal quantum dots; supraparticles; auger recombination; fluorescence resonance energy transfer

1. Introduction

Colloid quantum dots are widely studied light-emitting materials with unique structures and excellent optoelectronic properties such as a wide excitation spectrum, narrow emission spectrum, good color purity, and high photoluminescence quantum yield [1–6], which make them ideal building blocks for optoelectronic devices with collective characteristics. Their emission spectrum can be tuned across the entire visible wavelength band by simply adjusting their size and type, which is attractive for applications in displays [7,8], light-emitting diodes [9–11], solar cells [12,13], and lasers [14–17]. However, the stability and collective properties required for device applications cannot be achieved by monodisperse CQDs. To build a research platform for complex multiparticle systems with collective properties and to promote the development of CQD devices with novel optoelectronic properties, researchers often use self-assembly methods to create CQD superstructures with coupling effects [18–24]. Such superstructures are expected to exhibit new physical and optical collective properties due to the coupling between CQDs and have attracted considerable attention, and have great application prospects in catalysis [25,26], photonic materials [27], solar cells [28], and drug delivery [29]. Therefore, clarifying the superstructure construction mechanism of self-assembled CQDs and their unique collective optical properties is of great significance for the development of next generation optoelectronic devices [30–39].

In recent years, researchers have devoted tremendous efforts to the study of the optical properties of superstructured systems [40]. The microemulsion method is commonly used to control the self-assembly of a large number of CQDs into three-dimensional superstructures with spherical shapes, known as supraparticles. Previous studies on the preparation of CQD superstructures by the microemulsion method mainly focused on the relationship between experimental parameters and the final products, including the influence of the concentration of CQDs and the type of surfactant on the surface tension and stability of microemulsion droplets, as well as the control of the final morphology of the SPs by changing the volatilization conditions [41–44]. Another main direction is clarifying the mechanism of superstructure assembly by monitoring the dynamical properties of supraparticle nucleation in the process of CQD assembly [45–49]. The success of the synthesis method has enabled the use of CQDs as building blocks for multifunctional SPs. For instance, the combination of red-, green-, and blue-emitting QDs into a single SP enables white-light generation [50,51]. Additionally, the formation of micrometer-scale spherical SPs effectively combines the QDs into microcavities that support whispering gallery modes to achieve SP lasers [52–56]. However, the collective luminescence properties and spectral dynamics of quantum dot supraparticles are rarely studied [57,58]. Therefore, it is urgent and necessary to clarify the ultrafast optical processes in QD-SPs.

In this work, we synthesized spherical quantum dot supraparticles with a regular shape and high packing density using the microemulsion template method. This self-assembled SP structure enabled energy transfer between CQDs through fluorescence resonance energy transfer, resulting in a redshift in the steady-state fluorescence spectra of the SPs. Moreover, we investigated the dynamics of the energy transfer process of individual SPs by time-resolved fluorescence spectroscopy. The fast FRET process promotes rapid energy transfer between excitons, which then released energy by emitting photons. This resulted in significantly lower energy dissipation in the form of thermal energy due to Auger recombination and potentially improved stability at high temperatures. In addition, we achieved SPs with smaller particle spacing by short-chain ligand exchange, as evidenced by the faster spectral redshift rate and the faster FRET rate. Therefore, SP structures with short-chain ligands can better suppress Auger recombination, leading to better temperature stability of the SPs. Our results show that the CQD self-assembled supraparticle structure is an ideal platform for the research of multiparticle systems, and its novel FRET effect and temperature-insensitive fluorescence emission characteristics could promote the development of new types of optoelectronic devices.

2. Self-Assembly of CQDs into Supraparticles

Figure 1 shows a schematic of the synthesis of self-assembled supraparticles using CQDs as building blocks. The surface of the CQDs was passivated by long-chain organic ligands such as oleic acid and oleylamine, which spatially stabilized the colloid and limited the interaction between individual CQDs. The fabrication of self-assembled SPs from dispersed CQDs was achieved by evaporating the non-polar phases of the oil-in-water emulsions. Nearly monodisperse droplets of the CQD solution (oil phase) were formed using microfluidic chips, and specific size droplets could be prepared by accurately controlling the shear force and relative flow rate of the oil and water phases. These formed droplets were dispersed in an aqueous phase containing surfactants, which imparted spatial stability to the droplets and prevented them from fusing or breaking. The CQDs were confined to the microemulsion drop by the hydrophobic interaction between the surface ligand and the surfactant. After the formation of the microemulsions, the CQDs began to assemble with the evaporation of the low-boiling solvent in the droplet. The concentration of CQDs increased with the evaporation process, and as the volume fraction of CQDs increased to 20%, the CQDs began to aggregate through hydrophobic interactions [46]. Evaporation time is related to droplet size, oil phase type, and many other factors affecting the evaporation rate. After the oil phase evaporated completely, solid SPs were formed, which were bound together by van der Waals forces and no longer dispersed in polar or non-polar solvents [45,59]. The slow evaporation rate of the oil phase resulted in the formation of SPs with a regular spherical structure and smoother surface. These SPs were stabilized in the aqueous phase through the hydrophobic interaction of surfactants, and their optical properties did not change significantly even after being stored in water for months. By combining the microemulsion method with microfluidics, we can precisely control the size of the SPs by adjusting the size of microemulsion droplets and the initial concentration of CQDs in the oil phase.

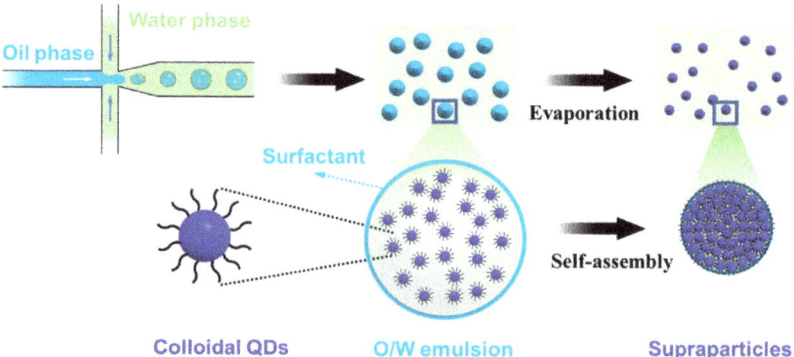

Figure 1. Schematic of the fabrication of colloidal quantum dot supraparticles through the microemulsion method.

3. Structural Characterizations of Supraparticles

The surface morphology and spatial structure of the synthesized SPs were characterized by scanning electron microscopy (SEM) and transmission electron microscopy (TEM). As shown in Figure 2a, most of the SP samples were spherical with a size distribution of 1.0 ± 0.2 μm due to interfacial tension during the volatilization of the oil phase. The inset of Figure 2a presents a high-resolution image of an individual SP microsphere, illustrating its regular spherical structure and smooth surface. To further investigate the internal structure of SP microspheres, TEM analysis was performed. Figure 2b shows a high-angle annular dark field (HAADF) TEM image of an individual SP microsphere, indicating that the SP microsphere was composed of a large number of CQDs. Due to the presence of hydrophobic surface ligands, the spacing of CQDs in SPs was about 3 nm, which is less than 2 times the length of oleic acid ligands. This indicates that the cross-linking between ligands in the SPs made it difficult to redisperse in solvents, and the smooth surface of the SPs can be attributed to the high filling factor of the cross-linked long-chain ligands. In Figure 2c, the fast Fourier transform of the TEM image clearly shows that the CQDs were disordered in the SPs, forming an amorphous glassy structure as expected. Previous studies have shown that if the basic constituent particles have regular spherical shapes and the same size, and their interactions can be well approximated with a hard sphere model, the SPs will form superlattice structures with icosahedral or face-centered cubic structures [41,45,60,61]. In our experiment, the initial CQDs building blocks were irregularly spherical and polydisperse, resulting a random arrangement of CQDs in SPs rather than superlattice structures despite the high packing factor (as show in Figure S1). These SPs were tightly packed via cross-linked ligands, resulting in an interparticle spacing of 3 nm, less the twice distance between ligand layers (Figure S1). Furthermore, as shown in Figure 2d, energy dispersive X-ray spectroscopy (EDS) confirmed the uniform distribution of CQDs in SPs. These results show that the as-prepared QD-SP samples have a regular shape, and the CQDs are tightly packed in SP microspheres without agglomeration and rupture. In addition, the SPs can be stored in water and had good solution operability, making them promising for further integration into functional materials and devices.

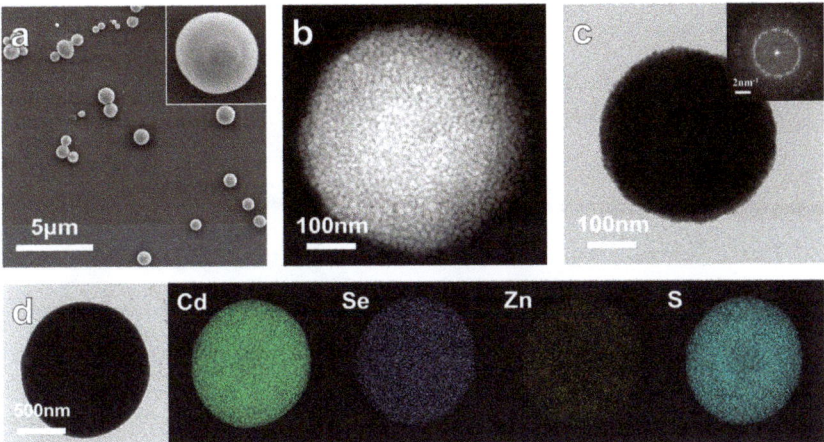

Figure 2. Structural characterization of supraparticles. (**a**) Typical SEM image of the supraparticles. Inset: magnified image of an individual supraparticle. (**b**) Representative high-angle annular dark field scanning transmission electron microscopy (HAADF-STEM) image of a supraparticle. (**c**) TEM image of a single QD-SP. Inset: the fast Fourier transform of the QD-SP. (**d**) EDS elemental mapping for cadmium, selenium, zinc, and sulfur to show the composition of the microspheres.

4. Single Supraparticle Spectroscopy and Analysis

The optical properties of the prepared QD-SP microspheres were characterized by a confocal micro-photoluminescence spectrometer. Figure 3a shows the absorption and emission spectra of dispersed CQD solution, as well as the emission spectrum of an individual SP microsphere. The photoluminescence (PL) emission center of dispersed CQD solutions was 1964 meV (631 nm) and the absorption spectrum showed the lowest exciton transition at 2025 meV (612 nm), and the emission spectrum of an individual SP was centered at 1944 meV (637 nm). We observed that the emission spectrum was slightly redshifted (20 meV) after the CQDs were assembled into SPs, which can be attributed to the fluorescence resonance energy transfer by short-range dipole–dipole coupling interactions between adjacent CQDs [62]. CQDs exhibit quantum size effects, and their size in colloidal solutions is not ideally uniform but follows a Gaussian distribution, resulting in CQDs having different exciton energies. When the distance between adjacent CQDs is less than 10 nm, small CQDs (with larger exciton energies) can act as donors while large CQDs (with smaller exciton energies) act as acceptors, activating FRET from small to large CQDs. We further investigated the energy dynamic processes of the SP samples using time-resolved and spectrum-resolved fluorescence spectroscopy to provide a more systematic demonstration of FRET. To quantify the energy transfer dynamics between CQDs, we excited the CQD solutions and SPs with 80 fs pulses (1 kHz repetition rate) using a wavelength of 400 nm. Figure 3c,d and Figure S2 show the emission intensity as a function of emission wavelength and time that was obtained using a streak camera. After exciting CQD solutions and SPs with femtosecond pulsed laser, we measured the time-resolved PL spectra on a time scale of 100 ps (Figure 3e,f). The results showed that the PL spectra of dispersed CQD solutions had a time-independent peak energy of 2037 meV. In contrast, the PL spectra of QD-SPs had an initial emission peak at 2035 meV, which matches the emission peak of dispersed CQD solution (this spectrum reflects the actual size distribution of CQDs). However, the emission peak rapidly redshifted within the next 100 ps and stabilized at 2020 meV after more than 100 ps, indicating that the energy transfer process had stopped.

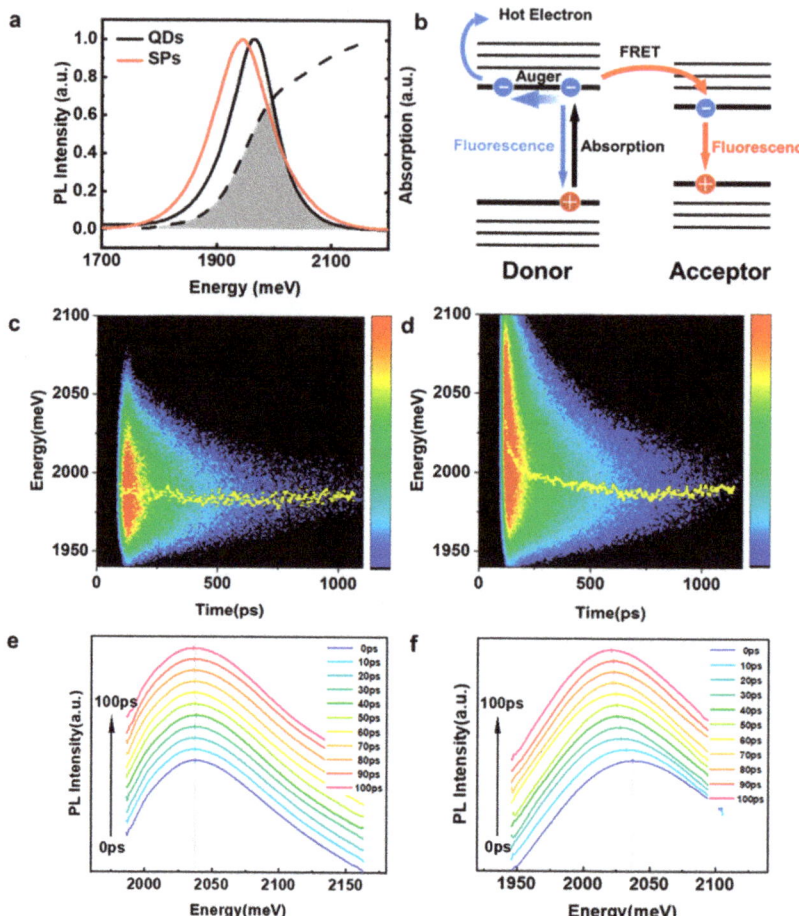

Figure 3. (**a**) Absorption (black dashed line) and emission (black curve) spectra for the CQD solution and emission spectrum for the SPs (red curve). (**b**) The excited-state pathways in a CQDs supraparticle considered in our model. (**c**,**d**) Spectrally resolved transient photoluminescence of (**c**) CdSe/ZnS CQD solution and (**d**) CQD supraparticles. (**e**,**f**) Emission spectra of CdSe/ZnS CQD solution (**e**) and CQD supraparticles (**f**) at 100 ps delay time after an excitation pulse. The short vertical bars are the peak energies. The gray vertical line is the PL peak energy of the first spectrum at 0 ps.

In order to elucidate the physical mechanism of PL spectra shift in self-assembled CQDs, we further analyzed the time-resolved PL spectra. First, we compare the transient PL decay curves of the dispersed CQD solution and a single SP at different emission wavelengths, as shown in Figure 4a,b. When the filter for wavelength selection was used, PL decay in the CQD solutions was almost wavelength-independent, which indicates that the CQDs in solution were independent emitters without any coupling or interaction between them. In contrast, PL decay at different wavelengths in SPs clearly revealed the time-dependent dynamics of FRET between CQDs. At short wavelengths, PL decayed faster at the initial stage, and the PL lifetime gradually increased with the increase in wavelength.

Furthermore, we fitted the lifetime data in Figure 4b with the bi-exponential equation $y(t) = A_0 + A_1\exp(-t/\tau_1) + A_2\exp(-t/\tau_2)$, where τ_1 and τ_2 represent fast and slow decay times, and A_1 and A_2 represent their contribution percentages, respectively. The fitting parameters are summarized in Table S1 and plotted in Figure 4c,d. As wavelength increased, we found that the fast decay time (52.5–64.8 ps) and slow decay time (308.6–355.4 ps) increased, but the fast component ratio decreased from 60% to 35%, while the slow component ratio increased from 40% to 65%. Since the fast and slow decay components come from non-radiative and radiative recombination processes, the fast component fraction decreased rapidly with wavelength from 595 to 620 nm, indicating that exciton energy was transferred from small to large CQDs due to dipole–dipole coupling interactions. In smaller CQDs, most excitons transferred energy to lower bandgap CQDs rather than radiative recombination. In contrast, excitons in larger CQDs received energy from FRET and radiated photons, which produced a longer PL lifetime than smaller CQDs. Data analysis of the ultra-fast time-resolved PL decay clearly showed the fast FRET process between different CQDs in SPs from high energy excitons to low energy excitons; indirect coupling via photon reabsorption was ruled out because it would not change the PL lifetime decay rate of the CQDs.

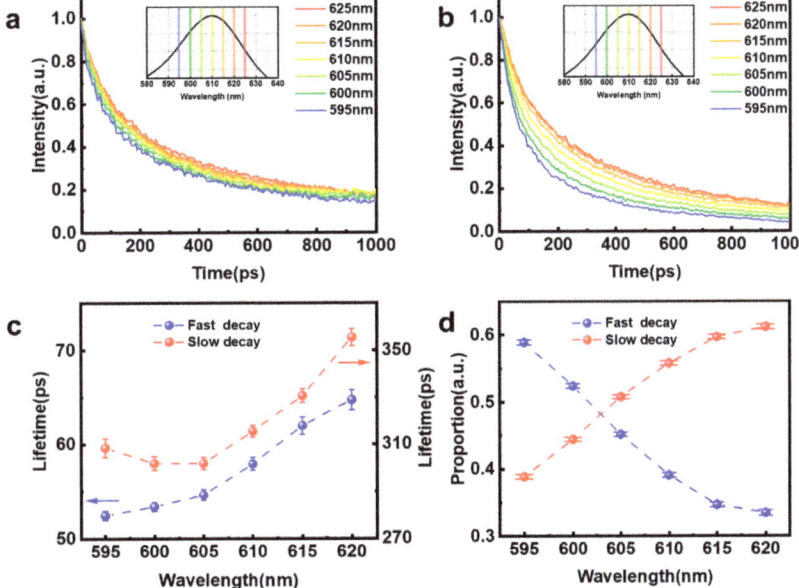

Figure 4. The energy transfer in CQD solution and QD-SPs. (**a**,**b**) Typical time-resolved photoluminescence decay curves of CQD solution (**a**) and QD-SPs (**b**) at various wavelengths. (**c**,**d**) The fast decay time and slow decay time of CQD solution (**c**) and SPs (**d**) PL decay curves with the emission peak in the range of 595 to 620 nm.

5. Suppression of Auger Recombination

The fluorescence resonance energy transfer has been demonstrated to be an effective means of enhancing the performance of sensing and light-harvesting functions [63]. In contrast to FRET, the non-radiative Auger recombination process can lead to the loss of energy in the form of heat, leading to reduced efficiency for CQD-based applications [64,65]. We noted that the FRET has an energy transfer mechanism similar to Auger recombination, but Auger recombination releases energy in the form of heat. Therefore, we suspect that when FRET occurs between CQDs, excitons transfer energy rapidly to neighboring excitons and then recombine to emit photons, reducing the probability of Auger recombination and thus inhibiting Auger recombination within CQDs. At higher excitation fluences, the transient PL of SPs showed an increased radiation lifetime compared with the CQD solution (Figure 5a), with a slow lifetime component of 9.6 ns for the CQD solution and 12.2 ns for the QD-SPs. Moreover, temperature-dependent PL showed that the PL intensity of the CQD solution decreased more rapidly with increasing temperature than QD-SPs (Figure 5b). The QD-SP structure was better at withstanding temperature increases (Figure 5b shows a lower slope), and considering the high correlation between Auger recombination and temperature [66], this difference suggests that the QD-SP structure effectively suppressed this non-radiative recombination process. To further verify the relevance of FRET to AR, we used a short-chain Octylamine (OctA) ligand exchange strategy to replace the long-chain organic oleic acid (OA) ligand, reducing the inter-dot distances in SPs (as show in Figure S2). Considering the strong distance-sensitivity of FRET, reducing the dot spacing will improve FRET efficiency, which can be manifested by an increase in spectral redshift [62]. Figure 5c shows the PL spectra of the dispersed CQD solution, OA-SPs, and OctA-SPs at room temperature. The emission centers of the three CQD structures were 1964 meV, 1944 meV, and 1930 meV, respectively. The spectrum of OctA-SPs was redshifted by 30 meV compared to 20 meV in the spectrum of OA-SPs. In addition, the PL emission centers of the dispersed CQD solution did not change after the ligand exchange, suggesting that the short-chain ligand increased the rate of FRET. Furthermore, we measured the temperature-dependent PL spectra of the dispersed CQD solution, OA-SPs, and OctA-SPs, as shown in Figure 5d. Under the same excitation conditions, the PL intensities of the three different CQD structures decreased to 10% of the initial value when the temperature increased to 310 K, 360 K, and 400 K, respectively. These results show that the SP structure achieved the acceleration of FRET process and the suppression of Auger recombination, improving the efficiency of CQD-based devices and has a promising application prospect in carrier-multiplication-enhanced photovoltaics and electrically pumped lasers [67].

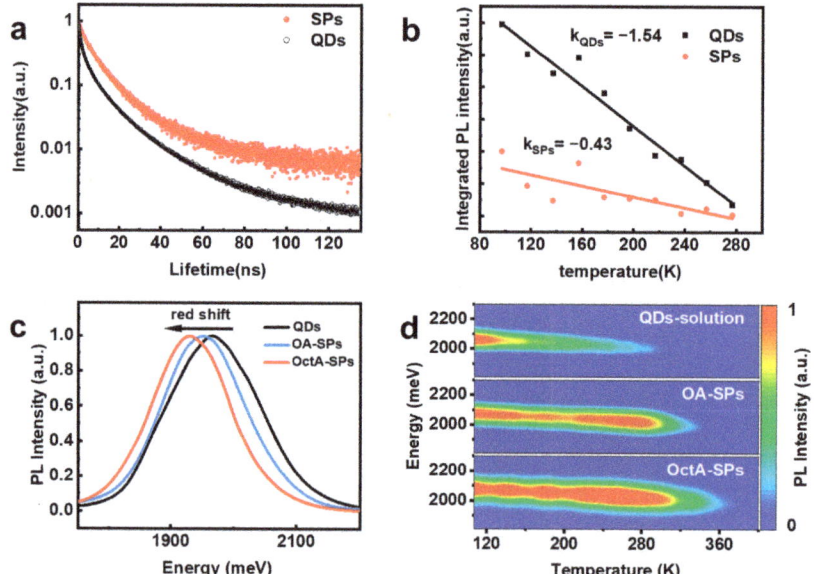

Figure 5. Auger recombination suppression using supraparticles. (**a**) PL decay of CQD solution and supraparticles. (**b**) Temperature-dependent PL intensity of CQD solution and supraparticles. (**c**) PL spectra of CQD solution, oleic acid ligand SPs, octylamine ligand SPs. (**d**) Contour plots of the temperature-dependent emission from different ligands supraparticles with the temperature varying from 110 to 400 K under the same excitation conditions.

6. Materials and Methods

6.1. Synthesis of Supraparticles Structures

Colloidal CdSe/ZnS quantum dots were obtained commercially (Xingzi (Shanghai, China) New Material Technology Development Co., Ltd.). We assembled QDs into supraparticles using an emulsion-based, bottom-up, self-assembly process following previously reported methods with slight modifications. In this work, we used microfluidic chips instead of ultrasonic methods to produce monodisperse oil-in-water microemulsion droplets. In a typical experiment, we prepared a solution of CdSe/ZnS QDs in hexane at a concentration of ~20 mg/mL as the dispersed phase and a solution of sodium dodecyl sulfate (SDS) in deionized water at a concentration of ~6 mg/mL as the continuous phase. We then connected a commercial microfluidic chip to a multichannel pressure regulator to generate monodisperse hexane microdroplets, controlling their size by adjusting the flow rates of the dispersed and continuous phases. We collected the resulting microemulsion in a glass vial covered with parafilm pierced by several 0.5 mm holes to slow down evaporation and stirred it at room temperature for 12–18 h until the hexane in the oil phase had fully evaporated. After evaporation, we washed the resulting QD SPs by three rounds of centrifugation (5000 rpm, 2 min) and redispersed them in deionized water to remove the residual surfactants. For ligand exchange of QD-SPs, we centrifuged the originally prepared SPs (5000 rpm, 2 min), added 1 wt% octylamine in 1 mL methanol, and mixed the solution at 500 rpm under magnetic stirring for 20 min. After 20 min, we washed the ligand-exchanged QD-SPs by centrifugation (5000 rpm, 2 min) and redispersed them in methanol and water to remove the residual ligands. We then drop-cast the dispersion of QD-SPs onto the desired substrate (silicon for SEM imaging and glass for spectroscopy) and dried it under a vacuum.

6.2. Structural Characterization

The SEM measurements were performed using field-emission scanning electron microscopy (FE-SEM; Auriga S40, Zeiss, Oberkochen, Germany) operated at 1 kV. The TEM measurements were performed using the Tecnai G2 F20 S-TWIN (FEI, Hillsboro, OR, USA) operated at 200 KV. The samples were prepared on a clean silicon wafer and then transferred onto a 300-mesh copper TEM grid by slightly touching the sample to the mesh.

6.3. Optical Characterization

Photoluminescence spectra were measured using a confocal microphotoluminescence system (LabRAM HR Evolution) with a high-numerical-aperture microscopy objective (N.A. = 0.5, 50×). The PL spectra were excited by a femtosecond laser (Libra, Coherent, B40 fs, 10 kHz, Santa Clara, CA, USA). Absorption spectra were measured using a PerkinElmer UV/VIS/NIR spectrometer (Lambda 750, Villeneuve-d'Ascq, France). The time-resolved PL measurements were performed using a streak camera with picosecond-order time resolution (Optronis, SC-10, Kehl, Germany). Low-temperature measurements were performed using a cryostat (80–475 K, Janis ST-500, Woburn, MA, USA) with a temperature controller (cryocon 22C, USA) and liquid N_2 for cooling.

7. Conclusions

In summary, we demonstrated a self-assembly method for preparing spherical SPs with a regular structure and smooth surface. The high packing density made the SP structure very stable in both water and air, and had good solution processability. PL spectroscopy and time-resolved experiments showed that the decay rate of PL intensity gradually increased with the increase in pump energy due to FRET, and the PL spectrum gradually redshifted with time. As the FRET rate increased, non-radiative Auger recombination of the CQDs was suppressed, and temperature-dependent PL spectra confirmed that the SP structure was more effective in suppressing Auger recombination than dispersed CQDs. Through short-chain ligand exchange, we further verified that the highly packed SPs had better high temperature optical properties, which can be attributed to the increased

FRET rate and suppressed Auger recombination in the SPs with smaller dot spacings. These properties make SP structures very attractive for applications in biosensors and light-emitting devices.

Supplementary Materials: The following supporting information can be downloaded at: https://www.mdpi.com/article/10.3390/inorganics11050218/s1, Figure S1: High resolution transmission electron microscopy image of SPs; Figure S2: Fourier transform in-frared spectroscopy spectra of OctA-SPs and OA-SPs; Table S1: Life decay curve fitting parameters at different wavelengths of 595–620 nm.

Author Contributions: Conceptualization, X.T. and H.C.; methodology, X.T. and H.C.; validation, X.T. and H.C.; formal analysis, X.T., H.C., H.D., C.Z. and L.Z.; investigation, X.T. and H.C.; resources, H.D., C.Z. and L.Z.; data curation, X.T. and H.C.; writing—original draft preparation, X.T. and H.C.; writing—review and editing, H.D., C.Z. and L.Z.; supervision, H.D. and L.Z.; funding acquisition, H.C., H.D. and L.Z. All authors have read and agreed to the published version of the manuscript.

Funding: This research was funded by the China Postdoctoral Science Foundation (No. 2022M723267), Shanghai Sailing Program (No. 23YF1453900), the Natural Science Foundation of Shanghai (Nos. 20JC1414605, 23ZR1471500), the National Natural Science Foundation of China (No. 61925506), Hangzhou Science and Technology Bureau of Zhejiang Province (No. TD2020002), and Academic/Technology Research Leader Program of Shanghai (23XD1404500).

Data Availability Statement: All data are available in the manuscript. Correspondence and requests for materials should be addressed to H.D. (hongxingd@siom.ac.cn), C.Z. (chizhang@ucas.ac.cn), and L.Z. (lzhang@siom.ac.cn).

Conflicts of Interest: The authors declare that they have no conflicts of interest.

References

1. Barak, Y.; Meir, I.; Shapiro, A.; Jang, Y.; Lifshitz, E. Fundamental Properties in Colloidal Quantum Dots. *Adv. Mater.* **2018**, *30*, e1801442. [CrossRef] [PubMed]
2. Chen, O.; Zhao, J.; Chauhan, V.P.; Cui, J.; Wong, C.; Harris, D.K.; Wei, H.; Han, H.-S.; Fukumura, D.; Jain, R.K.; et al. Compact high-quality CdSe-CdS core-shell nanocrystals with narrow emission linewidths and suppressed blinking. *Nat. Mater.* **2013**, *12*, 445–451. [CrossRef] [PubMed]
3. Yin, Y.; Alivisatos, A.P. Colloidal nanocrystal synthesis and the organic-inorganic interface. *Nature* **2005**, *437*, 664–670. [CrossRef] [PubMed]
4. Liu, M.; Chen, Y.; Tan, C.-S.; Quintero-Bermudez, R.; Proppe, A.H.; Munir, R.; Tan, H.; Voznyy, O.; Scheffel, B.; Walters, G.; et al. Lattice anchoring stabilizes solution-processed semiconductors. *Nature* **2019**, *570*, 96–101. [CrossRef]
5. Cargnello, M.; Johnston-Peck, A.C.; Diroll, B.T.; Wong, E.; Datta, B.; Damodhar, D.; Doan-Nguyen, V.V.T.; Herzing, A.A.; Kagan, C.R.; Murray, C.B. Substitutional doping in nanocrystal superlattices. *Nature* **2015**, *524*, 450–453. [CrossRef]
6. Peng, X.; Lv, L.; Liu, S.; Li, J.; Lei, H.; Qin, H. Synthesis of Weakly Confined, Cube-Shaped, and Monodisperse Cadmium Chalcogenide Nanocrystals with Unexpected Photophysical Properties. *J. Am. Chem. Soc.* **2022**, *144*, 16872–16882. [CrossRef]
7. Shirasaki, Y.; Supran, G.J.; Bawendi, M.G.; Bulovic, V. Emergence of colloidal quantum-dot light-emitting technologies. *Nat. Photonics* **2013**, *7*, 13–23. [CrossRef]
8. Chang, H.; Dong, H.; Zhao, J.; Zhang, L. Efficient and stable solid state luminophores with colloidal quantum dots-based silica monolith. *Solid State Commun.* **2020**, *305*, 113765. [CrossRef]
9. Yang, Y.; Zheng, Y.; Cao, W.; Titov, A.; Hyvonen, J.; Manders, J.R.; Xue, J.; Holloway, P.H.; Qian, L. High-efficiency light-emitting devices based on quantum dots with tailored nanostructures. *Nat. Photonics* **2015**, *9*, 259–266. [CrossRef]
10. Kagan, C.R.; Lifshitz, E.; Sargent, E.H.; Talapin, D.V. Building devices from colloidal quantum dots. *Science* **2016**, *353*, aac5523. [CrossRef]
11. Zhao, B.; Yao, Y.; Gao, M.; Sun, K.; Zhang, J.; Li, W. Doped quantum dot@silica nanocomposites for white light-emitting diodes. *Nanoscale* **2015**, *7*, 17231–17236. [CrossRef] [PubMed]
12. Swarnkar, A.; Marshall, A.R.; Sanehira, E.M.; Chernomordik, B.D.; Moore, D.T.; Christians, J.A.; Chakrabarti, T.; Luther, J.M. Quantum dot-induced phase stabilization of alpha-CsPbI3 perovskite for high-efficiency photovoltaics. *Science* **2016**, *354*, 92–95. [CrossRef] [PubMed]
13. Robel, I.; Subramanian, V.; Kuno, M.; Kamat, P.V. Quantum dot solar cells. Harvesting light energy with CdSe nanocrystals molecularly linked to mesoscopic TiO2 films. *J. Am. Chem. Soc.* **2006**, *128*, 2385–2393. [CrossRef] [PubMed]
14. Cooney, R.R.; Sewall, S.L.; Sagar, D.M.; Kambhampati, P. Gain Control in Semiconductor Quantum Dots via State-Resolved Optical Pumping. *Phys. Rev. Lett.* **2009**, *102*, 127404. [CrossRef] [PubMed]
15. Wang, Y.; Yu, D.; Wang, Z.; Li, X.; Chen, X.; Nalla, V.; Zeng, H.; Sun, H. Solution-Grown CsPbBr3/Cs4PbBr6 Perovskite Nanocomposites: Toward Temperature-Insensitive Optical Gain. *Small* **2017**, *13*, 1587.

16. Park, Y.-S.; Roh, J.; Diroll, B.T.; Schaller, R.D.; Klimov, V.I. Colloidal quantum dot lasers. *Nat. Rev. Mater.* **2021**, *6*, 382–401. [CrossRef]
17. Taghipour, N.; Dalmases, M.; Whitworth, G.L.; Dosil, M.; Othonos, A.; Christodoulou, S.; Liga, S.M.; Konstantatos, G. Colloidal Quantum Dot Infrared Lasers Featuring Sub-Single-Exciton Threshold and Very High Gain. *Adv. Mater.* **2023**, *35*, 2207678. [CrossRef]
18. Chen, O.; Riedemann, L.; Etoc, F.; Herrmann, H.; Coppey, M.; Barch, M.; Farrar, C.T.; Zhao, J.; Bruns, O.T.; Wei, H.; et al. Magneto-fluorescent core-shell supernanoparticles. *Nat. Commun.* **2014**, *5*, 5093. [CrossRef]
19. Wu, C.; Lu, Z.; Li, Z.; Yin, Y. Assembly of Colloidal Nanoparticles into Hollow Superstructures by Controlling Phase Separation in Emulsion Droplets. *Small Struct.* **2021**, *2*, 2100005. [CrossRef]
20. Kim, K.-H.; Dannenberg, P.H.; Yan, H.; Cho, S.; Yun, S.-H. Compact Quantum-Dot Microbeads with Sub-Nanometer Emission Linewidth. *Adv. Funct. Mater.* **2021**, *31*, 2103413. [CrossRef]
21. He, X.; Jia, K.; Bai, Y.; Chen, Z.; Liu, Y.; Huang, Y.; Liu, X. Quantum dots encoded white-emitting polymeric superparticles for simultaneous detection of multiple heavy metal ions. *J. Hazard. Mater.* **2021**, *405*, 124263. [CrossRef] [PubMed]
22. Lee, T.; Ohshiro, K.; Watanabe, T.; Hyeon-Deuk, K.; Kim, D. Temperature-Dependent Exciton Dynamics in CdTe Quantum Dot Superlattices Fabricated via Layer-by-Layer Assembly. *Adv. Opt. Mater.* **2022**, *10*, 2102781. [CrossRef]
23. Lan, X.; Chen, M.; Hudson, M.H.; Kamysbayev, V.; Wang, Y.; Guyot-Sionnest, P.; Talapin, D.V. Quantum dot solids showing state-resolved band-like transport. *Nat. Mater.* **2020**, *19*, 323–329. [CrossRef] [PubMed]
24. Crisp, R.W.; Schrauben, J.N.; Beard, M.C.; Luther, J.M.; Johnson, J.C. Coherent Exciton Delocalization in Strongly Coupled Quantum Dot Arrays. *Nano Lett.* **2013**, *13*, 4862–4869. [CrossRef]
25. Hou, K.; Han, J.; Tang, Z. Formation of Supraparticles and Their Application in Catalysis. *ACS Mater. Lett.* **2020**, *2*, 95–106. [CrossRef]
26. Wang, B.; Li, R.; Guo, G.; Xia, Y. Janus and core@shell gold nanorod@Cu2-xS supraparticles: Reactive site regulation fabrication, optical/catalytic synergetic effects and enhanced photothermal efficiency/photostability. *Chem. Commun.* **2020**, *56*, 8996–8999. [CrossRef]
27. Lin, C.H.; Lafalce, E.; Jung, J.; Smith, M.J.; Malak, S.T.; Aryal, S.; Yoon, Y.J.; Zhai, Y.; Lin, Z.; Vardeny, Z.V.; et al. Core/Alloyed-Shell Quantum Dot Robust Solid Films with High Optical Gains. *ACS Photonics* **2016**, *3*, 647–658. [CrossRef]
28. Gur, I.; Fromer, N.A.; Geier, M.L.; Alivisatos, A.P. Air-stable all-inorganic nanocrystal solar cells processed from solution. *Science* **2005**, *310*, 462–465. [CrossRef]
29. Yao, C.; Wang, P.; Li, X.; Hu, X.; Hou, J.; Wang, L.; Zhang, F. Near-Infrared-Triggered Azobenzene-Liposome/Upconversion Nanoparticle Hybrid Vesicles for Remotely Controlled Drug Delivery to Overcome Cancer Multidrug Resistance. *Adv. Mater.* **2016**, *28*, 9341–9348. [CrossRef]
30. Klimov, V.I.; Mikhailovsky, A.A.; McBranch, D.W.; Leatherdale, C.A.; Bawendi, M.G. Quantization of multiparticle Auger rates in semiconductor quantum dots. *Science* **2000**, *287*, 1011–1013. [CrossRef]
31. Talapin, D.V.; Lee, J.-S.; Kovalenko, M.V.; Shevchenko, E.V. Prospects of Colloidal Nanocrystals for Electronic and Optoelectronic Applications. *Chem. Rev.* **2010**, *110*, 389–458. [CrossRef] [PubMed]
32. Zhao, Y.; Shang, L.; Cheng, Y.; Gu, Z. Spherical Colloidal Photonic Crystals. *Acc. Chem. Res.* **2014**, *47*, 3632–3642. [CrossRef] [PubMed]
33. Park, J.-G.; Kim, S.-H.; Magkiriadou, S.; Choi, T.M.; Kim, Y.-S.; Manoharan, V.N. Full-Spectrum Photonic Pigments with Non-iridescent Structural Colors through Colloidal Assembly. *Angew. Chem. Int. Ed.* **2014**, *53*, 2899–2903. [CrossRef] [PubMed]
34. Vogel, N.; Utech, S.; England, G.T.; Shirman, T.; Phillips, K.R.; Koay, N.; Burgess, I.B.; Kolle, M.; Weitz, D.A.; Aizenberg, J. Color from hierarchy: Diverse optical properties of micron-sized spherical colloidal assemblies. *Proc. Natl. Acad. Sci. USA* **2015**, *112*, 10845–10850. [CrossRef]
35. Marino, E.; Sciortino, A.; Berkhout, A.; MacArthur, K.E.; Heggen, M.; Gregorkiewicz, T.; Kodger, T.E.; Capretti, A.; Murray, C.B.; Koenderink, A.F.; et al. Simultaneous Photonic and Excitonic Coupling in Spherical Quantum Dot Supercrystals. *ACS Nano* **2020**, *14*, 13806–13815. [CrossRef]
36. Chen, X.; Fu, Z.; Gong, Q.; Wang, J. Quantum entanglement on photonic chips: A review. *Adv. Photonics* **2021**, *3*, 064002. [CrossRef]
37. Jiang, B.; Zhu, S.; Ren, L.; Shi, L.; Zhang, X. Simultaneous ultraviolet, visible, and near-infrared continuous-wave lasing in a rare-earth-doped microcavity. *Adv. Photonics* **2022**, *4*, 046003. [CrossRef]
38. Shuklov, I.A.; Toknova, V.F.; Lizunova, A.A.; Razumov, V.F. Controlled aging of PbS colloidal quantum dots under mild conditions. *Mater. Today Chem.* **2020**, *18*, 100357. [CrossRef]
39. Wu, J.; Su, R.; Fieramosca, A.; Ghosh, S.; Zhao, J.; Liew, T.C.H.; Xiong, Q. Perovskite polariton parametric oscillator. *Adv. Photonics* **2021**, *3*, 055003. [CrossRef]
40. Bai, F.; Wang, D.; Huo, Z.; Chen, W.; Liu, L.; Liang, X.; Chen, C.; Wang, X.; Peng, Q.; Li, Y. A versatile bottom-up assembly approach to colloidal spheres from nanocrystals. *Angew. Chem. Int. Ed.* **2007**, *46*, 6650–6653. [CrossRef]
41. De Nijs, B.; Dussi, S.; Smallenburg, F.; Meeldijk, J.D.; Groenendijk, D.J.; Filion, L.; Imhof, A.; van Blaaderen, A.; Dijkstra, M. Entropy-driven formation of large icosahedral colloidal clusters by spherical confinement. *Nat. Mater.* **2015**, *14*, 56–60. [CrossRef] [PubMed]
42. Lacava, J.; Ouali, A.-A.; Raillard, B.; Kraus, T. On the behaviour of nanoparticles in oil-in-water emulsions with different surfactants. *Soft Matter* **2014**, *10*, 1696–1704. [CrossRef] [PubMed]

43. Plunkett, A.; Eldridge, C.; Schneider, G.A.; Domenech, B. Controlling the Large-Scale Fabrication of Supraparticles. *J. Phys. Chem. B* **2020**, *124*, 11263–11272. [CrossRef] [PubMed]
44. Koshkina, O.; Raju, L.T.; Kaltbeitzel, A.; Riedinger, A.; Lohse, D.; Zhang, X.; Landfester, K. Surface Properties of Colloidal Particles Affect Colloidal Self-Assembly in Evaporating Self-Lubricating Ternary Droplets. *ACS Appl. Mater. Interfaces* **2022**, *14*, 2275–2290. [CrossRef]
45. Marino, E.; Kodger, T.E.; Wegdam, G.H.; Schall, P. Revealing Driving Forces in Quantum Dot Supercrystal Assembly. *Adv. Mater.* **2018**, *30*, 1803433. [CrossRef]
46. Montanarella, F.; Geuchies, J.J.; Dasgupta, T.; Prins, P.T.; van Overbeek, C.; Dattani, R.; Baesjou, P.; Dijkstra, M.; Petukhov, A.V.; van Blaaderen, A.; et al. Crystallization of Nanocrystals in Spherical Confinement Probed by in Situ X-ray Scattering. *Nano Lett.* **2018**, *18*, 3675–3681. [CrossRef]
47. Schmitt, J.; Hajiw, S.; Lecchi, A.; Degrouard, J.; Salonen, A.; Imperor-Clerc, M.; Pansu, B. Formation of Superlattices of Gold Nanoparticles Using Ostwald Ripening in Emulsions: Transition from fcc to bcc Structure. *J. Phys. Chem. B* **2016**, *120*, 5759–5766. [CrossRef]
48. Marino, E.; Keller, A.W.; An, D.; van Dongen, S.; Kodger, T.E.; MacArthur, K.E.; Heggen, M.; Kagan, C.R.; Murray, C.B.; Schall, P. Favoring the Growth of High-Quality, Three-Dimensional Supercrystals of Nanocrystals. *J. Phys. Chem. C* **2020**, *124*, 11256–11264. [CrossRef]
49. Wang, D.; van der Wee, E.B.; Zanaga, D.; Altantzis, T.; Wu, Y.; Dasgupta, T.; Dijkstra, M.; Murray, C.B.; Bals, S.; van Blaaderen, A. Quantitative 3D real-space analysis of Laves phase supraparticles. *Nat. Commun.* **2021**, *12*, 3980. [CrossRef]
50. Montanarella, F.; Altantzis, T.; Zanaga, D.; Rabouw, F.T.; Bals, S.; Baesjou, P.; Vanmaekelbergh, D.; van Blaaderen, A. Composite Supraparticles with Tunable Light Emission. *ACS Nano* **2017**, *11*, 9136–9142. [CrossRef]
51. Zou, H.; Wang, D.; Gong, B.; Liu, Y. Preparation of CdTe superparticles for white light-emitting diodes without Forster resonance energy transfer. *RSC Adv.* **2019**, *9*, 30797–30802. [CrossRef] [PubMed]
52. Vanmaekelbergh, D.; van Vugt, L.K.; Bakker, H.E.; Rabouw, F.T.; de Nijs, B.; van Dijk-Moes, R.J.A.; van Huis, M.A.; Baesjou, P.J.; van Blaaderen, A. Shape-Dependent Multiexciton Emission and Whispering Gallery Modes in Supraparticles of CdSe/Multishell Quantum Dots. *ACS Nano* **2015**, *9*, 3942–3950. [CrossRef] [PubMed]
53. Montanarella, F.; Urbonas, D.; Chadwick, L.; Moerman, P.G.; Baesjou, P.J.; Mahrt, R.F.; van Blaaderen, A.; Stoferle, T.; Vanmaekelbergh, D. Lasing Supraparticles Self-Assembled from Nanocrystals. *ACS Nano* **2018**, *12*, 12788–12794. [CrossRef] [PubMed]
54. Neuhaus, S.J.; Marino, E.; Murray, C.B.; Kagan, C.R. Frequency Stabilization and Optically Tunable Lasing in Colloidal Quantum Dot Superparticles. *Nano Lett.* **2023**, *23*, 645–651. [CrossRef] [PubMed]
55. Chang, H.; Zhong, Y.; Dong, H.; Wang, Z.; Xie, W.; Pan, A.; Zhang, L. Ultrastable low-cost colloidal quantum dot microlasers of operative temperature up to 450 K. *Light Sci. Appl.* **2021**, *10*, 60. [CrossRef]
56. Gao, Z.; Zhang, W.; Yan, Y.; Yi, J.; Dong, H.; Wang, K.; Yao, J.; Zhao, Y.S. Proton-Controlled Organic Microlaser Switch. *ACS Nano* **2018**, *12*, 5734–5740. [CrossRef]
57. Blondot, V.; Bogicevic, A.; Coste, A.; Arnold, C.; Buil, S.; Quelin, X.; Pons, T.; Lequeux, N.; Hermier, J.-P. Fluorescence properties of self assembled colloidal supraparticles from CdSe/CdS/ZnS nanocrystals. *New J. Phys.* **2020**, *22*, 113026. [CrossRef]
58. Montanarella, F.; Biondi, M.; Hinterding, S.M.; Vanmaekelbergh, D.; Rabouw, F.T. Reversible Charge-Carrier Trapping Slows Forster Energy Transfer in CdSe/CdS Quantum-Dot Solids. *Nano Lett.* **2018**, *18*, 5867–5874. [CrossRef]
59. Kister, T.; Mravlak, M.; Schilling, T.; Kraus, T. Pressure-controlled formation of crystalline, Janus, and core-shell supraparticles. *Nanoscale* **2016**, *8*, 13377–13384. [CrossRef]
60. Wang, J.; Mbah, C.F.; Przybilla, T.; Zubiri, B.A.; Spiecker, E.; Engel, M.; Vogel, N. Magic number colloidal clusters as minimum free energy structures. *Nat. Commun.* **2018**, *9*, 5259. [CrossRef]
61. Kim, C.; Jung, K.; Yu, J.W.; Park, S.; Kim, S.-H.; Lee, W.B.; Hwang, H.; Manoharan, V.N.; Moon, J.H. Controlled Assembly of Icosahedral Colloidal Clusters for Structural Coloration. *Chem. Mater.* **2020**, *32*, 9704–9712. [CrossRef]
62. Chou, K.F.; Dennis, A.M. Forster Resonance Energy Transfer between Quantum Dot Donors and Quantum Dot Acceptors. *Sensors* **2015**, *15*, 13288–13325. [CrossRef] [PubMed]
63. Clapp, A.R.; Medintz, I.L.; Mauro, J.M.; Fisher, B.R.; Bawendi, M.G.; Mattoussi, H. Fluorescence resonance energy transfer between quantum dot donors and dye-labeled protein acceptors. *J. Am. Chem. Soc.* **2004**, *126*, 301–310. [CrossRef] [PubMed]
64. Hou, X.; Kang, J.; Qin, H.; Chen, X.; Ma, J.; Zhou, J.; Chen, L.; Wang, L.; Wang, L.-W.; Peng, X. Engineering Auger recombination in colloidal quantum dots via dielectric screening. *Nat. Commun.* **2019**, *10*, 1750. [CrossRef]
65. Vaxenburg, R.; Rodina, A.; Shabaev, A.; Lifshitz, E.; Efros, A.L. Nonradiative Auger Recombination in Semiconductor Nanocrystals. *Nano Lett.* **2015**, *15*, 2092–2098. [CrossRef]
66. Javaux, C.; Mahler, B.; Dubertret, B.; Shabaev, A.; Rodina, A.V.; Efros, A.L.; Yakovlev, D.R.; Liu, F.; Bayer, M.; Camps, G.; et al. Thermal activation of non-radiative Auger recombination in charged colloidal nanocrystals. *Nat. Nanotechnol.* **2013**, *8*, 206–212. [CrossRef]
67. Algar, W.R.; Kim, H.; Medintz, I.L.; Hildebrandt, N. Emerging non-traditional Forster resonance energy transfer configurations with semiconductor quantum dots: Investigations and applications. *Coord. Chem. Rev.* **2014**, *263*, 65–85. [CrossRef]

Disclaimer/Publisher's Note: The statements, opinions and data contained in all publications are solely those of the individual author(s) and contributor(s) and not of MDPI and/or the editor(s). MDPI and/or the editor(s) disclaim responsibility for any injury to people or property resulting from any ideas, methods, instructions or products referred to in the content.

Correction

Correction: Tian et al. Fluorescence Resonance Energy Transfer Properties and Auger Recombination Suppression in Supraparticles Self-Assembled from Colloidal Quantum Dots. *Inorganics* 2023, *11*, 218

Xinhua Tian [1,2,†], Hao Chang [1,†], Hongxing Dong [1,3,4,*], Chi Zhang [3,5,*] and Long Zhang [1,2,3,4,*]

1. Key Laboratory of Materials for High-Power Laser, Shanghai Institute of Optics and Fine Mechanics, Chinese Academy of Sciences, Shanghai 201800, China
2. School of Physical and Technology, ShanghaiTech University, Shanghai 201210, China
3. School of Physics and Optoelectronic Engineering, Hangzhou Institute for Advanced Study, University of Chinese Academy of Sciences, Hangzhou 310024, China
4. CAS Center for Excellence in Ultra-Intense Laser Science, Shanghai 201800, China
5. Department of Mechanical and Aerospace Engineering, University of Missouri, Columbia, MO 65211, USA

* Correspondence: hongxingd@siom.ac.cn (H.D.); chizhang@ucas.ac.cn (C.Z.); lzhang@siom.ac.cn (L.Z.)

† These authors contributed equally to this work.

Error in Figure

In the original publication [1], there was a mistake in two schematics (Figures 1 and 3b) as published. Both schematics may lead to misunderstanding. The corrected Figures 1 and 3b appear below. The authors apologize for any inconvenience caused and state that the scientific conclusions are unaffected. This correction was approved by the Academic Editor. The original publication has also been updated.

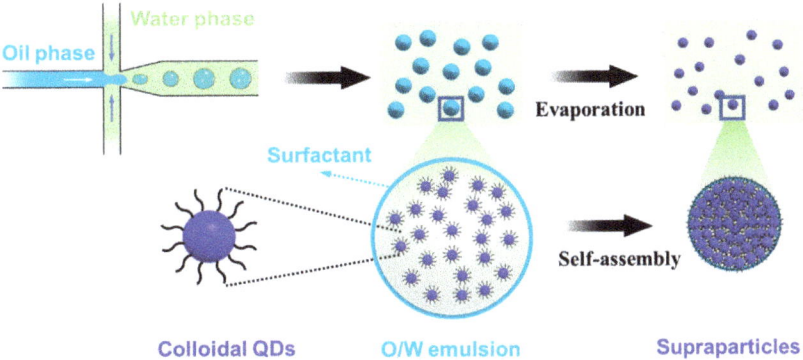

Figure 1. Schematic of the fabrication of colloidal quantum dot supraparticles through the microemulsion method.

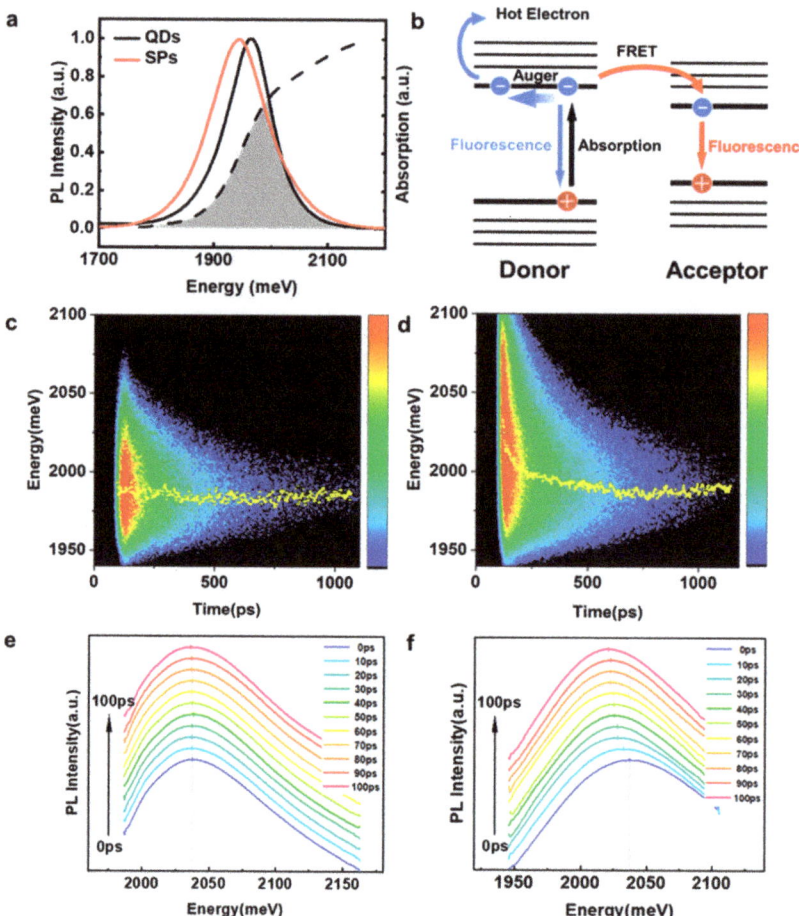

Figure 3. (**a**) Absorption (black dashed line) and emission (black curve) spectra for the CQD solution and emission spectrum for the SPs (red curve). (**b**) The excited-state pathways in a CQDs supraparticle considered in our model. (**c**,**d**) Spectrally resolved transient photoluminescence of (**c**) CdSe/ZnS CQD solution and (**d**) CQD supraparticles. (**e**,**f**) Emission spectra of CdSe/ZnS CQD solution (**e**) and CQD supraparticles (**f**) at 100 ps delay time after an excitation pulse. The short vertical bars are the peak energies. The gray vertical line is the PL peak energy of the first spectrum at 0 ps.

Reference

1. Tian, X.; Chang, H.; Dong, H.; Zhang, C.; Zhang, L. Fluorescence Resonance Energy Transfer Properties and Auger Recombination Suppression in Supraparticles Self-Assembled from Colloidal Quantum Dots. *Inorganics* **2023**, *11*, 218. [CrossRef]

Disclaimer/Publisher's Note: The statements, opinions and data contained in all publications are solely those of the individual author(s) and contributor(s) and not of MDPI and/or the editor(s). MDPI and/or the editor(s) disclaim responsibility for any injury to people or property resulting from any ideas, methods, instructions or products referred to in the content.

Article

Adsorption of Metal Atoms on SiC Monolayer

Lei Jiang *, Yanbo Dong and Zhen Cui *

School of Automation and Information Engineering, Xi'an University of Technology, Xi'an 710048, China
* Correspondence: jianglei@xaut.edu.cn (L.J.); zcui@xaut.edu.cn (Z.C.)

Abstract: The electronic, magnetic, and optical behaviors of metals (M = Ag, Al, Au, Bi, Ca, Co, Cr, Cu, Fe, Ga, K, Li, Mn, Na, Ni) adsorbed on the SiC monolayer have been calculated based on density functional theory (DFT). The binding energy results show that all the M-adsorbed SiC systems are stable. All the M-adsorbed SiC systems are magnetic with magnetic moments of 1.00 μ_B (Ag), 1.00 μ_B (Al), 1.00 μ_B (Au), 1.01 μ_B (Bi), 1.95 μ_B (Ca), 1.00 μ_B (Co), 4.26 μ_B (Cr), 1.00 μ_B (Cu), 2.00 μ_B (Fe), 1.00 μ_B (Ga), 0.99 μ_B (K), 1.00 μ_B (Li), 3.00 μ_B (Mn), and 1.00 μ_B (Na), respectively, except for the Ni-adsorbed SiC system. The Ag, Al, Au, Cr, Cu, Fe, Ga, Mn, and Na-adsorbed SiC systems become magnetic semiconductors, while Bi, Ca, Co, K, and Li-adsorbed SiC systems become semimetals. The Bader charge results show that there is a charge transfer between the metal atom and the SiC monolayer. The work function of the K-adsorbed SiC system is 2.43 eV, which is 47.9% lower than that of pristine SiC and can be used in electron-emitter devices. The Bi, Ca, Ga, and Mn-adsorbed SiC systems show new absorption peaks in the visible light range. These results indicate that M-adsorbed SiC systems have potential applications in the field of spintronic devices and solar energy conversion photovoltaic devices.

Keywords: adsorption; metal atoms; 2D SiC; magnetism; first-principles calculations

Citation: Jiang, L.; Dong, Y.; Cui, Z. Adsorption of Metal Atoms on SiC Monolayer. *Inorganics* 2023, 11, 240. https://doi.org/10.3390/inorganics11060240

Academic Editor: Antonino Gulino

Received: 15 May 2023
Revised: 27 May 2023
Accepted: 29 May 2023
Published: 30 May 2023

Copyright: © 2023 by the authors. Licensee MDPI, Basel, Switzerland. This article is an open access article distributed under the terms and conditions of the Creative Commons Attribution (CC BY) license (https://creativecommons.org/licenses/by/4.0/).

1. Introduction

Since the successful preparation of graphene [1], there has been a surge in research into two-dimensional (2D) materials, including 2D WS_2 [2,3], GaN [4–6], BN [7,8], black phosphorus [9,10], ZnO [11,12], SiC [13,14], etc. SiC is a third-generation semiconductor material with a wide band gap, high electron saturation drift rate, high breakdown field strength, high thermal conductivity, high radiation resistance, etc. It has a wide range of applications in solar cells, high-frequency high-power devices, and high-temperature electronic devices. Two-dimensional SiC has the advantages of high electron mobility, chemical stability, and high catalytic activity and is often used to make photocatalysts [15]. Based on the first-principles approach, 2D SiC has been predicted to have a graphene-like honeycomb structure and can exist stably as a semiconductor material with a direct band gap of 2.52–2.87 eV [16,17]. Chabi et al. have successfully prepared SiC nanosheets with an average thickness of 2–3 nm through a catalytic carbon thermal reduction method and ultrasonic pretreatment process [18]. Two-dimensional SiC has great potential in the field of nanoelectronic devices, but there are some problems in photocatalysis. Two-dimensional SiC is only responsive to partially visible light [19], so it is necessary to reduce the band gap and improve the absorption efficiency of visible light. Current methods to effectively modulate the band structure include doping [20,21], stacking [22,23], adsorption [24,25], heterojunctions [26–28], etc.

The adsorption of metal atoms is one of the most important means to modulate the properties of 2D materials. The adsorption of different atoms on the surface of 2D materials can modulate the optical, electrical, and magnetic properties of 2D materials. Nie et al. have studied the adsorption of 3D transition metals on the SnO monolayer [29]. They found that 3D transition metal adsorption induced magnetism and achieved *n*-type and *p*-type doping. Guo et al. have modulated the electronic properties of the

WSSe monolayer by adsorbing Fe, Co, and Ni atoms and developed its applications in gas sensors and single-atom catalysts [30]. Cui et al. have studied the adsorption of transition metals on the Pd_2Se_3 monolayer [31]. They found that the adsorption of transition metals improved light absorption in the ultraviolet, visible, and infrared regions. Xu et al. have predicted the magnetism of the $SnSe_2$ monolayer after the adsorption of transition metals and found that the adsorption of Ti atoms can endow the $SnSe_2$ monolayer with perpendicular magnetic anisotropy [32]. In this paper, the electronic structure, magnetic, and optical properties of 15 metal atoms adsorbed on SiC monolayer have been calculated using the first-principles approach. The influence of the M atoms on the properties of the SiC monolayer is analyzed according to the band structure, work function, and light absorption spectra, and the application prospects of M-adsorbed SiC systems in the field of spin devices and photovoltaic devices are explored.

2. Computational Details

The electronic, magnetic, and optical behaviors of M-adsorbed SiC systems have been investigated in the Vienna ab initio calculation simulation package (VASP) [33,34] using density functional theory (DFT) [35,36]. The electron-ion interactions are performed using the Perdew-Burke-Ernzerhof (PBE) form of the generalized gradient approximation (GGA) approach [37]. The exchange-correlation interactions are performed using the projector-enhanced wave (PAW) approach [38]. Dispersion corrections are considered by Grimme's DFT-D3 method [39]. The plane wave cutoff energy is 400 eV, the Monkhorst-Pack scheme [40] grid in the Brillouin zone is $4 \times 4 \times 1$, and the vacuum space is 20 Å. During structural relaxation, the convergences of the force and self-consistent energy are 1×10^{-2} eV Å^{-1} and 1×10^{-5} eV, respectively. The optical properties are considered according to the frequency-dependent dielectric response theory, including the local field effects in the random-phase-approximation (RPA) method [41].

3. Results and Discussion

The pristine SiC has a graphene-like structure with an alternating arrangement of C and Si atoms, and its lattice parameter is 3.1 Å with a bond length of 1.78 Å. From the band structure and density of state (DOS) of Figure 1b,c, it can be seen that pristine SiC is a nonmagnet semiconductor with a direct bandgap of 2.5 eV, and the conduction band minimum (CBM) is mainly contributed by the hybridization of the p-state of Si and C, while the valence band maximum (VBM) is mainly contributed by the $2p$-state of C. The VBM and CBM are not at the same high symmetry point, indicating that the pristine SiC is an indirect bandgap semiconductor. These results are consistent with previous reports [42], indicating that our computational method is reliable.

In order to study the stability of metal adsorption on the SiC system, we constructed four adsorption models for each type of metal, as shown in Figure 1a. The adsorption sites were located above a Si atom, above a C1 atom, above a C2 atom, and above the Si-C bond. Adsorption energy (E_{ads}) was used to characterize the stability of the adsorption system, which can be calculated using the following formula:

$$E_{ads} = E_{M-SiC} - E_{SiC} - \mu_M \quad (1)$$

where E_{M-SiC} is the total energy of the M-adsorbed SiC systems, including the interaction energy between the metal atom and the SiC monolayer; E_{SiC} is the energy of the pristine SiC monolayer; μ_M is the chemical potential of an isolated metal atom.

As listed in Table 1, the E_{ads} of the M-adsorbed SiC systems are −0.193 eV (Ag), −1.683 eV (Al), −1.706 eV (Au), −1.135 eV (Bi), −0.111 eV (Ca), −2.785 eV (Co), −2.545 eV (Cr), −1.388 eV (Cu), −2.104 eV (Fe), −1.493 eV (Ga), −5.357 eV (K), −1.026 eV (Li), −0.956 eV (Mg), −0.301 eV (Na), and −3.845 eV (Ni), respectively. It can be seen that the E_{ads} of all systems are negative, indicating that the systems are stable. Different metal adsorption has different adsorption sites that are the most stable. The most stable adsorption site for Ag is located at S_{C2}, while for Co, Cr, K, Li, Mn, Na, and Ni, it is at S_H.

The most stable adsorption site for Al, Au, Bi, Ca, Cu, Fe, and Ga is at S_{C1}. The interactions between different metals and the SiC monolayer are also different. The adsorption distances of the M-adsorbed SiC systems are 2.36 Å (Ag), 2.44 Å (Al), 2.04 Å (Au), 2.45 Å (Bi), 2.87 Å (Ca), 1.30 Å (Co), 1.42 Å (Cr), 2.06 Å (Cu), 1.36 Å (Fe), 2.54 Å (Ga), 3.07 Å (K), 1.61 Å (Li), 1.46 Å eV (Mg), 2.84 Å (Na), and 1.61 Å eV (Ni), respectively. Among them, Co, Cr, Fe, Li, Mn, and Ni-adsorbed SiC systems have smaller adsorption distances, indicating that the atoms of these systems have stronger interactions with the SiC monolayer. The subsequent calculations are based on the most stable structure.

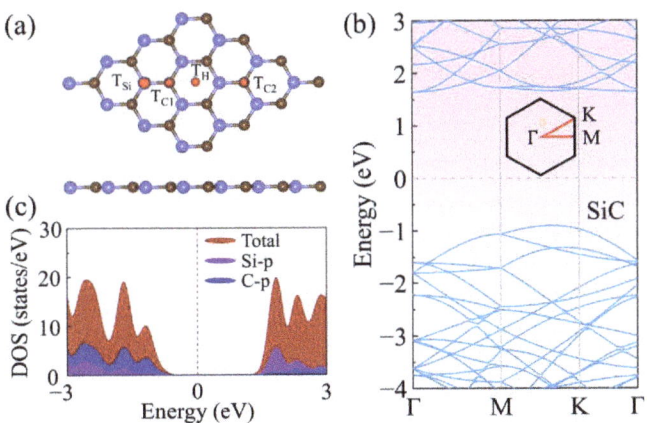

Figure 1. The (**a**) crystal structure, (**b**) band structure, and (**c**) DOS of pristine SiC monolayer. The Fermi level is shifted to zero. The blue ball represents the Si atom, and the brown ball represents the C atom.

Table 1. The adsorption position (S_x: x = H, Si, C1, C2), absorption energy (E_{ad}), adsorption height (*D*), band gap (E_g), total magnetic moments (*M*), and charge transfer (*C*) for M-adsorbed SiC systems.

Adsorption Style	S_x	E_{ad} (eV)	D (Å)	E_g (eV)	M (μ_B)	C (e)
Ag	S_{C2}	−0.193	2.36	0.521	1.00	+0.446
Al	S_{C1}	−1.683	2.44	0.659	1.00	−0.588
Au	S_{C1}	−1.706	2.04	0.837	1.00	+0.319
Bi	S_{C1}	−1.135	2.45	0	1.01	−0.109
Ca	S_{C1}	−0.111	2.87	0	1.95	−0.766
Co	S_H	−2.785	1.30	0	1.00	−0.110
Cr	S_H	−2.545	1.42	0.199	4.26	−0.560
Cu	S_{C1}	−1.388	2.06	0.705	1.00	+0.023
Fe	S_{C1}	−2.104	1.36	0.734	2.00	−0.280
Ga	S_{C1}	−1.493	2.54	0.640	1.00	−0.292
K	S_H	−5.357	3.07	0	0.99	+1.455
Li	S_H	−1.026	1.61	0	1.00	−0.867
Mn	S_H	−0.956	1.46	0.494	3.00	−0.468
Na	S_H	−0.301	2.84	0.442	1.00	−0.391
Ni	S_H	−3.845	1.61	1.754	0	+0.045

In order to investigate the effect of metal adsorption on the electronic properties of SiC systems, we studied the band structures of different metal-adsorbed silicon carbide systems, as shown in Figure 2. It can be seen that, except for the Ni-adsorbed SiC system, the spin-up and spin-down of other systems do not overlap, indicating that these systems all exhibit magnetism. Among them, the adsorption of Ag, Al, Au, Cr, Cu, Fe, Ga, Mn, and Na atoms on SiC systems result in a magnetic semiconductor, and the bandgaps of 0.521 eV (Ag), 0.659 eV (Al), 0.837 eV (Au), 0.199 eV (Cr), 0.705 eV (Cu), 0.734 eV (Fe), 0.640 eV (Ga), 0.494 eV (Mn), and 0.442 eV (Na), respectively. However, the adsorption of Ni on the SiC system leads to a non-magnetic semiconductor with a bandgap of 1.754 eV. The band gaps of the systems after adsorption are all smaller than those of the unadsorbed systems. Interestingly, the Bi, Ca, Co, K, and Li-adsorbed SiC systems exhibit semimetallic characteristics, indicating that they can be used as sensitive components in magnetic materials, electrodes, or electronic devices. Furthermore, Figure 3 describes the spin-polarized charge density of these magnetic systems. In addition to the magnetic distribution of Li and K-adsorbed SiC systems mainly distributed on the SiC monolayer, it can be clearly seen that the magnetic distribution of other systems mainly lies on the adsorbed metal and the atoms underneath it. The magnetic moments of the M-adsorbed SiC systems are 1.00 μ_B (Ag), 1.00 μ_B (Al), 1.00 μ_B (Au), 1.01 μ_B (Bi), 1.95 μ_B (Ca), 1.00 μ_B (Co), 4.26 μ_B (Cr), 1.00 μ_B (Cu), 2.00 μ_B (Fe), 1.00 μ_B (Ga), 0.99 μ_B (K), 1.00 μ_B (Li), 3.00 μ_B (Mn), and 1.00 μ_B (Na), respectively. This indicates that the adsorption of metal atoms can modulate the band structure and magnetic properties of SiC monolayers, so the M-adsorbed SiC systems can be applied to the production of spintronic devices.

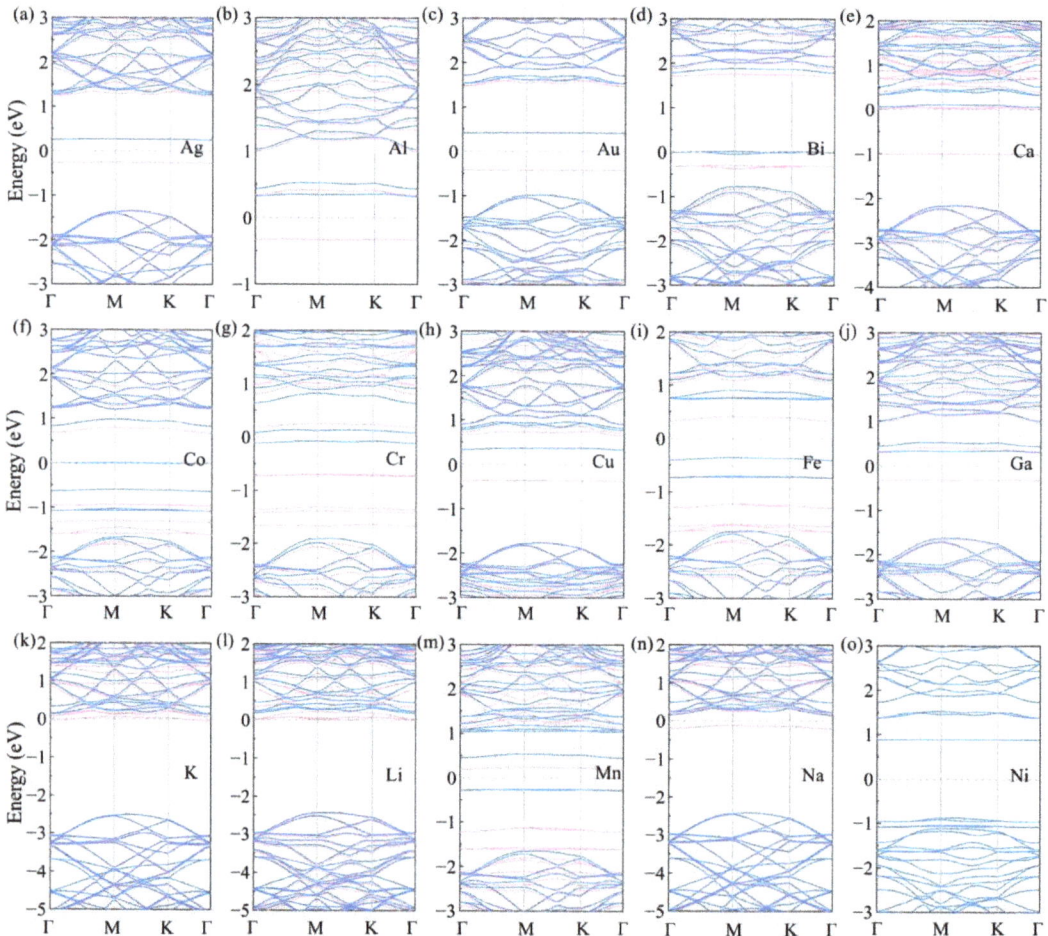

Figure 2. The band structures of (**a**) Ag, (**b**) Al, (**c**) Au, (**d**) Bi, (**e**) Ca, (**f**) Co, (**g**) Cr, (**h**) Cu, (**i**) Fe, (**j**) Ga, (**k**) K, (**l**) Li, (**m**) Mn, (**n**) Na, and (**o**) Ni atoms adsorbed on SiC. The pink lines represent the spin-up band structure, and the blue lines represent the spin-down band structure. The Fermi level is shifted to zero.

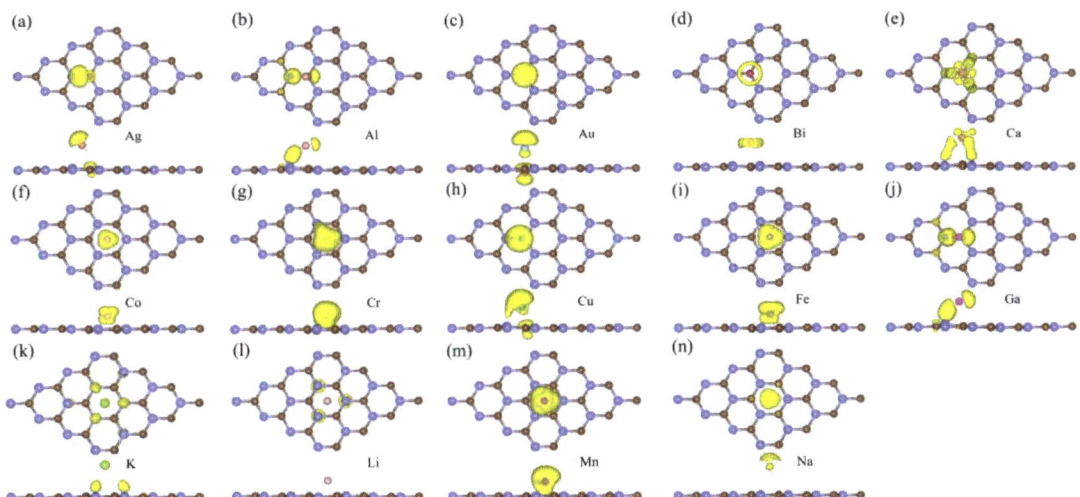

Figure 3. The spin-polarized charge density of (**a**) Ag, (**b**) Al, (**c**) Au, (**d**) Bi, (**e**) Ca, (**f**) Co, (**g**) Cr, (**h**) Cu, (**i**) Fe, (**j**) Ga, (**k**) K, (**l**) Li, (**m**) Mn, and (**n**) Na atoms adsorbed on SiC. The yellow region represents the spin-up magnetic state, and the green region represents the spin-down magnetic state.

Charge transfer is an important parameter for describing the interaction between the substrate material and the adsorbed atoms. The charge density difference (CDD) can clearly see the charge transfer and distribution, and the CDD of M-adsorbed SiC systems can be calculated using the following formula:

$$\Delta \rho = \rho_{M\text{-}SiC} - \rho_{SiC} - \rho_M \qquad (2)$$

where $\Delta \rho$ is the CDD; $\rho_{M\text{-}SiC}$ is the charge density of the M-adsorbed SiC systems; ρ_{SiC} is the charge density of the pristine SiC monolayer; and ρ_M is the charge density of an isolated metal atom. The CDD of M-adsorbed SiC systems is studied in Figure 4 of this section. It can be seen that there is a significant charge transfer between the metal atoms and the SiC monolayer. For the Ag, Au, Cu, K, and Ni-adsorbed SiC systems, the adsorbed atom is the acceptor, and the SiC monolayer is a donor. For other M-adsorbed SiC systems, the adsorbed atom is the donor, and the SiC monolayer is the acceptor. Bader charges [43–45] are used to accurately describe the amount of charge transfer. After calculation, the amount of charge transfer for various metals to the SiC monolayer are +0.446|e| (Ag), −0.588|e| (Al), +0.319|e| (Au), −0.109|e| (Bi), −0.766|e| (Ca), −0.110|e| (Co), −0.560|e| (Cr), +0.023|e| (Cu), −0.280|e| (Fe), −0.292|e| (Ga), +1.455|e| (K), −0.867|e| (Li), −0.468|e| (Mn), −0.391|e| (Na), and +0.045|e| (Ni), respectively.

The work function is a crucial parameter for evaluating the electron emission performance of optoelectronic materials, which can be calculated using the following formula:

$$\Phi = E_{vacuum} - E_{Fermi} \qquad (3)$$

where Φ, E_{vacuum}, and E_{Fermi} represent work function, vacuum level, and Fermi level, respectively. We have studied the work functions of various metals-adsorbed SiC monolayers and presented the results in Figure 5. It can be seen that the work function of pristine SiC is 4.8 eV, and the work function of the M-adsorbed SiC systems fluctuates after adsorption. Interestingly, apart from Bi-adsorbed SiC, the work functions of all other M-adsorbed SiC systems are lower than that of the pristine SiC of 3.58 eV (Ag), 4.23 eV (Al), 3.62 eV (Au), 3.10 eV (Ca), 3.40 eV (Co), 3.51 eV (Cr), 3.60 eV (Cu), 3.51 eV (Fe), 4.35 eV (Ga), 2.43 eV (K), 2.66 eV (Li), 2.61 eV (Mn), 4.66 eV (Na), and 3.76 eV (Ni), respectively. The K-adsorbed SiC system has a minimum work function of 2.43 eV, which is 47.9% lower than that of the pristine SiC. This suggests that most metal adsorption systems can be employed in electron-emitter devices.

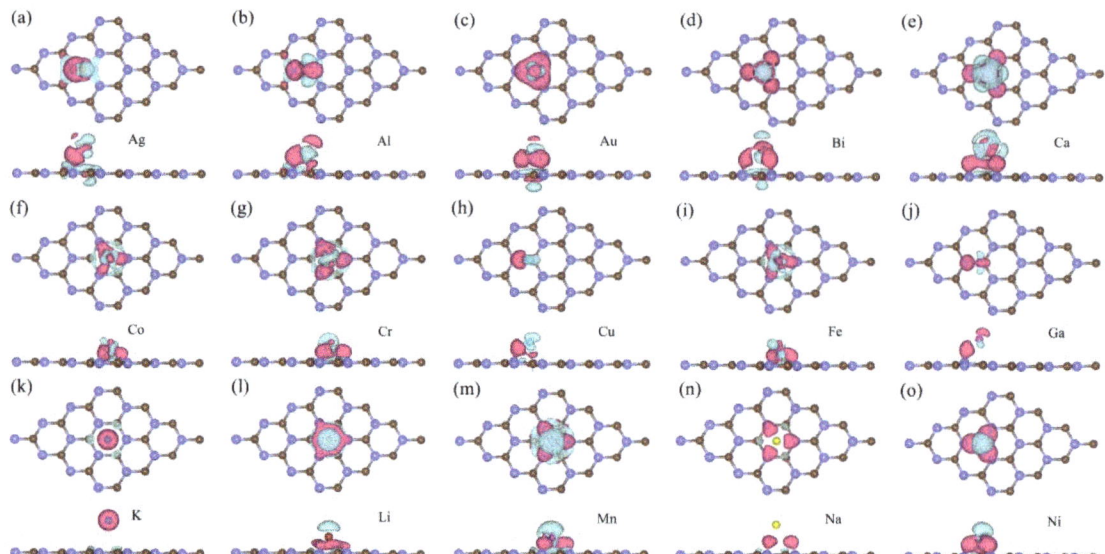

Figure 4. The CDD of (**a**) Ag, (**b**) Al, (**c**) Au, (**d**) Bi, (**e**) Ca, (**f**) Co, (**g**) Cr, (**h**) Cu, (**i**) Fe, (**j**) Ga, (**k**) K, (**l**) Li, (**m**) Mn, (**n**) Na, and (**o**) Ni atoms adsorbed on SiC monolayer. The fuchsia region represents the charge accumulation, and the cyan region represents the charge dissipation.

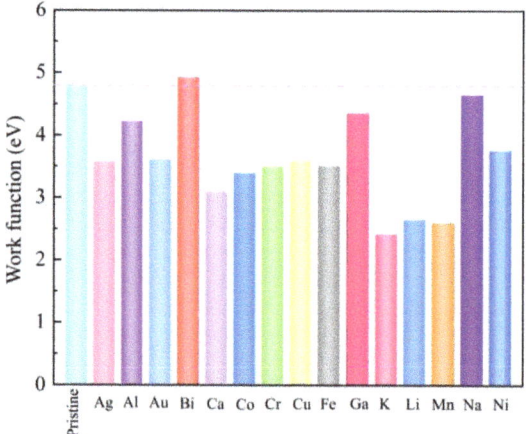

Figure 5. The work function of pristine and M-adsorbed SiC systems.

One of the important indicators for evaluating the performance of photoelectronic devices is light absorption. The optical properties of matter are represented by the transverse dielectric function $\varepsilon(\omega)$ [46,47].

$$\varepsilon(\omega) = \varepsilon_1(\omega) + i\varepsilon_2(\omega) \qquad (4)$$

where $\varepsilon_1(\omega)$ and $\varepsilon_2(\omega)$ are the real and imaginary parts of the dielectric function, and ω is the photon frequency. The $\varepsilon_2(\omega)$ can be obtained by dipole transition amplitude from the valence band (occupied states) to the conduction band (unoccupied states), while the

$\varepsilon_1(\omega)$ can be obtained from the Kramers-Kronig relationship. In additional, the absorption coefficient $\alpha(\omega)$ can be obtained from the $\varepsilon_1(\omega)$ and $\varepsilon_2(\omega)$ [48]:

$$\alpha(\omega) = \sqrt{2}\omega \left[\frac{\sqrt{\varepsilon_1^2(\omega) + \varepsilon_2^2(\omega)} - \varepsilon_1(\omega)}{2} \right]^{\frac{1}{2}} \quad (5)$$

Figure 6 shows the light absorption spectra of different metals adsorbed on the SiC monolayer. The pristine SiC mainly absorbs in the ultraviolet region and hardly absorbs in the visible light range, indicating that SiC can be used as a UV photodetector, but its application in the visible light range is limited. After metal adsorption, the absorption peak in the ultraviolet region is enhanced. The Bi, Ca, Ga, and Mn-adsorbed SiC systems show new absorption peaks in the visible light range. The Cu-adsorbed SiC system shows a strong absorption peak at 352.1 nm. These indicate that the systems can be used for solar energy conversion photovoltaic devices.

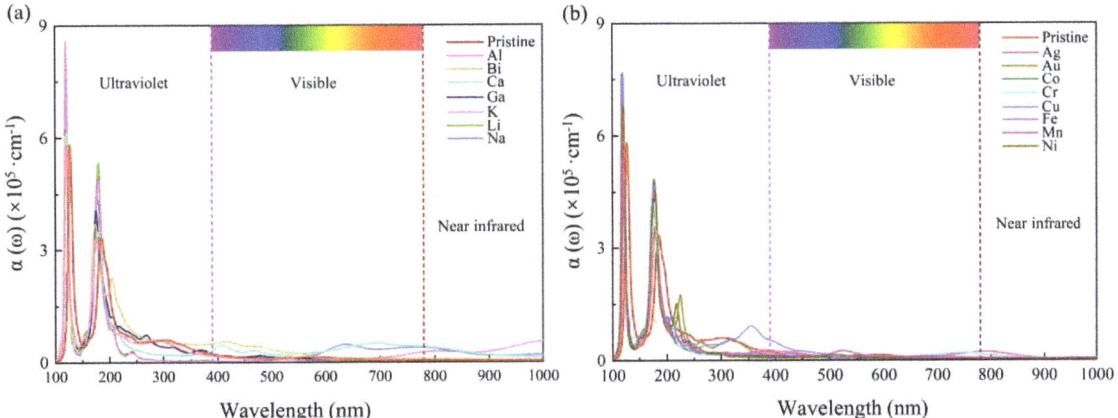

Figure 6. Light absorption spectra of pristine and M-adsorbed SiC systems: (**a**) non-transition metal adsorbed SiC monolayer and (**b**) transition metal adsorbed SiC monolayer.

4. Conclusions

The electronic, magnetic, and optical behaviors of the metals (M = Ag, Al, Au, Bi, Ca, Co, Cr, Cu, Fe, Ga, K, Li, Mn, Na, Ni) adsorbed SiC systems have been calculated based on the first-principles. The binding energy results show that the most stable adsorption sites are S_{C2} for Ag atoms, S_H for Co, Cr, K, Li, Mn, Na, and Ni atoms, and S_{C1} for Al, Au, Bi, Ca, Cu, Fe, and Ga atoms. All the M-adsorbed SiC systems are magnetic except for the Ni-adsorbed SiC system. The magnetic distribution of Li and K-adsorbed SiC systems is mainly distributed on the SiC monolayer, while the magnetic distribution of the other systems mainly lies on the adsorbed metal and the atoms underneath it. The band gap is smaller in the M-adsorbed SiC systems compared to the pristine SiC. The Ag, Al, Au, Cr, Cu, Fe, Ga, Mn, and Na-adsorbed SiC systems are magnetic semiconductors with band gaps of 0.521 eV (Ag), 0.659 eV (Al), 0.837 eV (Au), 0.199 eV (Cr), 0.705 eV (Cu), 0.734 eV (Fe), 0.640 eV (Ga), 0.494 eV (Mn), and 0.442 eV (Na), while SiC becomes semimetal after adsorption of Bi, Ca, Co, K, and Li atoms. The Bader charge results show that the adsorbed atom is more readily charged in the Ag, Au, Cu, K, and Ni-adsorbed SiC systems, while the SiC monolayer is more readily charged in the other M-adsorbed SiC systems. The work function of the K-adsorbed SiC system is 2.43 eV, which is 47.9% lower than the work function of pristine SiC and can be used in an electron emitter device. After metal atom adsorption, the absorption peak of the M-adsorbed SiC systems in the UV region is enhanced, and new absorption peaks in the visible range appeared for the Bi, Ca, Ga, and Mn-adsorbed SiC systems. These results show that the M-adsorbed SiC systems are expected to be used in spintronic devices and solar energy conversion photovoltaic devices.

Author Contributions: Conceptualization, L.J. and Z.C.; methodology, Z.C.; software, Z.C.; validation, L.J., Y.D. and Z.C.; investigation, Y.D.; resources, Z.C.; data curation, Y.D.; writing—original draft preparation, Y.D.; writing—review and editing, L.J. and Z.C. All authors have read and agreed to the published version of the manuscript.

Funding: This work was supported by the Key Research and Development Project of Shaanxi Province of China (No.2022QCY-LL-27).

Data Availability Statement: Not applicable.

Conflicts of Interest: The authors declare no conflict of interest.

References

1. Novoselov, K.S.; Geim, A.K.; Morozov, S.V.; Jiang, D.; Zhang, Y.; Dubonos, S.V.; Grigorieva, I.V.; Firsov, A.A. Electric field effect in atomically thin carbon films. *Science* **2004**, *306*, 666–669. [CrossRef]
2. Cui, Z.; Yang, K.; Shen, Y.; Yuan, Z.; Dong, Y.; Yuan, P.; Li, E. Toxic gas molecules adsorbed on intrinsic and defective WS_2: Gas sensing and detection. *Appl. Surf. Sci.* **2023**, *613*, 155978. [CrossRef]
3. Xia, S.; Wang, Y.; Shi, H.; Diao, Y.; Kan, C. Structural and electronic properties of nanoclusters (Xn, X = Au, Ag, Al; n = 1–4) adsorption on GaN/WS_2 van der Waals heterojunction: A first principle study. *Appl. Surf. Sci.* **2022**, *605*, 154716. [CrossRef]
4. Sanders, N.; Bayerl, D.; Shi, G.; Mengle, K.A.; Kioupakis, E. Electronic and optical properties of two-dimensional GaN from first-principles. *Nano Lett.* **2017**, *17*, 7345–7349. [CrossRef] [PubMed]
5. Li, H.; Dai, J.; Li, J.; Zhang, S.; Zhou, J.; Zhang, L.; Chu, W.; Chen, D.; Zhao, H.; Yang, J. Electronic structures and magnetic properties of GaN sheets and nanoribbons. *J. Phys. Chem. C* **2010**, *114*, 11390–11394. [CrossRef]
6. Dong, Y.; Li, E.; Cui, Z.; Shen, Y.; Ma, D.; Wang, F.; Yuan, Z.; Yang, K. Electronic properties and photon scattering of buckled and planar few-layer 2D GaN. *Vacuum* **2023**, *210*, 111861. [CrossRef]
7. Tang, Q.; Bao, J.; Li, Y.; Zhou, Z.; Chen, Z. Tuning band gaps of BN nanosheets and nanoribbons via interfacial dihalogen bonding and external electric field. *Nanoscale* **2014**, *6*, 8624–8634. [CrossRef] [PubMed]
8. Wang, C.; Wang, S.; Li, S.; Zhao, P.; Xing, S.; Zhuo, R.; Liang, J. Effects of strain and Al doping on monolayer h-BN: First-principles calculations. *Phys. E* **2023**, *146*, 115546. [CrossRef]
9. Ling, X.; Wang, H.; Huang, S.; Xia, F.; Dresselhaus, M.S. The renaissance of black phosphorus. *Proc. Nat. Acad. Sci. USA* **2015**, *112*, 4523–4530. [CrossRef] [PubMed]
10. Han, R.; Feng, S.; Sun, D.-M.; Cheng, H.-M. Properties and photodetector applications of two-dimensional black arsenic phosphorus and black phosphorus. *Sci. China Inform. Sci.* **2021**, *64*, 140402. [CrossRef]
11. Wang, S.; Ren, C.; Tian, H.; Yu, J.; Sun, M. MoS_2/ZnO van der Waals heterostructure as a high-efficiency water splitting photocatalyst: A first-principles study. *Phys. Chem. Chem. Phys.* **2018**, *20*, 13394–13399. [CrossRef]
12. Xia, S.; Diao, Y.; Kan, C. Electronic and optical properties of two-dimensional GaN/ZnO heterojunction tuned by different stacking configurations. *J. Colloid Interface Sci.* **2022**, *607*, 913–921. [CrossRef]
13. Zhang, L.; Cui, Z. Electronic, Magnetic, and Optical Performances of Non-Metals Doped Silicon Carbide. *Front. Chem.* **2022**, *10*, 898174. [CrossRef] [PubMed]
14. Li, S.; Sun, M.; Chou, J.-P.; Wei, J.; Xing, H.; Hu, A. First-principles calculations of the electronic properties of SiC-based bilayer and trilayer heterostructures. *Phys. Chem. Chem. Phys.* **2018**, *20*, 24726–24734. [CrossRef]
15. Zhan, J.; Yao, X.; Li, W.; Zhang, X. Tensile mechanical properties study of SiC/graphene composites based on molecular dynamics. *Comp. Mater. Sci.* **2017**, *131*, 266–274. [CrossRef]
16. Yan, W.-J.; Xie, Q.; Qin, X.-M.; Zhang, C.-H.; Zhang, Z.-Z.; Zhou, S.-Y. First-principle analysis of photoelectric properties of silicon-carbon materials with graphene-like honeycomb structure. *Comp. Mater. Sci.* **2017**, *126*, 336–343. [CrossRef]
17. Shi, Z.; Zhang, Z.; Kutana, A.; Yakobson, B.I. Predicting two-dimensional silicon carbide monolayers. *ACS Nano* **2015**, *9*, 9802–9809. [CrossRef] [PubMed]
18. Chabi, S.; Chang, H.; Xia, Y.; Zhu, Y. From graphene to silicon carbide: Ultrathin silicon carbide flakes. *Nanotechnology* **2016**, *27*, 075602. [CrossRef] [PubMed]
19. Lin, S. Light-emitting two-dimensional ultrathin silicon carbide. *J. Phys. Chem. C* **2012**, *116*, 3951–3955. [CrossRef]
20. Zhao, Q.; Xiong, Z.; Qin, Z.; Chen, L.; Wu, N.; Li, X. Tuning magnetism of monolayer GaN by vacancy and nonmagnetic chemical doping. *J. Phys. Chem. Solids* **2016**, *91*, 1–6. [CrossRef]
21. Zhao, Q.; Xiong, Z.; Luo, L.; Sun, Z.; Qin, Z.; Chen, L.; Wu, N. Design of a new two-dimensional diluted magnetic semiconductor: Mn-doped GaN monolayer. *Appl. Surf. Sci.* **2017**, *396*, 480–483. [CrossRef]
22. Chernozatonskii, L.A.; Katin, K.P.; Kochaev, A.I.; Maslov, M.M. Moiré and non-twisted sp3-hybridized structures based on hexagonal boron nitride bilayers: Ab initio insight into infrared and Raman spectra, bands structures and mechanical properties. *Appl. Surf. Sci.* **2022**, *606*, 154909. [CrossRef]

23. Cai, X.; Deng, S.; Li, L.; Hao, L. A first-principles theoretical study of the electronic and optical properties of twisted bilayer GaN structures. *J. Comput. Electron.* **2020**, *19*, 910–916. [CrossRef]
24. Mu, Y. Chemical functionalization of GaN monolayer by adatom adsorption. *J. Phys. Chem. C* **2015**, *119*, 20911–20916. [CrossRef]
25. Tang, W.; Sun, M.; Yu, J.; Chou, J.-P. Magnetism in non-metal atoms adsorbed graphene-like gallium nitride monolayers. *Appl. Surf. Sci.* **2018**, *427*, 609–612. [CrossRef]
26. Islam, S.; Lee, K.; Verma, J.; Protasenko, V.; Rouvimov, S.; Bharadwaj, S.; Xing, H.; Jena, D. MBE-grown 232–270 nm deep-UV LEDs using monolayer thin binary GaN/AlN quantum heterostructures. *Appl. Phys. Lett.* **2017**, *110*, 041108. [CrossRef]
27. Cui, Z.; Ren, K.; Zhao, Y.; Wang, X.; Shu, H.; Yu, J.; Tang, W.; Sun, M. Electronic and optical properties of van der Waals heterostructures of g-GaN and transition metal dichalcogenides. *Appl. Surf. Sci.* **2019**, *492*, 513–519. [CrossRef]
28. Ren, K.; Wang, S.; Luo, Y.; Xu, Y.; Sun, M.; Yu, J.; Tang, W. Strain-enhanced properties of van der Waals heterostructure based on blue phosphorus and g-GaN as a visible-light-driven photocatalyst for water splitting. *RSC Adv.* **2019**, *9*, 4816–4823. [CrossRef]
29. Nie, K.; Wang, X.; Mi, W. Electronic structure and magnetic properties of 3d transition-metal atom adsorbed SnO monolayers. *Appl. Surf. Sci.* **2019**, *493*, 404–410. [CrossRef]
30. Guo, J.-X.; Wu, S.-Y.; Zhong, S.-Y.; Zhang, G.-J.; Shen, G.-Q.; Yu, X.-Y. Janus WSSe monolayer adsorbed with transition-metal atoms (Fe, Co and Ni): Excellent performance for gas sensing and CO catalytic oxidation. *Appl. Surf. Sci.* **2021**, *565*, 150558. [CrossRef]
31. Cui, Z.; Zhang, S.; Wang, L.; Yang, K. Optoelectronic and magnetic properties of transition metals adsorbed Pd_2Se_3 monolayer. *Micro Nanostruct.* **2022**, *167*, 207260. [CrossRef]
32. Xu, B.; Chen, C.; Liu, X.; Ma, S.; Zhang, J.; Wang, Y.; Li, J.; Gu, Z.; Yi, L. Magnetic properties and electronic structure of 3d transition-metal atom adsorbed two-dimensional $SnSe_2$. *J. Magn. Magn. Mater.* **2022**, *562*, 169817. [CrossRef]
33. Kresse, G.; Furthmüller, J. Efficient iterative schemes for ab initio total-energy calculations using a plane-wave basis set. *Phys. Rev. B* **1996**, *54*, 11169. [CrossRef]
34. Hafner, J. Ab-initio simulations of materials using VASP: Density-functional theory and beyond. *J. Comput. Chem.* **2008**, *29*, 2044–2078. [CrossRef]
35. Hohenberg, P.; Kohn, W. Inhomogeneous Electron Gas. *Phys. Rev.* **1964**, *136*, B864–B871. [CrossRef]
36. Kohn, W.; Sham, L.J. Self-Consistent Equations Including Exchange and Correlation Effects. *Phys. Rev.* **1965**, *140*, A1133–A1138. [CrossRef]
37. Perdew, J.P.; Burke, K.; Ernzerhof, M. Generalized gradient approximation made simple. *Phys. Rev. Lett.* **1996**, *77*, 3865. [CrossRef]
38. Kresse, G.; Joubert, D. From ultrasoft pseudopotentials to the projector augmented-wave method. *Phys. Rev. B* **1999**, *59*, 1758. [CrossRef]
39. Grimme, S.; Antony, J.; Ehrlich, S.; Krieg, H. A consistent and accurate ab initio parametrization of density functional dispersion correction (DFT-D) for the 94 elements H-Pu. *J. Chem. Phys.* **2010**, *132*, 154104. [CrossRef]
40. Monkhorst, H.J.; Pack, J.D. Special points for Brillouin-zone integrations. *Phys. Rev. B* **1976**, *13*, 5188. [CrossRef]
41. Hybertsen, M.S.; Louie, S.G. Electron correlation in semiconductors and insulators: Band gaps and quasiparticle energies. *Phys. Rev. B* **1986**, *34*, 5390. [CrossRef] [PubMed]
42. Zhao, Z.; Yong, Y.; Zhou, Q.; Kuang, Y.; Li, X. Gas-sensing properties of the SiC monolayer and bilayer: A density functional theory study. *ACS Omega* **2020**, *5*, 12364–12373. [CrossRef] [PubMed]
43. Henkelman, G.; Arnaldsson, A.; Jónsson, H. A fast and robust algorithm for Bader decomposition of charge density. *Comp. Mater. Sci.* **2006**, *36*, 354–360. [CrossRef]
44. Sanville, E.; Kenny, S.D.; Smith, R.; Henkelman, G. Improved grid-based algorithm for Bader charge allocation. *J. Comput. Chem.* **2007**, *28*, 899–908. [CrossRef]
45. Tang, W.; Sanville, E.; Henkelman, G. A grid-based Bader analysis algorithm without lattice bias. *J. Phys. Condens. Matter* **2009**, *21*, 084204. [CrossRef]
46. Ehrenreich, H.; Cohen, M.H. Self-consistent field approach to the many-electron problem. *Phys. Rev.* **1959**, *115*, 786. [CrossRef]
47. Toll, J.S. Causality and the dispersion relation: Logical foundations. *Phys. Rev.* **1956**, *104*, 1760. [CrossRef]
48. Fox, M.; Bertsch, G.F. Optical properties of solids. *Am. Assoc. Phys. Teach.* **2002**, *70*, 1269–1270. [CrossRef]

Disclaimer/Publisher's Note: The statements, opinions and data contained in all publications are solely those of the individual author(s) and contributor(s) and not of MDPI and/or the editor(s). MDPI and/or the editor(s) disclaim responsibility for any injury to people or property resulting from any ideas, methods, instructions or products referred to in the content.

Article

Strain Modulation of Electronic Properties in Monolayer SnP$_2$S$_6$ and GeP$_2$S$_6$

Junlei Zhou [1], Yuzhou Gu [1], Yue-E Xie [1], Fen Qiao [1], Jiaren Yuan [2,*], Jingjing He [3], Sake Wang [4], Yangsheng Li [2,*] and Yangbo Zhou [2,*]

[1] School of Physics and Electronic Engineering, Jiangsu University, Zhenjiang 212013, China
[2] School of Physics and Materials Science, Nanchang University, Nanchang 330031, China
[3] College of Information Science and Technology, Nanjing Forestry University, Nanjing 210037, China
[4] School of Science, Jinling Institute of Technology, Nanjing 211169, China
* Correspondence: jryuan@ncu.edu.cn (J.Y.); ysli@ncu.edu.cn (Y.L.); yangbozhou@ncu.edu.cn (Y.Z.)

Abstract: In recent years, two-dimensional (2D) materials have attracted significant attention due to their distinctive properties, including exceptional mechanical flexibility and tunable electronic properties. Via the first-principles calculation, we investigate the effect of strain on the electronic properties of monolayer SnP$_2$S$_6$ and GeP$_2$S$_6$. We find that monolayer SnP$_2$S$_6$ is an indirect bandgap semiconductor, while monolayer GeP$_2$S$_6$ is a direct bandgap semiconductor. Notably, under uniform biaxial strains, SnP$_2$S$_6$ undergoes an indirect-to-direct bandgap transition at 4.0% biaxial compressive strains, while GeP$_2$S$_6$ exhibits a direct-to-indirect transition at 2.0% biaxial tensile strain. The changes in the conduction band edge can be attributed to the high-symmetry point Γ being more sensitive to strain than K. Thus, the relocation of the conduction band and valence band edges in monolayer SnP$_2$S$_6$ and GeP$_2$S$_6$ induces a direct-to-indirect and indirect-to-direct bandgap transition, respectively. Consequently, the strain is an effective band engineering scheme which is crucial for the design and development of next-generation nanoelectronic and optoelectronic devices.

Keywords: monolayer semiconductor; bandgap; electronic structure; biaxial strain; DFT calculations

Citation: Zhou, J.; Gu, Y.; Xie, Y.-E.; Qiao, F.; Yuan, J.; He, J.; Wang, S.; Li, Y.; Zhou, Y. Strain Modulation of Electronic Properties in Monolayer SnP$_2$S$_6$ and GeP$_2$S$_6$. *Inorganics* **2023**, *11*, 301. https://doi.org/10.3390/inorganics11070301

Academic Editor: Antonino Gulino

Received: 3 June 2023
Revised: 10 July 2023
Accepted: 13 July 2023
Published: 15 July 2023

Copyright: © 2023 by the authors. Licensee MDPI, Basel, Switzerland. This article is an open access article distributed under the terms and conditions of the Creative Commons Attribution (CC BY) license (https://creativecommons.org/licenses/by/4.0/).

1. Introduction

Two-dimensional (2D) materials, such as graphene [1,2], have attracted significant interest because of their exotic physical properties and potential application in nanoelectronic and optoelectronic devices [3]. However, graphene is a semimetal with zero bandgaps [4], and the absence of a bandgap will limit its application in nanoscale optoelectronic and cutting-edge ultra-fast electronic devices of the next generation [5]. Hence, the exploration of 2D materials with appropriate bandgaps is of great importance for device applications. The transition metal dichalcogenides (TMDCs) [6], hexagonal boron nitride (b-BN) [7], germanene (2D germanium) [8], and metal halogenides [9,10], among others, have garnered substantial interest in the realms of materials science, microelectronics, physics, and related fields. Among these encouraging candidates, 2D nanoporous metal chalcogen phosphates MP$_2$S$_6$ (M = metal, X = S, Se) including SnP$_2$S$_6$ and GeP$_2$S$_6$ have also captured considerable attention because of their moderate-to-wide bandgaps ranging from 1.2 eV to 3.5 eV [11]. Similar to TMDCs, SnP$_2$S$_6$ and GeP$_2$S$_6$ share properties of typical 2D materials [12], with a weak van der Waals interlayer interaction [13], a high surface-to-volume ratio and excellent optical properties [14]. Moreover, metal chalcogen phosphates MP$_2$X$_6$ have rich properties, including topological magnetism [15], ferroelectric ordering [16], photocatalytic properties [17], and H$_2$ storage and Li intercalation for batteries [11]. For example, monolayer SnP$_2$S$_6$ is an indirect bandgap semiconductor with high carrier mobility and with a bandgap of 1.1 eV, while monolayer GeP$_2$S$_6$ is a direct bandgap semiconductor with a bandgap of 1.06 eV, which shows promising potential in electronic and photoelectronic applications [18,19]. Furthermore, SnP$_2$S$_6$ and GeP$_2$S$_6$ are 2D porous materials that are important as membranes, adsorbents, catalysts, and in other chemical applications. The novel physical properties of 2D nanoporous metal chalcogen phosphate materials have the potential to expand the range of applications in nanoelectronics [11]. He et al. reported that SnP_2S_6 exhibits nonlinear optical characteristics with significant nonresonant second harmonic generation, which lays the foundation for developing novel optoelectronic devices [20].

The development of 2D materials has attracted considerable attention owing to its unique attributes, including tunable electronic properties. Strain is a promising avenue for tuning the electric properties of two-dimensional materials [21,22]. Theoretical studies have shown that transition metal disulfide compounds (TMDs) are sensitive to strain [23], and strain engineering can shift the conduction band minimum (CBM) and valence band maximum (VBM) [23–26]. It has been reported that the application of tensile strain can lead to a narrowing of the bandgap in MoS_2 in a monolayer system [27,28]. Moreover, the direct-to-indirect bandgap transition can occur in monolayer MoS_2 with 6% tensile strain [29]. The strain-tuned electronic properties in 2D materials have significant implications for the development of next-generation nanoelectronic and optoelectronic devices [30–32]. Furthermore, studies have also investigated the effects of applying pressure to layered SnP_2S_6 and GeP_2S_6. The transition from a semiconductor to a metal has been reported in layered SnP_2S_6 and GeP_2S_6 [33]. We will now shift our research focus towards investigating the effects of biaxial strain on monolayer SnP_2S_6 and GeP_2S_6. The metal chalcogen phosphates SnP_2S_6 and GeP_2S_6 with intrinsic nanoporous structures possess excellent mechanical performance, and are expected to be able to effectively tune the electronic properties under strain [34].

In this study, we investigated the ground-state monolayers of SnP_2S_6 and GeP_2S_6, revealing that monolayer SnP_2S_6 and GeP_2S_6 are thermodynamically stable. Monolayer SnP_2S_6 and GeP_2S_6 are indirect bandgap semiconductors with a 1.35 eV gap and a direct bandgap semiconductor with a gap of 1.06 eV in equilibrium, respectively. Additionally, we modulated their electronic structures by applying strain from 6% compression to 6% tension. When a 4% biaxial compression (BC) strain was applied to the monolayer SnP_2S_6, it transformed from an indirect bandgap semiconductor to a direct bandgap semiconductor. Similarly, when a 2% biaxial tension (BT) strain was applied to the monolayer GeP_2S_6, it transitioned from a direct bandgap semiconductor to an indirect bandgap semiconductor. These results provide valuable insights into the strain engineering in tunning the electronic structures of the monolayer systems.

2. Result and Discussion

SnP_2S_6 and GeP_2S_6 are members of the family of novel 2D metal thiophosphates. These monolayers are characterized by space group P312 and contain one metal cation (Sn or Ge) and one anionic $[P_2S_6]^{4-}$ unit, as illustrated in Figure 1a,b [34]. The metal cation undergoes coordination with six sulfur (S) atoms, thereby forming a hexahedral structure, whereas each phosphorus (P) atom is coordinated with three sulfur atoms in a tetrahedral arrangement, as depicted in Figure 1b. The monolayer SnP_2S_6 and GeP_2S_6 exhibit optimized lattice parameters of 6.13 Å and 5.99 Å, respectively, which is consistent with the previous result [35]. The optimized lattice parameters reflect the equilibrium distances between the atoms in the monolayers, providing insights into their structural stability and interatomic bonding characteristics. Undoubtedly, the stability of materials is a crucial consideration in materials science and applications. Therefore, we employed the finite displacement method to calculate the phonon spectra of monolayer SnP_2S_6 and monolayer GeP_2S_6 using a $4 \times 4 \times 1$ supercell via phonopy [36]. Notably, as shown in Figure 2, no imaginary phonon frequencies were found throughout the entire Brillouin zone for monolayer GeP_2S_6 and the imaginary frequency near Γ is negligible for monolayer SnP_2S_6, indicating that both structures are stable. In fact, monolayer SnP_2S_6 has been experimentally synthesized via chemical vapor transport technique along with the mechanical exfoliation method and used for preparing optoelectronic devices [37].

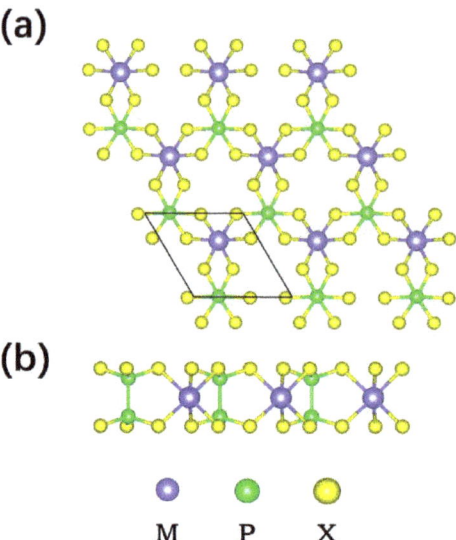

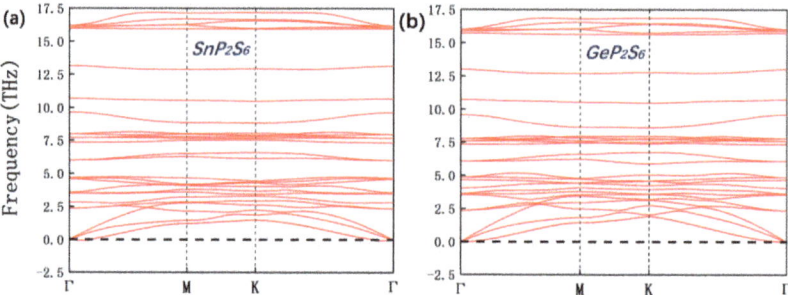

Figure 1. (a) Top and (b) side views of the monolayer MP$_2$S$_6$ are shown in the figure. The M (M = Sn, Ge) atoms, P atoms, and S atoms are represented by blue, green, and yellow spheres, respectively.

Figure 2. (a) Phonon spectrum of monolayer SnP$_2$S$_6$. (b) Phonon spectrum of monolayer GeP$_2$S$_6$.

Biaxial strain is employed in this study to investigate the impact of strain on the electronic properties of monolayer SnP$_2$S$_6$ and GeP$_2$S$_6$. Biaxial strain is defined as follows:

$$\delta = \frac{\Delta a}{a_0} \qquad (1)$$

where a_0 represents the optimized lattice constant when the structure is unstrained and Δa denotes the variation in the lattice constant after the application of a specific strain in the xy plane. Electronic property calculations are conducted over a range of δ values spanning from −6% to +6%, representing the percentage change in the lattice constant. The negative value of δ indicates compressive strain, where the lattice is compressed along the xy plane. Conversely, the positive sign refers to tensile strain, where the lattice is stretched along the xy plane. By systematically varying the strain within this range, we were able to investigate the influence of strain on the electronic properties of the monolayers. This analysis provides valuable insights into the strain-dependent behavior of the materials, including changes in band structure, bandgap, and other electronic characteristics. The comprehensive range of strain values allows us to explore the full spectrum of electronic responses in SnP$_2$S$_6$ and GeP$_2$S$_6$, enhancing our understanding of their potential applications in strain-engineered devices and electronic systems.

The electronic band structures of monolayer SnP_2S_6 and GeP_2S_6 structures under biaxial strain are investigated. Our calculations reveal that the total energy of the system exhibits a weak dependence on the applied biaxial strain, as depicted in Figure 3. The calculated total energy for the relaxed monolayer structures of SnP_2S_6 and GeP_2S_6 is determined to be −42.6808 eV and −42.4794806 eV, respectively. The maximum change in total energy is only around 0.5 eV and 0.6 eV within the considered strain range. While the impact of strain on total energy may appear to be minimal, it plays a significant role in determining the material properties. Even slight changes in total energy can result in noticeable modifications to the electronic band structure, bandgap, and other electronic characteristics of the monolayers. Thus, comprehending the influence of strain on total energy is vital for the precise prediction and control of the electronic behavior exhibited by SnP_2S_6 and GeP_2S_6 under varying strain conditions.

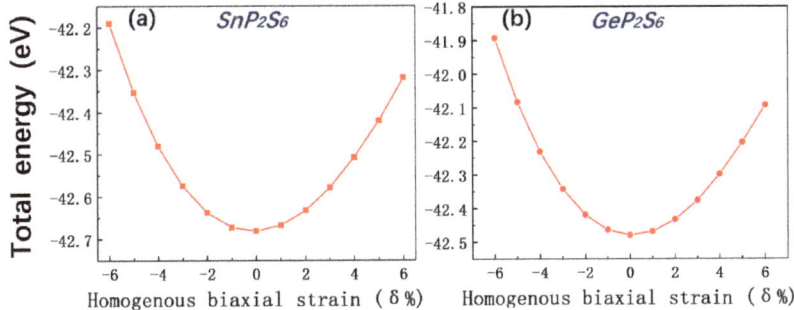

Figure 3. (**a**) Total energy vs. strain in monolayer SnP_2S_6. (**b**) Total energy vs. strain in monolayer GeP_2S_6.

The electronic band structure of the unstrained SnP_2S_6 is presented in Figure 4a, where the VBM is located at the K point, while the CBM is located at the Γ point, with an indirect bandgap of 1.346 eV, which is consistent with previous calculation results (1.35 eV) [34]. The contribution of different orbitals to the VBM and CBM are investigated by calculating the orbitals' resolved density of states. S-p orbitals contribute mostly to the VBM, while Sn-s orbitals have the highest contribution to the CBM, along with a minor contribution from S-p orbitals, as illustrated in Figure 4a. To further explore the effect of homogeneous biaxial strain on the electronic structure of SnP_2S_6, we examined the impact of a 4.0% BC strain. Under this strain condition, a notable shift in the CBM is observed from the Γ point to the K point, while the VBM remains localized at the K point. This shift results in an intriguing indirect-to-direct bandgap transition, as depicted in Figure 4b. The findings emphasize the significant role of strain engineering in modifying the electronic properties of SnP_2S_6. By manipulating the strain, it becomes possible to tailor the band structure and control the bandgap characteristics of the material. This newfound capability holds immense promise for the development of strain-tunable electronic devices, as well as for applications in fields such as optoelectronics and energy harvesting. Importantly, the information obtained from this study contributes to a deeper understanding of the underlying mechanisms governing the electronic behavior of SnP_2S_6 under strain, providing valuable insights for future research endeavors in this field.

Similarly, we investigated the electronic structure of the unstrained GeP_2S_6 and strained system, as illustrated in Figure 4c,d. It is worth noting that the VBM and CBM of the unstrained GeP_2S_6 monolayer are both located at the K point, indicating a direct bandgap semiconductor with a bandgap of 1.06 eV, which is consistent with the results of previous studies employing the same methodology (1.06 eV) [19]. As depicted in Figure 4c, the majority of the orbital contribution to the VBM results from S-p orbitals, whereas Ge-s orbital dominates the CBM along with a minor contribution from S-p orbitals. Under the influence of a 2.0% BT strain, the CBM experienced a significant shift from the K point to the Γ point, while the VBM remained unchanged at the K point. This strain-induced phenomenon led to an intriguing transition from a direct bandgap to an indirect bandgap in the material. By manipulating the strain, researchers can fine-tune the electronic properties of GeP$_2$S$_6$, enabling the design and development of strain-tailored devices with enhanced performance. In contrast, molybdenum disulfide (MoS$_2$) undergoes a semiconductor-to-metal transition under significantly large strains [38], exhibiting a pronounced shift in the valence band maximum from the high-symmetry point K to the Γ point when the tensile strain reaches 6%. This transition leads to a transformation of the initial direct bandgap into an indirect one [24]. Therefore, compared to MoS$_2$, both systems investigated in this study exhibit a stronger strain-tuning effect under a relatively small strain, which holds significant implications for researchers studying two-dimensional and nanoscale electronic materials.

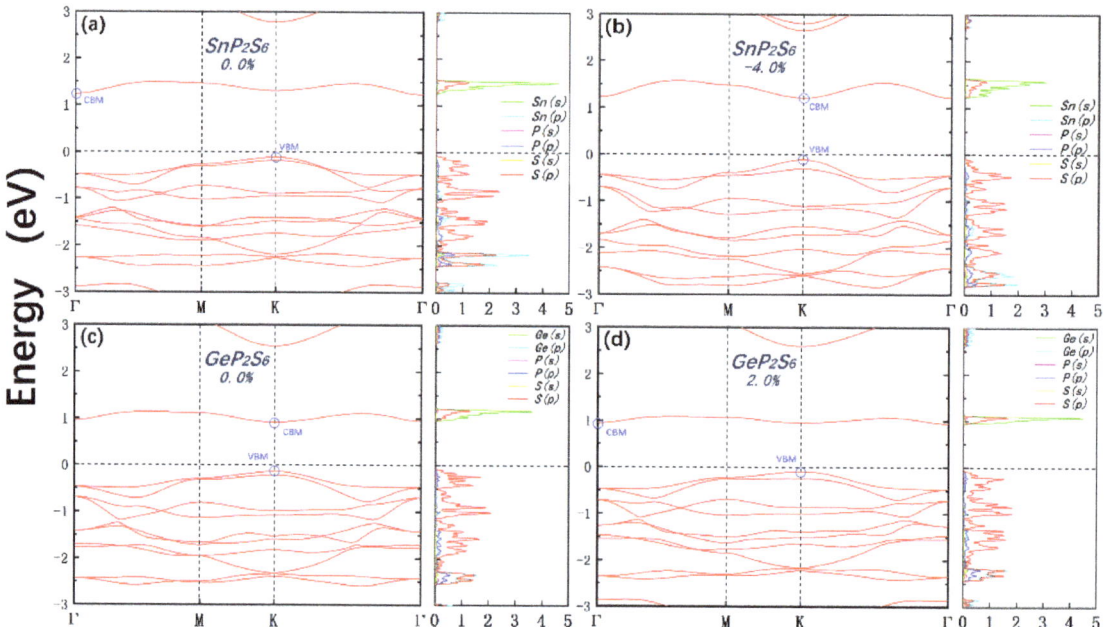

Figure 4. (**a**,**b**) Monolayer SnP$_2$S$_6$ electronic band structure and angular-momentum resolved density of states at 0.0% and 4.0% BC strain, respectively. (**c**,**d**) The monolayer GeP$_2$S$_6$ at 0.0% and 2.0% BT strain.

In order to gain deeper insights into the orbital contributions to the band structure, we conducted an analysis of the projected band structure in the absence of strain. Our calculations reveal distinct orbital contributions to the conduction and valence band edges in both materials. It is evident that the conduction band edge of system SnP$_2$S$_6$ is primarily contributed by Sn-s orbitals, with additional small contributions from S-p_z orbitals and a small contribution from S-p_x/p_y orbitals, while the valence band edge is predominantly attributed to the S-p_x orbital with a small contribution from the S-p_y orbital in Figure 5a–d. Similarly, the conduction band edge of the GeP$_2$S$_6$ monolayer is primarily influenced by the Ge-s orbital, while the valence band edge is predominantly affected by the S-p_x orbital with a small contribution from S-p_z orbitals, as depicted in Figure 6a–d. This analysis provides valuable insights into the orbital characteristics responsible for the electronic properties of SnP$_2$S$_6$ and GeP$_2$S$_6$. Understanding the specific orbital contributions aids the elucidation of the underlying mechanisms that govern the material's electronic behavior and facilitates the design and optimization of electronic devices based on these materials.

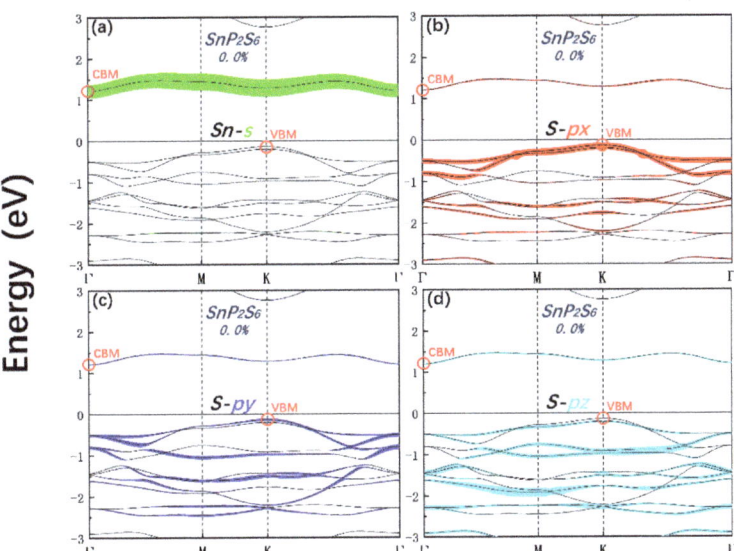

Figure 5. The orbital-resolved band structure of SnP$_2$S$_6$ under unstrained conditions for the Sn-s (**a**), S-px (**b**), S-py (**c**), and S-pz (**d**) orbitals is as follows: Red circles represent the corresponding VBM and CBM in the band structures.

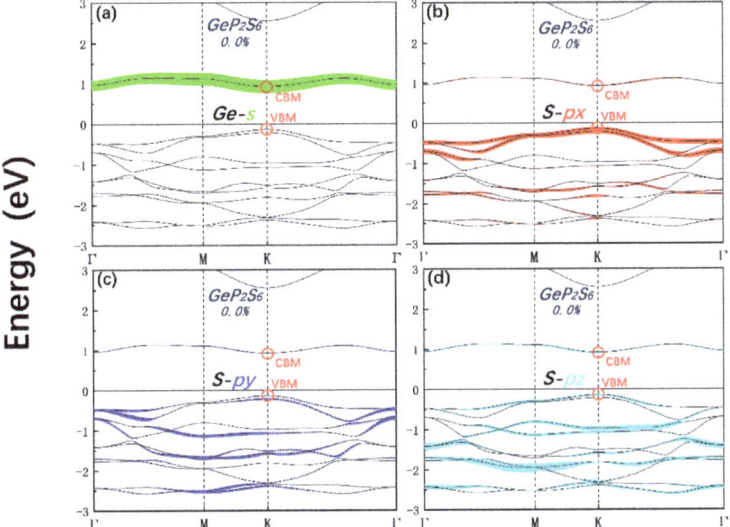

Figure 6. The orbital-resolved band structure of GeP$_2$S$_6$ under unstrained conditions for the Ge-s (**a**), S-px (**b**), S-py (**c**), and S-pz (**d**) orbitals is as follows: Red circles represent the corresponding VBM and CBM in the band structures.

According to Figure 4a–d, the conversion from an indirect to direct bandgap for monolayer SnP$_2$S$_6$ and the transformation from a direct to indirect bandgap for monolayer GeP$_2$S$_6$ is primarily caused by the shift in the conduction band minimum. To further investigate this phenomenon, we plotted the variation in the energy of the conduction band edge state at the Γ point (CB-Γ) and at the K point (CB-K) as a function of strain as depicted in Figure 7a,b. It is evident that both systems exhibit an approximately linear decrease in energy with increasing tensile strain and an increase in energy with increasing compressive strain, for both the Γ and K points. The high-symmetry point Γ exhibits a higher rate under strain as compared to the K point, indicating the band edge near Γ point is more sensitive to strain. Hence, the positional relationship the energy bands at Γ and K points reverses under certain strain. In the case of system SnP_2S_6, when the applied strain is less than −4%, $E_{CB-\Gamma}$ is smaller than E_{CB-K} located at K. However, when the applied strain exceeds −4%, $E_{CB-\Gamma}$ becomes larger than E_{CB-K}, and the CBM shifts to the Γ point. Conversely, in the case of GeP_2S_6, when the strain is less than 1%, $E_{CB-\Gamma}$ is larger than E_{CB-K}, and the CBM is located at the K point. However, when the strain exceeds 2%, $E_{CB-\Gamma}$ becomes smaller than E_{CB-K}, and the CBM shifts to the Γ point. These findings indicate that the strain-induced changes in the position of the CBM play a crucial role in determining the bandgap characteristics of SnP_2S_6 and GeP_2S_6. The precise control of strain allows for the manipulation of the electronic properties of these materials, providing opportunities for tailoring their behavior and optimizing their performance in various electronic applications.

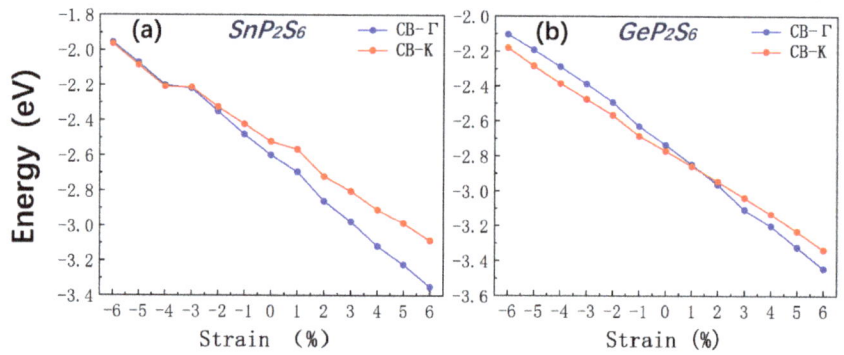

Figure 7. The variation in CBM energy at the Γ point (CB-Γ) and at the K point (CB-K) with strain for SnP_2S_6 (**a**) and GeP_2S_6 (**b**).

Figure 7 reveals that the two systems have the same energy level shift tendency. The energy level of the Γ point and K point for the conduction band in both materials increases with increasing compressive strain and decreases with increasing tensile strain. Furthermore, the energy level of the Γ point for the conduction band in both materials is more sensitive than that of the K point. The difference is that the energy band types of monolayer SnP_2S_6 and monolayer GeP_2S_6 at zero strain are inconsistent. Monolayer SnP_2S_6 with an indirect bandgap will transform into a direct bandgap semiconductor; since the Γ point is more sensitive, the energy of the Γ point rises faster than that of the K point, and the energy positions are reversed at 2% compressive strain. However, monolayer GeP_2S_6 with a direct bandgap will transform into an indirect bandgap semiconductor since the energy of the Γ point drops faster than that of the K point, and the energy positions are reversed at 4% tensile strain. Furthermore, we believe these insights can be applied to other 2D materials because other 2D systems also have similar phenomenon with different strain sensitivities at different k points. Therefore, these insights provide crucial references and guidance for researchers exploring strain modulation effects in other two-dimensional materials.

Furthermore, we investigated the modulation of the bandgap as a function of strain along distinct symmetry axes, namely K-K and K-Γ. Analyzing the bandgap behavior, we observed that in the case of SnP_2S_6, when the strain is less than −4%, the bandgaps along the K-K and K-Γ directions exhibit remarkable proximity, suggesting a nearly degenerate nature. This finding implies that the band structure is relatively insensitive to strain in this range. However, as the strain exceeds −4%, an intriguing trend emerges. The disparity between the bandgaps along K-K and K-Γ becomes increasingly prominent, as depicted in Figure 8a. Specifically, we found that the bandgap undergoes only subtle changes under compressive strain but exhibits significant variations under tensile strain. This strain-induced modification in the bandgap characteristics highlights the potential for tailoring the electronic properties of SnP_2S_6 through strain engineering. On the other hand, our investigation of monolayer GeP_2S_6 revealed a different pattern. Under compressive strain, the distinction between the bandgaps along the K-K and K-Γ directions is relatively more pronounced compared to that of SnP_2S_6. This implies that the band structure of GeP_2S_6 is more sensitive to compressive strain, leading to a significant difference in the bandgap behavior between these two symmetry directions. Figure 8b illustrates the substantial changes in the bandgap under both compressive and tensile strains, further highlighting the potential for strain engineering in manipulating the electronic properties of GeP_2S_6. By precisely controlling the strain, researchers can fine-tune the bandgap characteristics of SnP_2S_6 and GeP_2S_6, enabling the development of tailored electronic devices with enhanced performance. The ability to manipulate the band structure through strain engineering opens up new avenues for exploring novel device concepts and applications. Furthermore, our study contributes to the broader understanding of strain-dependent phenomena in 2D materials. By unraveling the intricate relationship between strain and bandgap variations, we gain deeper insights into the underlying mechanisms governing the electronic behavior of these materials. This knowledge serves as a valuable foundation for future research endeavors in the field of strain engineering and paves the way for the discovery and optimization of other 2D materials with tailored electronic properties. The distinct variations observed under different strain conditions offer exciting possibilities for bandgap engineering and the design of advanced electronic and optoelectronic devices.

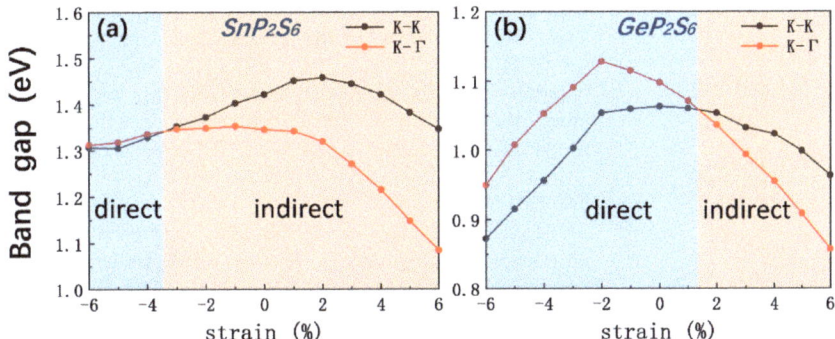

Figure 8. (a) Variations in bandgap along the high-symmetry directions K-K and K-Γ in SnP_2S_6. (b) Bandgap fluctuations along the high-symmetry directions K-K and K-Γ in GeP_2S_6.

In addition, we have also investigated the electronic structures considering spin–orbit coupling (SOC). When SOC is included as shown in Figure 9, the band splitting is observed for some bands, especially at high-symmetry points in both systems. However, no noticeable change in the electronic properties near the band edges is observed. The band structures with SOC are highly similar to those without SOC. Specifically, an indirect bandgap semiconductor with a bandgap of 1.35 eV remains for SnP_2S_6 with SOC, while for GeP_2S_6 with SOC, a direct bandgap semiconductor with a bandgap of 1.06 eV remains. Moreover, the positions of the conduction band minimum and valence band maximum remain the same as those without SOC. Based on these results, we conclude that the inclusion of SOC does not significantly affect the electronic structures as discussed in the previous sections. Therefore, SOC was not considered in the previous analysis.

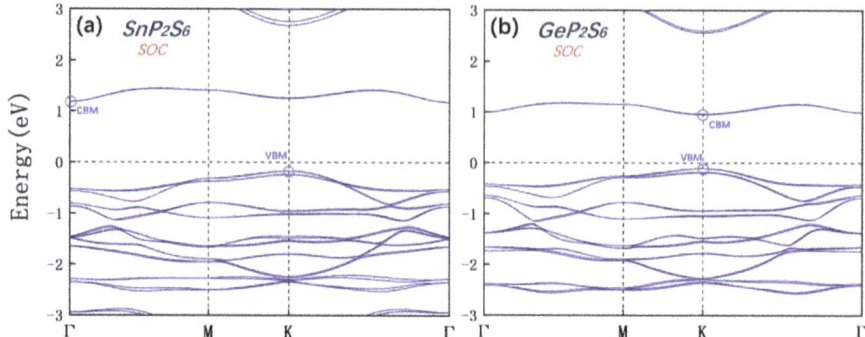

Figure 9. (**a**) Monolayer SnP_2S_6 electronic band structure with SOC. (**b**) Monolayer SnP_2S_6 electronic band structure with SOC.

3. Calculation Method

Density functional theory (DFT) [39] calculation was performed to investigate the electronic properties of monolayer metal chalcogen phosphates of SnP_2S_6 and GeP_2S_6 using the Vienna ab initio simulation package (VASP) [40]. The projector-augmented wave method was adopted with a cutoff energy of 450 eV [40]. The Perdew–Burke–Ernzerhof (PBE) [41] exchange–correlation function, based on the generalized gradient approximation (GGA), was employed in this study. To minimize the number of interactions between vertically periodic layers, a vacuum layer of 25 Å was set in the z direction for the monolayer structures. A $15 \times 15 \times 1$ Monkhorst–Pack k-grid mesh was employed for Brillouin zone sampling. The phonon spectra of monolayer SnP_2S_6 and monolayer GeP_2S_6 were calculated using the finite displacement method with a low repetition rate of a $4 \times 4 \times 1$ supercell. The atomic positions were iteratively optimized until the magnitude of the Hellmann–Feynman forces acting on all atoms was below 0.01 eV/Å, ensuring the convergence of the system.

4. Conclusions

In summary, we have conducted a comprehensive investigation on the electronic properties of monolayer SnP_2S_6 and GeP_2S_6 under biaxial strain. The results demonstrate that the monolayer SnP_2S_6 with an indirect bandgap of 1.346 eV undergoes a transition from an indirect to a direct bandgap under 4.0% uniform BC strain. Additionally, we report a direct-to-indirect bandgap transition in the monolayer semiconductor GeP_2S_6 under a uniform BT strain of 2.0%. The transition occurs because the high-symmetry point Γ is more sensitive to strain than K. Overall, our findings highlight the significant role of strain engineering in modulating the electronic properties of monolayer SnP_2S_6 and GeP_2S_6. The ability to induce bandgap transitions through strain opens up avenues for designing innovative devices with enhanced performance and functionality. Further exploration of strain effects and the development of strain engineering techniques will undoubtedly contribute to the advancement of next-generation electronic and optoelectronic devices based on these materials.

Author Contributions: Conceptualization, J.Z.; methodology, Y.G.; validation, Y.-E.X. and F.Q.; formal analysis, J.Z. and Y.G.; investigation, J.Z. and Y.G.; resources, J.H.; data curation, J.Z.; writing—original draft, J.Z. and J.Y.; writing—review and editing, S.W., Y.L. and Y.Z.; supervision, J.Y., Y.L. and Y.Z. All authors have read and agreed to the published version of the manuscript.

Funding: This work was supported by the National Natural Science Foundation of China (Nos. 62264010, 12264026, 12004142, 62201268), the Natural Science Foundation of Jiangxi Province, China (Nos. 20212BAB211023, 20224BAB211013) and the Innovation and Entrepreneurship Leading Talent Plan of Jiangxi Province (Nos. Jxsq2023101068).

Data Availability Statement: Not applicable.

Conflicts of Interest: The authors declare no conflict of interest.

References

1. Geim, A.K.; Novoselov, K.S. The rise of graphene. *Nat. Mater.* **2007**, *6*, 183–191. [CrossRef] [PubMed]
2. Novoselov, K.S.; Geim, A.K.; Morozov, S.V.; Jiang, D.; Katsnelson, M.I.; Grigorieva, I.V.; Dubonos, S.V.; Firsov, A.A. Two-dimensional gas of massless Dirac fermions in graphene. *Nature* **2005**, *438*, 197–200. [CrossRef] [PubMed]
3. Tang, Q.; Zhou, Z.; Chen, Z. Innovation and discovery of graphene-like materials via density-functional theory computations. *Wiley Interdiscip. Rev. Comput. Mol. Sci.* **2015**, *5*, 360–379. [CrossRef]
4. Neto, A.C.; Guinea, F.; Peres, N.M.; Novoselov, K.S.; Geim, A.K. The electronic properties of graphene. *Rev. Mod. Phys.* **2009**, *81*, 109. [CrossRef]
5. Novoselov, K.S.; Geim, A.K.; Morozov, S.V.; Jiang, D.; Zhang, Y.; Dubonos, S.V.; Grigorieva, I.V.; Firsov, A.A. Electric field effect in atomically thin carbon films. *Science* **2004**, *306*, 666–669. [CrossRef] [PubMed]
6. Kanoun, M.B.; Goumri-Said, S. Tailoring optoelectronic properties of monolayer transition metal dichalcogenide through alloying. *Materialia* **2020**, *12*, 100708. [CrossRef]
7. Chettri, B.; Patra, P.K.; Lalmuanchhana; Lalhriatzuala; Verma, S.; Rao, B.K.; Verma, M.L.; Thakur, V.; Kumar, N.; Hieu, N.N.; et al. Induced magnetic states upon electron–hole injection at B and N sites of hexagonal boron nitride bilayer: A density functional theory study. *Int. J. Quantum Chem.* **2021**, *121*, e26680. [CrossRef]
8. Raya, S.S.; Ansari, A.S.; Shong, B. Adsorption of gas molecules on graphene, silicene, and germanene: A comparative first-principles study. *Surf. Interfaces* **2021**, *24*, 101054. [CrossRef]
9. Lu, F.; Wang, W.; Luo, X.; Xie, X.; Cheng, Y.; Dong, H.; Liu, H.; Wang, W.-H. A class of monolayer metal halogenides MX_2: Electronic structures and band alignments. *Appl. Phys. Lett.* **2016**, *108*, 132104. [CrossRef]
10. Liu, P.; Lu, F.; Wu, M.; Luo, X.; Cheng, Y.; Wang, X.-W.; Wang, W.; Wang, W.-H.; Liu, H.; Cho, K. Electronic structures and band alignments of monolayer metal trihalide semiconductors MX_3. *J. Mater. Chem. C* **2017**, *5*, 9066–9071. [CrossRef]
11. Susner, M.A.; Chyasnavichyus, M.; McGuire, M.A.; Ganesh, P.; Maksymovych, P. Metal thio-and selenophosphates as multifunctional van der Waals layered materials. *Adv. Mater.* **2017**, *29*, 1602852. [CrossRef]
12. Li, X.; Zhu, H. Two-dimensional MoS_2: Properties, preparation, and applications. *J. Mater.* **2015**, *1*, 33–44. [CrossRef]
13. Gurarslan, A.; Jiao, S.; Li, T.D.; Li, G.; Yu, Y.; Gao, Y.; Riedo, E.; Xu, Z.; Cao, L. Van der waals force isolation of monolayer MoS_2. *Adv. Mater.* **2016**, *28*, 10055–10060. [CrossRef]
14. Molina-Sánchez, A.; Hummer, K.; Wirtz, L. Vibrational and optical properties of MoS_2: From monolayer to bulk. *Surf. Sci. Rep.* **2015**, *70*, 554–586. [CrossRef]
15. Sugita, Y.; Miyake, T.; Motome, Y. Multiple dirac cones and topological magnetism in honeycomb-monolayer transition metal trichalcogenides. *Phys. Rev. B* **2018**, *97*, 035125. [CrossRef]
16. Chyasnavichyus, M.; Susner, M.A.; Ievlev, A.V.; Eliseev, E.A.; Kalinin, S.V.; Balke, N.; Morozovska, A.N.; McGuire, M.A.; Maksymovych, P. Size-effect in layered ferrielectric $CuInP_2S_6$. *Appl. Phys. Lett.* **2016**, *109*, 172901. [CrossRef]
17. Zhang, X.; Zhao, X.; Wu, D.; Jing, Y.; Zhou, Z. $MnPSe_3$ monolayer: A promising 2d visible-light photohydrolytic catalyst with high carrier mobility. *Adv. Sci.* **2016**, *3*, 1600062. [CrossRef]
18. Jing, Y.; Zhou, Z.; Zhang, J.; Huang, C.; Li, Y.; Wang, F. SnP_2S_6 monolayer: A promising 2D semiconductor for photocatalytic water splitting. *Phys. Chem. Chem. Phys.* **2019**, *21*, 21064–21069. [CrossRef]
19. Liu, J.; Shen, Y.; Lv, L.; Wang, X.; Zhou, M.; Zheng, Y.; Zhou, Z. Rational design of porous GeP_2S_6 monolayer for photocatalytic water splitting under the irradiation of visible light. *Flatchem* **2021**, *30*, 100296. [CrossRef]
20. He, J.; Lee, S.H.; Naccarato, F.; Brunin, G.; Zu, R.; Wang, Y.; Miao, L.; Wang, H.; Alem, N.; Hautier, G.; et al. SnP_2S_6: A Promising Infrared Nonlinear Optical Crystal with Strong Nonresonant Second Harmonic Generation and Phase-Matchability. *ACS Photonics* **2022**, *9*, 1724–1732. [CrossRef]
21. Jacobsen, R.S.; Andersen, K.N.; Borel, P.I.; Fage-Pedersen, J.; Frandsen, L.H.; Hansen, O.; Kristensen, M.; Lavrinenko, A.V.; Moulin, G.; Ou, H.; et al. Strained silicon as a new electro-optic material. *Nature* **2006**, *441*, 199–202. [CrossRef] [PubMed]
22. Falvo, M.R.; Clary, G.J.; Taylor, R.M., II; Chi, V.; Brooks, F.P., Jr.; Washburn, S.; Superfine, R. Bending and buckling of carbon nanotubes under large strain. *Nature* **1997**, *389*, 582–584. [CrossRef]
23. Johari, P.; Shenoy, V.B. Tuning the electronic properties of semiconducting transition metal dichalcogenides by applying mechanical strains. *ACS Nano* **2012**, *6*, 5449–5456. [CrossRef]
24. Shi, H.; Pan, H.; Zhang, Y.W.; Yakobson, B.I. Quasiparticle band structures and optical properties of strained monolayer MoS_2 and WS_2. *Phys. Rev. B* **2013**, *87*, 155304. [CrossRef]
25. Horzum, S.; Sahin, H.; Cahangirov, S.; Cudazzo, P.; Rubio, A.; Serin, T.; Peeters, F.M. Phonon softening and direct to indirect band gap crossover in strained single-layer $MoSe_2$. *Phys. Rev. B* **2013**, *87*, 125415. [CrossRef]
26. Lee, Y.; Cho, S.B.; Chung, Y.C. Tunable indirect to direct band gap transition of monolayer Sc_2CO_2 by the strain effect. *ACS Appl. Mater. Interfaces* **2014**, *6*, 14724–14728. [CrossRef]
27. He, K.; Poole, C.; Mak, K.F.; Shan, J. Experimental demonstration of continuous electronic structure tuning via strain in atomically thin MoS_2. *Nano Lett.* **2013**, *13*, 2931–2936. [CrossRef] [PubMed]
28. Conley, H.J.; Wang, B.; Ziegler, J.I.; Haglund, R.F., Jr.; Pantelides, S.T.; Bolotin, K.I. Bandgap engineering of strained monolayer and bilayer MoS_2. *Nano Lett.* **2013**, *13*, 3626–3630. [CrossRef]

29. Yun, W.S.; Han, S.W.; Hong, S.C.; Kim, I.G.; Lee, J.D. Thickness and strain effects on electronic structures of transition metal dichalcogenides: 2H-MX_2 semiconductors (M = Mo, W; X = S, Se, Te). *Phys. Rev. B* **2012**, *85*, 033305. [CrossRef]
30. Zhong, F.; Wang, H.; Wang, Z.; Wang, Y.; He, T.; Wu, P.; Peng, M.; Wang, H.; Xu, T.; Wang, F.; et al. Recent progress and challenges on two-dimensional material photodetectors from the perspective of advanced characterization technologies. *Nano Res.* **2021**, *14*, 1840–1862. [CrossRef]
31. Wang, L.; Boutilier, M.S.H.; Kidambi, P.R.; Jang, D.; Hadjiconstantinou, N.G.; Karnik, R. Fundamental transport mechanisms, fabrication and potential applications of nanoporous atomically thin membranes. *Nat. Nanotechnol.* **2017**, *12*, 509–522. [CrossRef]
32. Zhao, J.; Ma, D.; Wang, C.; Guo, Z.; Zhang, B.; Li, J.; Nie, G.; Xie, N.; Zhang, H. Recent advances in anisotropic two-dimensional materials and device applications. *Nano Res.* **2021**, *14*, 897–919. [CrossRef]
33. Haborets, V.; Glukhov, K.; Banys, J.; Vysochanskii, Y. Layered GeP_2S_6, GeP_2Se_6, GeP_2Te_6, SnP_2S_6, SnP_2Se_6, and SnP_2Te_6 Polar Crystals with Semiconductor–Metal Transitions Induced by Pressure or Chemical Composition. *Integr. Ferroelectr.* **2021**, *220*, 90–99. [CrossRef]
34. Lin, M.; Liu, P.; Wu, M.; Cheng, Y.; Liu, H.; Cho, K.; Wang, W.H.; Lu, F. Two-dimensional nanoporous metal chalcogenophosphates MP_2X_6 with high electron mobilities. *Appl. Surf. Sci.* **2019**, *493*, 1334–1339. [CrossRef]
35. Whangbo, M.H.; Brec, R.; Ouvrard, G.; Rouxel, J. Reduction sites of transition-metal phosphorus trichalcogenides MPX_3. *Inorg. Chem.* **1985**, *24*, 2459–2461. [CrossRef]
36. Togo, A.; Tanaka, I. First principles phonon calculations in materials science. *Scr. Mater.* **2015**, *108*, 1–5. [CrossRef]
37. Zhang, Y.; Wang, F.; Feng, X.; Sun, Z.; Su, J.; Zhao, M.; Wang, S.; Hu, X.; Zhai, T. Inversion symmetry broken 2D SnP_2S_6 with strong nonlinear optical response. *Nano Res.* **2022**, *15*, 2391–2398. [CrossRef]
38. Scalise, E.; Houssa, M.; Pourtois, G.; Afanas'ev, V.; Stesmans, A. Strain-induced semiconductor to metal transition in the two-dimensional honeycomb structure of MoS_2. *Nano Res.* **2012**, *5*, 43–48. [CrossRef]
39. Kohn, W.; Sham, L.J. Self-consistent equations including exchange and correlation effects. *Phys. Rev.* **1965**, *140*, A1133. [CrossRef]
40. Kresse, G.; Furthmüller, J. Efficient iterative schemes for ab initio total-energy calculations using a plane-wave basis set. *Phys. Rev. B* **1996**, *54*, 11169. [CrossRef]
41. Perdew, J.P.; Burke, K.; Ernzerhof, M. Generalized gradient approximation made simple. *Phys. Rev. Lett.* **1996**, *77*, 3865. [CrossRef] [PubMed]

Disclaimer/Publisher's Note: The statements, opinions and data contained in all publications are solely those of the individual author(s) and contributor(s) and not of MDPI and/or the editor(s). MDPI and/or the editor(s) disclaim responsibility for any injury to people or property resulting from any ideas, methods, instructions or products referred to in the content.

Article

A First-Principle Study of Two-Dimensional Boron Nitride Polymorph with Tunable Magnetism

Liping Qiao [1], Zhongqi Ma [1], Fulong Yan [1], Sake Wang [2] and Qingyang Fan [3,*]

[1] Team of Micro & Nano Sensor Technology and Application in High-Altitude Regions, School of Information Engineering, Xizang Minzu University, Xianyang 712082, China; lpqiao@126.com (L.Q.); mazhongqi_my@163.com (Z.M.); yanfulong_xzmu@163.com (F.Y.)
[2] College of Science, Jinling Institute of Technology, Nanjing 211169, China; isaacwang@jit.edu.cn
[3] College of Information and Control Engineering, Xi'an University of Architecture and Technology, Xi'an 710055, China
* Correspondence: qyfan_xidian@163.com

Citation: Qiao, L.; Ma, Z.; Yan, F.; Wang, S.; Fan, Q. A First-Principle Study of Two-Dimensional Boron Nitride Polymorph with Tunable Magnetism. *Inorganics* **2024**, *12*, 59. https://doi.org/10.3390/inorganics12020059

Academic Editor: Jean-François Halet

Received: 25 December 2023
Revised: 30 January 2024
Accepted: 30 January 2024
Published: 15 February 2024

Copyright: © 2024 by the authors. Licensee MDPI, Basel, Switzerland. This article is an open access article distributed under the terms and conditions of the Creative Commons Attribution (CC BY) license (https://creativecommons.org/licenses/by/4.0/).

Abstract: Using the first-principles calculation, two doping two-dimensional (2D) BN (boron nitride) polymorphs are constructed in this work. The two doping 2D BN polymorphs B_5N_6Al and B_5N_6C sheets are thermally stable under 500 K. All the B_6N_6, B_5N_6Al, and B_5N_6C sheets are semiconductor materials with indirect band gaps on the basis of a hybrid functional. The anisotropic calculation results indicate that Young's modulus (E) and Poisson's ratio (v) of the B_6N_6, B_5N_6Al, and B_5N_6C sheets are anisotropic in the xy plane. In addition, the magnetic properties of the B_6N_6, B_5N_6Al, and B_5N_6C sheets have also been investigated. According to the calculation of the magnetic properties, B_6N_6 sheet does not exhibit magnetism, while it shows weak magnetism after doping carbon atom to the BN sheet. This paper explores the influence mechanism of doping different atoms on the basic physical properties of two-dimensional BN sheets. It not only constructs a relationship between structure and performance but also provides theoretical support for the performance regulation of BN materials.

Keywords: two-dimensional (2D) BN sheet; electronic band structure; density of states; anisotropic properties; magnetism

1. Introduction

Graphene is a planar 2D carbon allotrope with single atomic thickness. Due to its special electronic and magnetic characteristics, graphene is considered to be a revolutionary material for multiple facilities, such as high-speed electronic devices, thermal and conductivity-enhancing composite materials, sensors, RF logic devices, transparent electrodes, etc. [1–6]. In recent years, novel theoretical two-dimensional materials based on first-principles calculations have shown many interesting physical and chemical properties [7–12]. As isoelectronic bodies, BN and carbon also have rich and colorful physical properties and polymorphs [13–18].

Research on low-dimensional boron nitride nanomaterials is important in the field of materials science [19–24], especially because their excellent chemical properties and thermal stability have been widely studied [25]. They can enhance the mechanical enhancement and thermal conductivity of various crystal structures, for instance, polymers, ceramics, and metals [26–28]. Research has shown that graphene-like h-BN sheets have remarkable electronic and optical characteristics [29] and show a wide band gap and intense absorption capacity in the ultraviolet region. In other studies in the literature, it was shown that at low doping rates, B-N prefers to replace *sp* hybrid carbon on the chain than hexagonal. At high doping rates, it first attacks the hexagonal structure and then the chain.

Qi et al. [30] predicted 2D BxNy ($1 < x/y \leq 2$) sheets using the density functional theory (DFT) [31,32]. B_5N_3 and B_7N_5 sheets possess enough low enthalpy of formation and outstanding dynamic stability, which makes them possible to be found in experiments. Unlike previous BN, both B_5N_3 and B_7N_5 sheets exhibit narrow band gaps of 1.99 eV and 2.40 eV, respectively. Two-dimensional orthorhombic boron nitride crystal, named o-B_2N_2, was designed by Demirci et al. [23]. The stability of o-B_2N_2 at room temperature and ambient pressure has been confirmed. o-B_2N_2 is a semiconductor with a direct band gap of 1.70 eV. An appropriate band gap makes this structure exhibit higher light absorption in the visible light range along the armchair direction. The in-plane stiffness of o-B_2N_2 is also very close to that of hexagonal BN. Based on the DFT, Fan et al. [33] first proposed and studied in detail a new 2D *Pmma*-BN sheet. The stability of *Pmma*-BN sheet was demonstrated by phonon spectroscopy and ab initio molecular dynamics simulations at 300 and 500 K. Uniaxial strain has weakened the ZT of the *Pmma*-BN sheet and led to a decrease in thermoelectric conversion efficiency. Anota et al. [34] reported boron nitride nanosheets containing homonuclear boron bonds and found that the proportion of boron atoms in nanosheets is related to conductivity.

By inserting *sp*-hybridized BN bonds into a monolayer h-BN structure, Li et al. [24] designed BNyne, Bndiyne, and BNtriyne. To explore the influence mechanism on the physical properties of doped 2D boron nitride, B_6N_6 sheets (called BNyne in ref. [24]) with doping Al or C atoms are proposed in this work, named B_5N_6Al sheet and B_5N_6C sheet, respectively. Both carbon and aluminum have the advantages of being cheap and easy to obtain. The doping of C element is selected to improve the mechanical properties of B_6N_6, and the selection of Al element is to observe the influence of metal elements on the electrical properties of semiconductor materials. We verify the structural stability of B_5N_6Al sheet and B_5N_6C sheet from mechanical and thermal perspectives and study their elementary physical characteristics based on first-principles calculations. B_5N_6Al and B_5N_6C are both indirect band gap semiconductor materials, which are the same as B_6N_6. The band gap width of B_5N_6C is significantly reduced. More importantly, the spin-up and spin-down electronic band structures indicate that B_6N_6 sheet does not exhibit magnetism. But after adding carbon to the BN sheet, it changes from a nonmagnetic material to a magnetic material.

2. Results and Discussion

Two-dimensional boron nitride structures and the crystal structure model doped with aluminum and carbon atoms are shown in Figure 1, together with the configurations of h-BN, o-B_2N_2 [23], B_5N_3 [30], and B_7N_5 [30]. Here, light blue, light pink, green, and light red represent boron atoms, nitrogen atoms, aluminum atoms, and carbon atoms, respectively. h-BN is a hexagonal network-layered crystal composed of nitrogen atoms and boron atoms; its layered structure is similar to graphite. oB_2N_2 is a two-dimensional monolayer of boron nitride in an orthorhombic structure with a B-B and N-N double-atom structure. h-BN and oB_2N_2 are only composed of six-membered rings, while the crystal structure of B_5N_3 is composed of seven-membered rings and five-membered rings. B_7N_5 contains seven-membered rings, six-membered rings, and five-membered rings. By inserting B-N bonds into h-BN sheet, Li et al. [24] proposed two-dimensional B_6N_6, as shown in Figure 1. After being fully optimized, all of them—B_6N_6 sheet, B_5N_6Al sheet, and B_5N_6C sheet—retain their planar structure. All the unit cells of three B_6N_6, B_5N_6Al, and B_5N_6C sheets contain twelve atoms. The boron and nitrogen atoms in B_6N_6 sheet are half the same, and the other two doping models replace one B atom.

The lattice parameters of the three B_6N_6, B_5N_6Al, and B_5N_6C sheets are listed in Table 1. The optimized lattice parameter of B_6N_6 is a = 6.20 Å. After doping with Al and C atoms, the symmetry of the crystal structure changes, and the lattice parameters a and b of B_5N_6Al and B_5N_6C are no longer equal. In B_6N_6 sheet structure, due to the larger radius of aluminum atoms, the lattice parameters of B_5N_6Al increase, while the lattice parameters of B_5N_6C do not change much. However, the radius difference between carbon atoms, boron atoms, and nitrogen atoms is not significant.

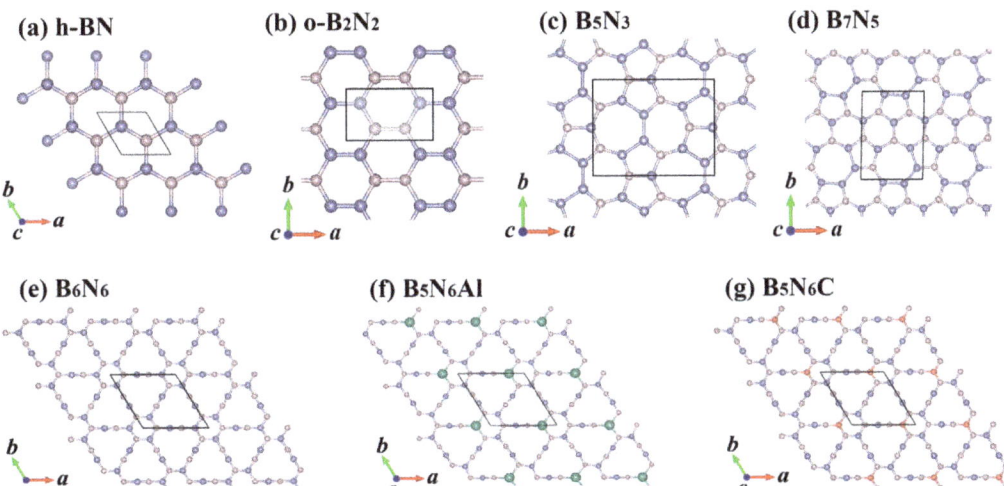

Figure 1. Crystal structures of the h-BN (**a**), o-B$_2$N$_2$ (**b**), B$_5$N$_3$ (**c**), B$_7$N$_5$ (**d**), B$_6$N$_6$ (**e**), B$_5$N$_6$Al (**f**), and B$_5$N$_6$C (**g**). Light blue, light pink, green, and light red represent boron atoms, nitrogen atoms, aluminum atoms, and carbon atoms, respectively.

Table 1. Lattice constant (Å) and bond length (Å) of B$_6$N$_6$, B$_5$N$_6$Al, and B$_5$N$_6$C.

	a	b	b1	b2	b3	b4	b5	b6
B$_6$N$_6$	6.20		1.275	1.278	1.094			
B$_5$N$_6$Al	7.40	7.18	1.796	1.451	1.465	1.390	1.264	1.374
B$_5$N$_6$C	6.86	5.93	1.381	1.451	1.441	1.380	1.260	1.391

The decrease in lattice parameters is caused by structural distortion in the six-membered ring in the structure, and the details of this are shown in Figure 2. The bond lengths of B$_6$N$_6$, B$_5$N$_6$Al, and B$_5$N$_6$C sheets are also listed in Table 1. The doping of atoms makes the bond lengths in both B$_5$N$_6$Al and B$_5$N$_6$C sheets increase to six different bond lengths, which is three more than in B$_6$N$_6$ sheet, and the angles of the six-membered ring have also become irregular. In B$_6$N$_6$ sheet, the bond length of the six-membered ring is uniform, the bond angle is 120°, the maximum bond length is 1.278 Å, and the shortest bond length is 1.094 Å. For B$_5$N$_6$Al sheet, the six-membered ring has undergone serious distortion, and the bond angle and bond length have changed. The maximum bond length of B$_5$N$_6$Al is 1.796 Å, which is approximately 41% higher than that of B$_6$N$_6$. The shortest bond length is 1.264 Å. Regarding B$_5$N$_6$C sheet, the six-membered ring has a small degree of distortion, and the maximum bond length is 1.451 Å, which is approximately 14% longer than that of B$_6$N$_6$, and the shortest bond length is 1.260 Å.

Ab initio molecular dynamic (AIMD) simulations were performed to evaluate the thermal stabilities of B$_5$N$_6$Al and B$_5$N$_6$C sheets under 500 K. The supercell of B$_5$N$_6$Al and B$_5$N$_6$C sheets before and after 5 ps simulation at a temperature of 500 K is also shown in Figure 3. The supercells of B$_5$N$_6$Al and B$_5$N$_6$C did not change significantly, and the B-N, N-Al and B-C bonds were not broken. The total energy fluctuation was also maintained at a stable level, indicating that B$_5$N$_6$Al and B$_5$N$_6$C have thermal stability at 500 K.

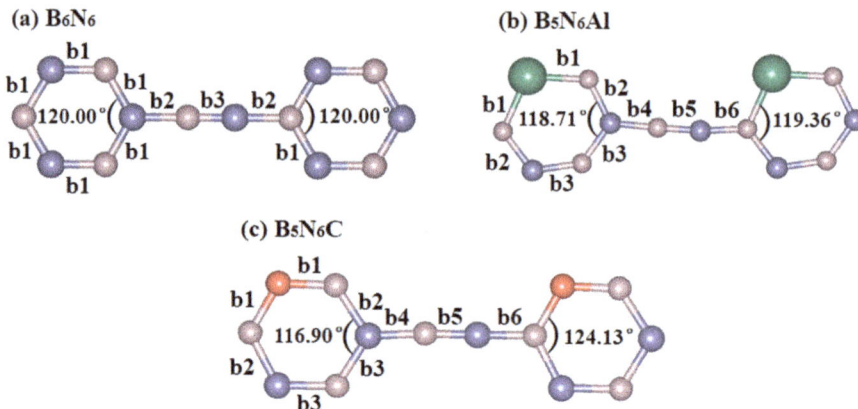

Figure 2. The bond length and bond angle of B_6N_6 (**a**), B_5N_6Al (**b**), and B_5N_6C (**c**).

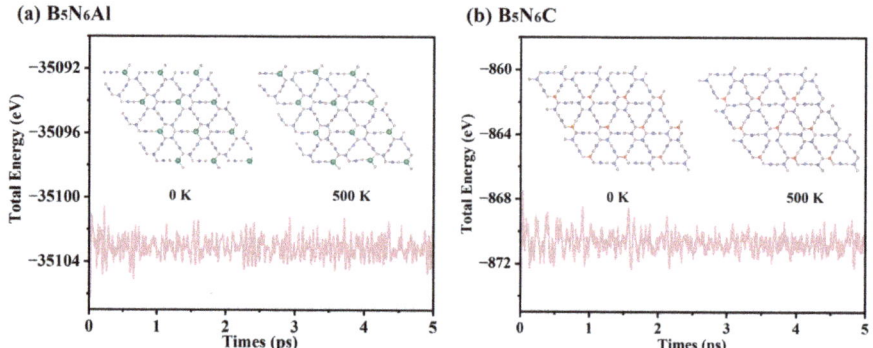

Figure 3. Total energy fluctuations of B_5AlN_6 and B_5N_6C as a function of the AIMDs simulation at 500 K.

The mechanical stability of B_6N_6, B_5N_6Al, and B_5N_6C sheets was also estimated, and the mechanical stability was evaluated by estimating the elastic constant. For 2D materials, the elastic constants can be obtained by the energy–strain method. That is, by applying different strains to the structure, the total energy of the system relative to the ground state energy change can be calculated. The relationship between the strain and the resulting energy change is expressed as [35]:

$$E(\varepsilon) - E_0 = \frac{1}{2}C_{11}\varepsilon_{xx}^2 + \frac{1}{2}C_{22}\varepsilon_{yy}^2 + C_{12}\varepsilon_{xx}\varepsilon_{yy} + 2C_{44}\varepsilon_{xy}^2 \quad (1)$$

where E_0 and $E(\varepsilon)$ represent the ground state configuration and total energy after strain application, respectively. ε denotes the strain; xx and yy denote the direction. By fitting the energy curve corresponding to the strain, C_{11}, C_{22}, C_{12}, and C_{44} as elastic constants can be obtained. The Born–Huang criteria [36] are a necessary condition for the elastic constants of 2D materials with mechanical stability, that is, $C_{11}C_{22} - C_{12}^2 > 0$ and $C_{44} > 0$. The elastic constants of B_6N_6, B_5N_6Al, and B_5N_6C sheets are shown in Table 2, which obviously meet the mechanical stability conditions. The $C_{11} = C_{22} = 180.98$ N/m of B_6N_6, which is lower than that of h-BN ($C_{11} = C_{22} = 290$ N/m [37]) and *pmma* BN ($C_{11} = 195$ N/m, $C_{22} = 256$ N/m).

Table 2. Elastic constants C_{ij} (N/m), Young's modulus E (N/m), and Poisson's ratio v of B_6N_6, B_5N_6Al, and P-$62m$ B_5N_6C.

	C_{11}	C_{12}	C_{22}	C_{66}	E_x	E_y	v_x	v_y
B_6N_6	180.98	82.09	180.98	49.45	141.18	141.18	0.46	0.46
B_5N_6Al	152.14	86.84	126.11	44.78	97.83	76.19	0.69	0.57
B_5N_6C	201.78	103.48	204.66	53.35	149.46	151.59	0.50	0.51

Compared with B_6N_6, the C_{11} of B_5N_6Al is reduced by approximately 16%, and the C_{22} is reduced by approximately 30%, which indicates that the ability to resist deformation along the x and y directions is weakened. That is, the Al atom plays a weakening role in the mechanical properties of B_6N_6. However, the addition of carbon atoms leads to an improvement in mechanical properties. The elastic constants C_{11} and C_{22} of B_5N_6C are increased by approximately 12~13%. Young's modulus along the x and y directions E_x and E_y of B_5N_6C is also greater than that of B_6N_6, which may be due to the fact that carbon atoms are more likely to form stable covalent bonds. With the addition of aluminum and carbon atoms, B_5N_6Al and B_5N_6C sheets have lower and higher elastic constants than B_6N_6 sheets, respectively. Thus, the doping of carbon and aluminum atoms has different influence mechanisms on the mechanical properties of B_6N_6 sheets. The former significantly improves the elastic constants and elastic modulus, while the latter weakens the ability of B_6N_6 sheets to resist deformation and makes them easier to compress. Poisson's ratios of B_5N_6Al and B_5N_6C sheets are 0.69 and 0.50 along the x-axis and 0.57 and 0.51 along the y-axis, respectively, which are higher than those of B_6N_6 sheets.

On the basis of the elastic constants, the in-plane E and v along any direction θ can be taken as [35]:

$$E(\theta) = \frac{C_{11}C_{22} - C_{12}^2}{C_{11}\alpha^4 + C_{22}\beta^4 + (\frac{C_{11}C_{22} - C_{12}^2}{C_{44}} - 2C_{12})\alpha^2\beta^2} \quad (2)$$

$$v(\theta) = -\frac{(C_{11} + C_{22} - \frac{C_{11}C_{22} - C_{12}^2}{C_{44}})\alpha^2\beta^2 - C_{12}(\alpha^4 + \beta^4)}{C_{11}\alpha^4 + C_{22}\beta^4 + (\frac{C_{11}C_{22} - C_{12}^2}{C_{44}} - 2C_{12})\alpha^2\beta^2} \quad (3)$$

where $\alpha = \sin\theta$; $\beta = \cos\theta$. In order to further explore the effect of carbon and aluminum atoms doping on the elastic anisotropy of B_6N_6 sheets, the angle-dependent in-plane E and v of B_6N_6, B_5N_6Al, and B_5N_6C sheets are illustrated in Figure 4. Further, 0° and 90° represent the directions of orthorhombic unit cells along the x- and y-axes, respectively. With the addition of aluminum, the maximum value of the E of B_5N_6Al sheet is lower than that of B_6N_6 sheet, but the minimum value is greater than that of B_6N_6 sheet, while the E of B_5N_6C sheet is larger than that of B_6N_6 sheet, and it shows an in-plane stiffness superior to B_6N_6 sheet. For Poisson's ratio, B_6N_6 sheet has a higher Poisson ratio than that of B_5N_6Al and B_5N_6C sheet, indicating that B_6N_6 sheet is more likely to expand laterally than B_5N_6Al and B_5N_6C sheet when tension is applied. The smallest in-plane Poisson's ratio of B_6N_6, B_5N_6Al, and B_5N_6C sheets occurs along the x-axis. When they are stretched diagonally, a maximum value occurs.

The orientation dependence of 2D Young's modulus is closer to the sphere, indicating that the weaker the anisotropy, the smaller the difference between the maximum and minimum values. On the contrary, the more it deviates from the spherical shape, the stronger the anisotropy. The minimum value of Young's modulus of B_6N_6 is 2.36 N/m, and the maximum value is 144 N/m, showing strong anisotropy. The minimum value of Young's modulus of B_5N_6Al is 4.63 N/m, and the maximum value is 92 N/m; anisotropy is weakened compared with B_6N_6. The minimum Young's modulus of B_5N_6C is 59 N/m, and the maximum is 152 N/m, showing the weakest Young's modulus anisotropy. The ratio of the maximum to minimum Young's modulus of B_6N_6, B_5N_6Al, and B_5N_6C sheets is 60.91, 19.94, and 2.57, respectively. These values clearly and intuitively show the influence of atomic doping on the anisotropy of Young's modulus. Both carbon atoms and aluminum atoms can weaken anisotropy, especially when the effect of carbon atoms is more significant. The ratio of the 2D extreme values of Poisson's ratio of B_6N_6, B_5N_6Al, and B_5N_6C sheets is 2.2, 1.72, and 1.60, respectively, indicating that the order of Poisson's ratio anisotropy is $B_6N_6 > B_5N_6Al > B_5N_6C$, which is consistent with the order of anisotropy of Young's modulus.

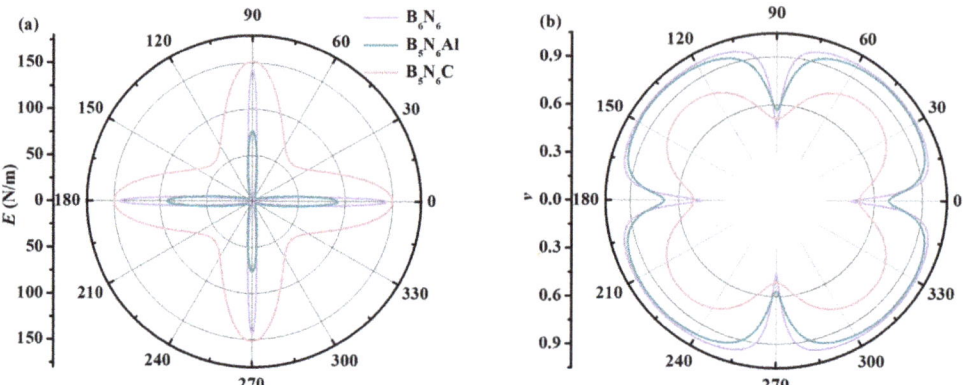

Figure 4. Orientation dependencies of Young's modulus (**a**) and Poisson's ratio (**b**) for B_6N_6, B_5N_6Al, and B_5N_6C.

In addition to in-plane stiffness, the stress neutralization of B_6N_6, B_5N_6Al, and B_5N_6C sheets under uniaxial strain is further analyzed, and the detailed results are presented in Figure 5. The ultimate strength of B_6N_6 sheet is 18.49 N/m loaded along the x (y)-axis with 19.5% uniaxial strain, while with the addition of aluminum atoms, although the ultimate strength has decreased slightly, the strain along the x-axis is not significantly different (18.5%). The x- and y-axis of B_5N_6C sheet have similar and good uniaxial strain limitations. The ultimate strength along the x- and y-axis of B_5N_6C sheet under the calculated maximum strength is 17.0%, respectively. For B_5N_6C sheet, the ultimate strength is 18.14 N/m and 18.50 N/m along the x or y direction, respectively. Therefore, B_6N_6 sheet and B_6N_6 sheet doped with carbon atom have good mechanical properties compared to B_5N_6Al sheet, which may be suitable for nanomechanical applications.

The band structures of the B_6N_6, B_5N_6Al, and B_5N_6C sheets are shown in Figure 6. The related results show that all the B_6N_6, B_5N_6Al, and B_5N_6C sheets exhibit semiconductor characters, and the band gaps of B_6N_6 and B_5N_6Al sheets are 5.684 and 5.418 eV, respectively. After doping with Al atoms, the variation in the band gap of B_5N_6Al sheet is relatively small, while after doping the C atoms, the change in the band gap of B_5N_6C sheet is significant. To study the magnetism of these materials, we also calculated spin-up and spin-down band structures of B_6N_6, B_5N_6Al, and B_5N_6C sheets. From the band structures analysis, there is no difference between spin-up and spin-down for B_6N_6 and B_5N_6Al sheets, which indicates that they do not conform to the characteristics of magnetic materials. Nevertheless, after doping with carbon atoms, there is a significant difference in the band structure of spin-up and spin-down. The conduction band minimum (CBM) and valence band maximum (VBM) of the spin-up band structure are located at the X and Γ points, respectively, with a wide indirect band gap of 5.181 eV, while the CBM and VBM of the -up appear at the Y and X points, with the narrow band gap being only 1.062 eV.

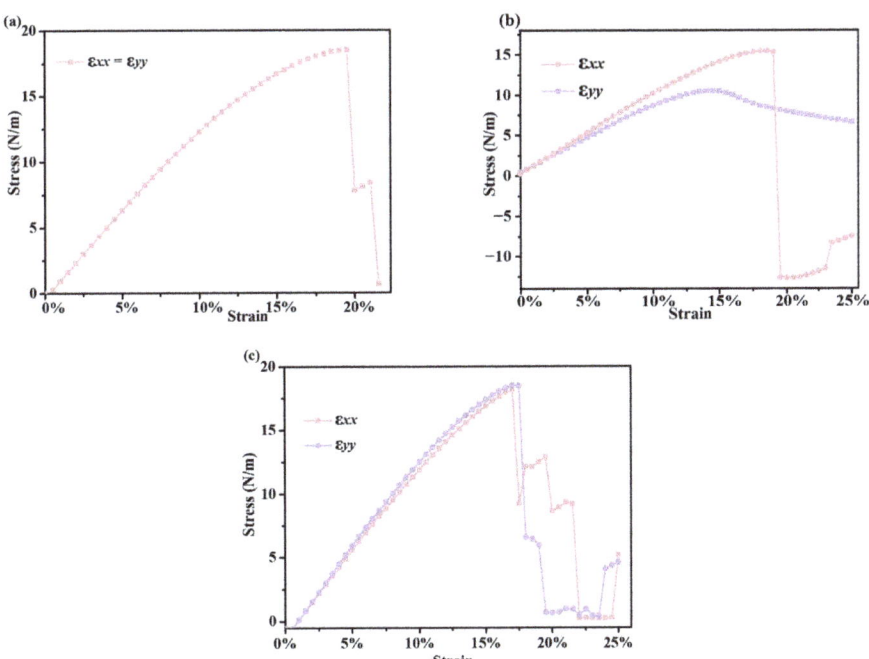

Figure 5. Strain–stress curves for the uniaxial tensile strains of B_6N_6 (**a**), B_5N_6Al (**b**), and B_5N_6C (**c**).

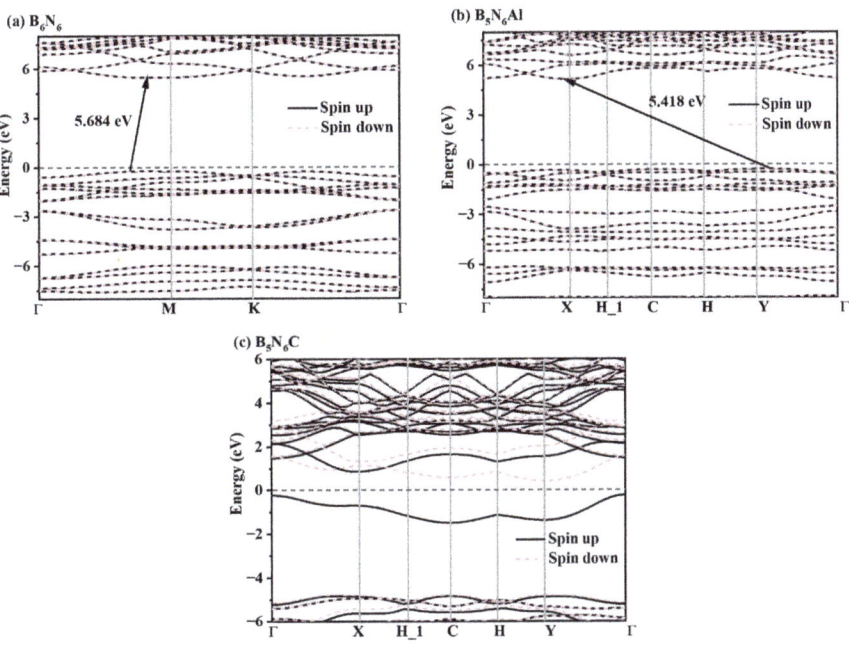

Figure 6. Electronic band structures of B_6N_6 (**a**), B_5N_6Al (**b**), and B_5N_6C (**c**). Black and red lines in the band structure present spin-up electrons and spin-down electrons, respectively.

We then explored the electronic properties of B_6N_6, B_5N_6Al, and B_5N_6C sheets by using the density of states (DOSs) and band decomposition charge density (BDCD) at the CBM and VBM. To verify the magnetism of these materials, their spin-up and spin-down

DOSs were simulated and are shown in Figure 7. For B_6N_6 sheet, the spin-up and spin-down density of states of N and B atoms are the same. In the energy range of 0~−20 eV, the contribution of N atoms is more than that of B atoms. In 5~10 eV, the contribution of the energy band is mainly from B atoms, that is, the conduction band is mainly contributed by N atoms, while the valence band is mainly contributed by B atoms. After doping with Al atoms, the spin-up and spin-down DOSs of nitrogen, boron, and aluminum atoms are also the same. It can be seen that the electrons contributed by Al atoms have always been the least, which may be due to the smallest proportion of Al atoms. At 0~−20 eV, N atoms contribute the most, which is similar to B_6N_6. At 5~8 eV, electrons mainly come from B atoms, while after doping with carbon atoms, both boron, nitrogen, and carbon atoms in B_5N_6C sheet exhibit differences, which indicates that B_5N_6C has a modicum of magnetism. For B_5N_6C sheet, the spin-up electrons come from B and C atoms in the energy window of −2~0 eV, which indicates that the narrowing of the spin-up band gap is mainly due to B and C atoms, independent of N atoms. In the energy ranges of −3~−27 eV and 0~5 eV, the N atom contributes the most to the electronic band, while at 5~10 eV, and the electrons are mainly derived from the B atom.

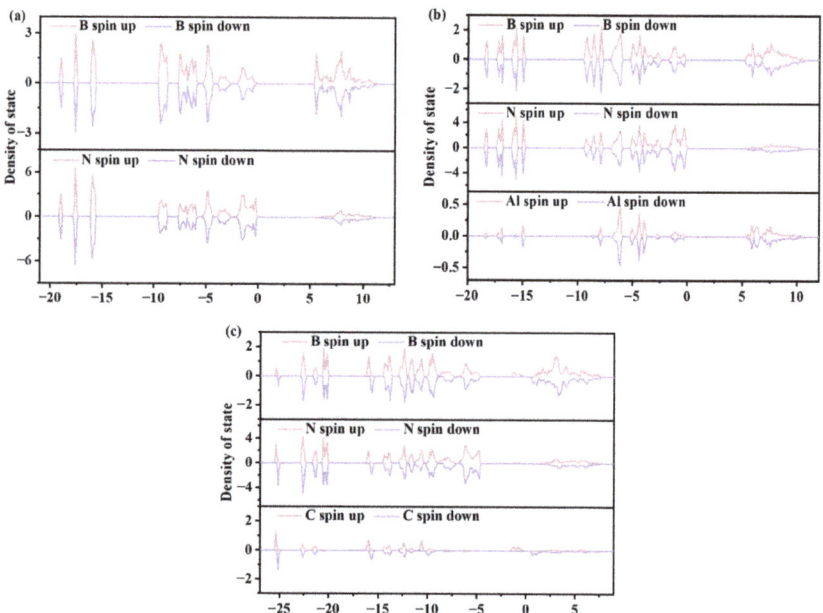

Figure 7. Spin-up and spin-down density of states for B_6N_6 (**a**), B_5N_6Al (**b**), and B_5N_6C (**c**).

The band decomposed charge densities (BDCDs) at the CBM and VBM of B_6N_6, B_5N_6Al, and B_5N_6C sheets are shown in Figure 8. As shown in the BDCDs, the electrons at the VBM of B_6N_6 sheet are mainly from the B atom, while the electrons at the CBM are mainly from the N atom. With the doping of Al atoms, the main contribution atoms of the CBM and VBM have reversed from before doping, while the Al atoms have no contribution to the CBM and VBM. For B_5N_6Al sheet, the VBM of B_5N_6Al sheet mainly comes from N atoms, and the CBM of B_6N_6 sheet mainly comes from B atoms. However, with the addition of the C atom, the distribution of electrons near the B_5N_6C sheet's CBM and VBM is irregular compared to that of the previous B_6N_6 sheet and B_5N_6Al sheet. After doping the C atom, the main contribution of the CBM and VBM is participated by doped atoms, but this also occurs mainly by nitrogen and boron atoms. In addition, the charge density at the spin-up VBM and CBM is also significantly different from that at the spin-down VBM and CBM. The charge at the spin-up VBM is mainly contributed by B and C atoms, while the charge at the spin-down VBM is mainly contributed by N atoms. More rarely, negative charge appears at the spin-down CBM, as shown in the green part of Figure 8c.

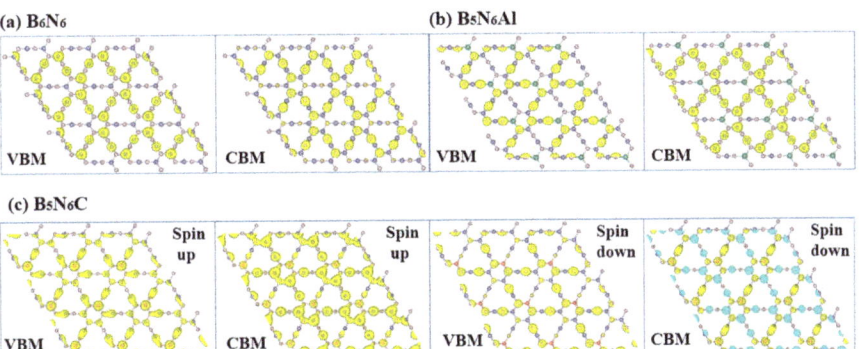

Figure 8. The band decomposed charge densities at the CBM and VBM of B_6N_6 (**a**), B_5N_6Al (**b**), and B_5N_6C (**c**).

Figure 9 illustrates the electron localization function (ELF) of B_6N_6, B_5N_6Al, and B_5N_6C. The electrons are well localized around the B-N bond, Al-N bond, and C-N bond. The strength of the covalent bond in B_6N_6 and B_5N_6Al sheets is similar, indicating that the doping of Al does not have a significant effect on the ELF. Compared with B_6N_6, the electron localization function of B_5N_6C is weaker. For B_5N_6C, the difference between the spin-up and spin-down ELF is less evident, but they are not exactly the same. Careful observation shows that the spin-down charge is slightly more than the spin-up charge. The magnetization direction for each atom of B_5N_6C is shown in Figure 10, where the blue arrow represents spin-up, and the red arrow represents spin-down. Both B and C atoms are spin-up, while N atoms are spin-down.

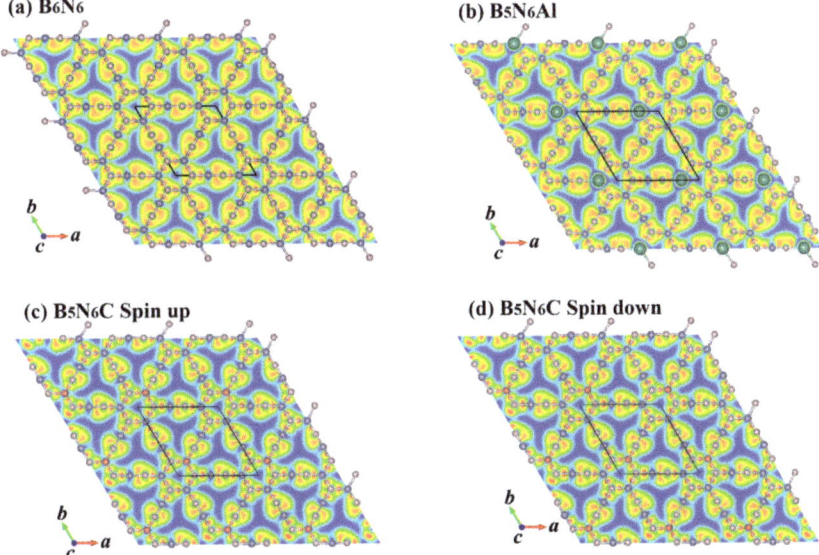

Figure 9. The electron localization function (ELF) of B_6N_6 (**a**), B_5N_6Al (**b**), and B_5N_6C (**c**,**d**).

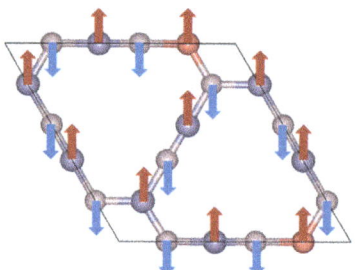

Figure 10. The magnetization direction for each atom of B_5N_6C; arrows denote spins.

3. Materials and Methods

The prediction of all the geometric optimization and properties of B_6N_6, B_5N_6Al, and B_5N_6C in this article was made on the basis of the first-principles calculation of the density functional theory (DFT) implemented in Mede A and VASP (3.3) [38,39]. The cutoff energy was selected to be 500 eV for plane waves. Electron–ion interactions were represented with the projector augmented wave (PAW) [40] pseudopotentials. The generalized gradients approximation proposed by the Perdew–Burke–Ernzerhof (GGA-PBE) functional is employed for the exchange correlation potential [41]. In order to fully optimize the geometry of the unit cell, the total energy convergence was set to 1×10^{-8} eV, and the atomic convergence force was 0.001 eV/Å. The Brillouin zone was sampled with an $8 \times 8 \times 1$ Monkhorst–Pack (MP) [42] special k-point grid for geometric optimization and properties prediction. The hybrid Heyd–Scuseria–Ernzerhof functional (HSE06) [43] was used to simulate the band structures. In order to verify the mechanical stability, the elastic constants of B_6N_6, B_5N_6Al, and B_5N_6C sheets were estimated in vaspkit (1.4.0) [44] using the energy–strain method.

4. Conclusions

Based on density functional theory, we studied the electronic properties, mechanical properties, anisotropy properties, and magnetic properties of B_6N_6, B_5N_6Al, and B_5N_6C sheets. B_6N_6, B_5N_6Al, and B_5N_6C sheets are all composed of six-membered rings and twelve-membered rings. All the B_6N_6, B_5N_6Al, and B_5N_6C sheets are anisotropic materials, and B_6N_6 sheet has the largest anisotropy in terms of Young's modulus and Poisson's ratio. B_6N_6 and B_5N_6C sheets have good mechanical properties compared to B_5N_6Al sheet, making them suitable for nanomechanical applications. B_6N_6 and B_5N_6Al sheets have indirect band gaps of 5.684 and 5.418 eV, respectively. Therefore, by calculating the spin-up and spin-down band structures, it was found that B_6N_6 sheet did not exhibit magnetic properties. With the insertion of aluminum atoms, B_5N_6Al sheet also did not show magnetism. However, with the doping of carbon atoms, B_5N_6C sheet shows a modicum of magnetism, and all three atoms (B atom, N atom, and C atom) that were not magnetic atoms themselves showed magnetism.

Author Contributions: Conceptualization, L.Q.; methodology, L.Q.; software, L.Q.; validation, Z.M.; formal analysis, Z.M.; investigation, F.Y.; resources, Q.F.; data curation, F.Y.; writing—original draft preparation, Q.F.; writing—review and editing, S.W.; visualization, S.W.; supervision, L.Q.; project administration, Q.F.; funding acquisition, L.Q. All authors have read and agreed to the published version of the manuscript.

Funding: This work was supported by the National Natural Science Foundation of China (Nos. 62164011 and 61804120), China Postdoctoral Science Foundation (Nos. 2019TQ0243, 2019M663646), the Natural Science Basic Research Program of Shaanxi Province (No. 2023-JC-YB-567), Key Science and Technology Innovation Team of Shaanxi Province (2022TD-34), the Natural Science Foundation of Jiangsu Province (No. BK20211002), as well as Qinglan Project of Jiangsu Province of China.

Data Availability Statement: Data are contained within the article.

Conflicts of Interest: The authors declare no conflicts of interest.

References

1. Bi, J.X.; Du, Z.Z.; Sun, J.M.; Liu, Y.H.; Wang, K.; Du, H.F.; Ai, W.; Huang, W. On the Road to the Frontiers of Lithium-Ion Batteries: A Review and Outlook of Graphene Anodes. *Adv. Mater.* **2023**, *35*, e2210734. [CrossRef]
2. Olabi, A.G.; Abdelkareem, M.A.; Wilberforce, T.; Sayed, E.T. Application of graphene in energy storage device—A review. *Renew. Sust. Energ. Rev.* **2021**, *135*, 110026. [CrossRef]
3. Xiao, Y.Q.; Pang, Y.X.; Yan, Y.X.; Qian, P.; Zhao, H.T.; Manickam, S.; Wu, T.; Pang, C.H. Synthesis and Functionalization of Graphene Materials for Biomedical Applications: Recent Advances, Challenges, and Perspectives. *Adv. Sci.* **2023**, *10*, e2205292. [CrossRef]
4. Yang, H.B.; Zheng, H.J.; Duan, Y.X.; Xu, T.; Xie, H.X.; Du, H.S.; Si, C.L. Nanocellulose-graphene composites: Preparation and applications in flexible electronics. *Int. J. Biol. Macromol.* **2023**, *253*, 126903. [CrossRef]
5. Ghosa, S.; Mondal, N.S.; Chowdhury, S.; Jana, D. Two novel phases of germa-graphene: Prediction, electronic and transport applications. *Appl. Surf. Sci.* **2023**, *614*, 156107. [CrossRef]
6. Asim, N.; Badiei, M.; Samsudin, N.A.; Mohammad, M.; Razali, H.; Soltani, S.; Amin, N. Application of graphene-based materials in developing sustainable infrastructure: An overview. *Compos. B. Eng.* **2022**, *245*, 110188. [CrossRef]
7. Sun, M.; Chou, J.P.; Hu, A.; Schwingenschlögl, U. Point Defects in Blue Phosphorene. *Chem. Mater.* **2019**, *31*, 8129–8135. [CrossRef]
8. Ma, Y.; Yan, Y.; Luo, L.; Pazos, S.; Zhang, C.; Lv, X.; Chen, M.; Liu, C.; Wang, Y.; Chen, A.; et al. High-performance van der Waals antiferroelectric CuCrP2S6-based memristors. *Nat. Commun.* **2023**, *14*, 7891. [CrossRef]
9. Zhang, C.; Ren, K.; Wang, S.; Luo, Y.; Tang, W.; Sun, M. Recent progress on two-dimensional van der Waals heterostructures for photocatalytic water splitting: A selective review. *J. Phys. D Appl. Phys.* **2023**, *56*, 483001. [CrossRef]
10. Zhang, W.; Chai, C.; Fan, Q.; Yang, Y.; Sun, M.; Palummo, M.; Schwingenschlögl, U. Two-dimensional borocarbonitrides for photocatalysis and photovoltaics. *J. Mater. Chem. C* **2023**, *11*, 3875–3884. [CrossRef]
11. Ren, K.; Yan, Y.; Zhang, Z.; Sun, M.; Schwingenschlogl, U. A family of LixBy monolayers with a wide spectrum of potential applications. *Appl. Surf. Sci.* **2022**, *604*, 154317. [CrossRef]
12. Sun, M.; Re Fiorentin, M.; Schwingenschlögl, U.; Palummo, M. Excitons and light-emission in semiconducting MoSi$_2$X$_4$ two-dimensional materials. *NPJ 2D Mater. Appl.* **2022**, *6*, 81. [CrossRef]
13. Wang, Y.; Miao, M.; Lv, J.; Zhu, L.; Yin, K.; Liu, H.; Ma, Y. An Effective Structure Prediction Method for Layered Materials Based on 2D Particle Swarm Optimization Algorithm. *J. Chem. Phys.* **2012**, *137*, 224108–224114. [CrossRef]
14. Pakdel, A.; Bando, Y.; Golberg, D. Nano boron nitride flatland. *Chem. Soc. Rev.* **2014**, *43*, 934–959. [CrossRef]
15. Entani, S.; Larionov, K.V.; Popov, Z.I.; Takizawa, M.; Mizuguchi, M.; Watanabe, H.; Li, S.T.; Naramoto, H.; Sorokin, P.B.; Sakai, S. Non-chemical fluorination of hexagonal boron nitride by high-energy ion irradiation. *Nanotechnology* **2020**, *31*, 125705. [CrossRef] [PubMed]
16. Wang, J.G.; Mu, X.J.; Wang, X.X.; Wang, N.; Ma, F.C.; Liang, W.J.; Sun, M.T. The thermal and thermoelectric properties of in-plane C-BN hybrid structures and graphene/h-BN van der Waals heterostructures. *Mater. Today Phys.* **2018**, *5*, 29–57. [CrossRef]
17. Sun, M.; Tang, W.; Ren, Q.; Wang, S.; Yu, J.; Du, Y. A first-principles study of light non-metallic atom substituted blue phosphorene. *Appl. Surf. Sci.* **2015**, *356*, 110–114. [CrossRef]
18. Wang, Z.H.; Zhou, X.F.; Zhang, X.M.; Zhu, Q.; Dong, H.F.; Zhao, M.W.; Oganov, A.R. Phagraphene: A Low-Energy Graphene Allotrope Composed of 5–6–7 Carbon Rings with Distorted Dirac Cones. *Nano Lett.* **2015**, *15*, 6182–6186. [CrossRef] [PubMed]
19. Singh, N.B.; Bhattacharya, B.; Sarkar, U. A first principle study of pristine and BN-doped graphyne family. *Struct. Chem.* **2014**, *25*, 1695–1710. [CrossRef]
20. Enyashin, A.N.; Ivanovskii, A.L. Graphene-like BN allotropes: Structural and electronic properties from DFTB calculations. *Chem. Phys. Lett.* **2011**, *509*, 143–147. [CrossRef]
21. Bu, H.; Zhao, M.; Zhang, H.; Wang, X.; Xi, Y.; Wang, Z. Isoelectronic Doping of Graphdiyne with Boron and Nitrogen: Stable Configurations and Band Gap Modification. *J. Phys. Chem. A* **2012**, *116*, 3934–3939. [CrossRef]
22. Shahrokhi, M.; Mortazavi, B.; Berdiyorov, R.G. New two-dimensional Boron Nitride allotropes with attractive electronic and optical properties. *Solid State Commun.* **2017**, *253*, 51–56. [CrossRef]
23. Demirci, S.; Rad, S.E.; Kazak, S.; Nezir, S.; Jahangirov, S. Monolayer diboron dinitride: Direct band-gap semiconductor with high absorption in the visible range. *Phys. Rev. B* **2020**, *101*, 125408. [CrossRef]
24. Li, X.D.; Cheng, X.L. Predicting the structural and electronic properties of two-dimensional single layer boron nitride sheets. *Chem. Phys. Lett.* **2018**, *694*, 102–106. [CrossRef]
25. Chen, Y.; Zou, J.; Campbell, S.J.; Le Caer, G. Boron nitride nanotubes: Pronounced resistance to oxidation. *Appl. Phys. Lett.* **2004**, *84*, 2430–2432. [CrossRef]
26. Wang, X.; Zhi, C.; Li, L.; Zeng, H.; Li, C.; Mitome, M.; Golberg, D.; Bando, Y. "Chemical Blowing" of Thin-Walled Bubbles: High-Throughput Fabrication of Large-Area, Few-Layered BN and C$_x$-BN Nanosheets. *Adv. Mater.* **2011**, *23*, 4072–4076. [CrossRef] [PubMed]
27. Zhi, C.; Bando, Y.; Tang, C.; Kuwahara, H.; Golberg, D. Large-Scale Fabrication of Boron Nitride Nanosheets and Their Utilization in Polymeric Composites with Improved Thermal and Mechanical Properties. *Adv. Mater.* **2009**, *21*, 2889–2893. [CrossRef]
28. Anota, E.C.; Hernández, A.B.; Morales, A.E.; Castro, M. Design of the magnetic homonuclear bonds boron nitride nanosheets using DFT methods. *J. Mol. Graph. Model.* **2017**, *74*, 135–142. [CrossRef]

29. Cao, X.; Li, Y.; Cheng, X.; Zhang, Y. Structural analogues of graphyne family: New types of boron nitride sheets with wide band gap and strong UV absorption. *Chem. Phys. Lett.* **2011**, *502*, 217–221. [CrossRef]
30. Qi, J.; Wang, S.; Wang, J.; Umezawa, N.; Blatov, V.A.; Hosono, H. B_5N_3 and $B7N5$ Monolayers with High Carrier Mobility and Excellent Optical Performance. *J. Phys. Chem. Lett.* **2021**, *12*, 4823–4832. [CrossRef]
31. Hohenberg, P.; Kohn, W. Inhomogeneous electron gas. *Phys. Rev.* **1964**, *136*, B864. [CrossRef]
32. Kohn, W.; Sham, L.J. Self-consistent equations including exchange and correlation effects. *Phys. Rev.* **1956**, *140*, A1133. [CrossRef]
33. Fan, Q.; Zhou, H.; Zhao, Y.; Yun, S. Predicting a novel two-dimensional BN material with a wide band gap. *Energy Mater.* **2022**, *2*, 200022. [CrossRef]
34. Anota, E.C. 2D boron nitride incorporating homonuclear boron bonds: Stabilized in neutral, anionic and cationic charge. *SN Appl. Sci.* **2022**, *4*, 295. [CrossRef]
35. Cadelano, E.; Palla, P.L.; Giordano, S.; Colombo, L. Elastic properties of hydrogenated graphene. *Phys. Rev. B* **2010**, *82*, 235414. [CrossRef]
36. Ding, Y.; Wang, Y. Density functional theory study of the silicene-like SiX and XSi_3 (X = B, C, N, Al, P) honeycomb lattices: The various buckled structures and versatile electronic properties. *J. Phys. Chem. C* **2013**, *117*, 18266–18278. [CrossRef]
37. Andrew, R.C.; Mapasha, R.E.; Ukpong, A.M.; Chetty, N. Mechanical properties of graphene and boronitrene. *Phys. Rev. B* **2012**, *85*, 125428. [CrossRef]
38. Kresse, G.; Furthmüller, J. Efficiency of ab initio total energy calculations for metals and semiconductors using a plane-wave basis set. *Comput. Mater. Sci.* **1996**, *6*, 15–50. [CrossRef]
39. Kresse, G.; Furthmüller, J. Efficient iterative schemes for ab initio total-energy calculations using a plane-wave basis set. *Phys. Rev. B* **1996**, *54*, 11169–11186. [CrossRef] [PubMed]
40. Blöchl, P.E. Projector augmented-wave method. *Phys. Rev. B* **1994**, *50*, 17953. [CrossRef] [PubMed]
41. Perdew, J.P.; Burke, K.; Ernzerhof, M. Generalized gradient approximation made simple. *Phys. Rev. Lett.* **1996**, *77*, 3865. [CrossRef] [PubMed]
42. Monkhorst, H.J.; Pack, J.D. Special points for brillouinzone integrations. *Phys. Rev. B* **1976**, *13*, 5188. [CrossRef]
43. Krukau, A.V.; Vydrov, O.A.; Izmaylov, A.F.; Scuseria, G.E. Influence of the exchange screening parameter on the performance of screened hybrid functionals. *J. Chem. Phys.* **2006**, *125*, 224106. [CrossRef] [PubMed]
44. Wang, V.; Liu, J.C.; Tang, G.; Geng, W.T. VASPKIT: A user-friendly interface facilitating high-throughput computing and analysis using VASP code. *Comput. Phys. Commun.* **2021**, *267*, 108033. [CrossRef]

Disclaimer/Publisher's Note: The statements, opinions and data contained in all publications are solely those of the individual author(s) and contributor(s) and not of MDPI and/or the editor(s). MDPI and/or the editor(s) disclaim responsibility for any injury to people or property resulting from any ideas, methods, instructions or products referred to in the content.

Article

Dependence of Ge/Si Avalanche Photodiode Performance on the Thickness and Doping Concentration of the Multiplication and Absorption Layers

Hazem Deeb, Kristina Khomyakova, Andrey Kokhanenko, Rahaf Douhan and Kirill Lozovoy *

Department of Quantum Electronics and Photonics, Faculty of Radiophysics, National Research Tomsk State University, 634050 Tomsk, Russia
* Correspondence: lozovoymailbox@gmail.com

Abstract: In this article, the performance and design considerations of the planar structure of germanium on silicon avalanche photodiodes are presented. The dependences of the breakdown voltage, gain, bandwidth, responsivity, and quantum efficiency on the reverse bias voltage for different doping concentrations and thicknesses of the absorption and multiplication layers of germanium on the silicon avalanche photodiode were simulated and analyzed. The study revealed that the gain of the avalanche photodiode is directly proportional to the thickness of the multiplication layer. However, a thicker multiplication layer was also associated with a higher breakdown voltage. The bandwidth of the device, on the other hand, was inversely proportional to the product of the absorption layer thickness and the carrier transit time. A thinner absorption layer offers a higher bandwidth, but it may compromise responsivity and quantum efficiency. In this study, the dependence of the photodetectors' operating characteristics on the doping concentration used for the multiplication and absorption layers is revealed for the first time.

Keywords: optoelectronics; avalanche photodiode; Ge/Si heterojunction; avalanche multiplication; photodetector; optical fiber telecommunication

Citation: Deeb, H.; Khomyakova, K.; Kokhanenko, A.; Douhan, R.; Lozovoy, K. Dependence of Ge/Si Avalanche Photodiode Performance on the Thickness and Doping Concentration of the Multiplication and Absorption Layers. *Inorganics* **2023**, *11*, 303. https://doi.org/10.3390/inorganics11070303

Academic Editors: Sake Wang, Minglei Sun and Nguyen Tuan Hung

Received: 26 June 2023
Revised: 13 July 2023
Accepted: 14 July 2023
Published: 15 July 2023

Copyright: © 2023 by the authors. Licensee MDPI, Basel, Switzerland. This article is an open access article distributed under the terms and conditions of the Creative Commons Attribution (CC BY) license (https://creativecommons.org/licenses/by/4.0/).

1. Introduction

Infrared photo-electronics is one of the most technologically advanced and rapidly developing areas of modern optoelectronics. Of particular interest are studies on the creation of highly sensitive and high-speed detectors for the fields of fiber communications [1–7], spectroscopy [8], and imaging systems [9,10]. Thus, avalanche photodetectors (APDs) have been very attractive with respect to high-sensitivity systems due to their intrinsic ability to enhance receiver sensitivity through internal avalanche gain [11].

The separate-absorption-charge-multiplication (SACM) germanium on silicon avalanche photodiode (Ge-on-Si APD) is an advanced photodetector structure that combines the properties of germanium and silicon to achieve efficient light detection and signal amplification [12–17]. In the SACM Ge-on-Si APD, the separate absorption and multiplication regions are designed to optimize the absorption of incident light and the multiplication of charge carriers, respectively. The absorption region is responsible for efficiently converting photons into electron–hole pairs, while the multiplication region provides the electric field required to accelerate the charge carriers and initiate the avalanche multiplication process. The integration of germanium with a silicon substrate enables compatibility with standard silicon fabrication processes, thereby facilitating large-scale production and integration with existing technology [18]. Moreover, the SACM Ge-on-Si APD benefits from the unique properties of germanium, such as its band gap (0.66 eV), which provides effective absorption at wavelengths in the entire visible and infrared ranges up to a maximum wavelength of approximately 1600 nm, while the fast mobility of electrons and holes offers the potential for fast response times.

Ongoing research and development efforts focus on further improving the performance of Ge-on-Si APDs. This includes enhancing charge multiplication efficiency, reducing dark current levels, and optimizing the device's response time [19–22]. Several studies have made notable contributions in this field. Kim et al. [23] demonstrated enhanced performance of a vertical illumination-type Ge-on-Si APD by leveraging internal RF-gain effects. The fabricated APD exhibited a negative differential resistance (NDR) beyond the avalanche breakdown voltage. This NDR allowed the device to achieve eye-opening results, corresponding to the minimal signal distortion, at up to 50 Gb/s accompanied by improved signal-to-noise ratios and signal amplitudes. Huang et al. [24] reported the development of a waveguide Ge/Si APD with 3 dB bandwidths of 56 GHz and 36 GHz, with responsivities of 1.08 A/W and 6 A/W at a wavelength of 1310 nm, respectively. Zeng et al. [25] focused on the effects of bias voltage and incident power on the bandwidth of Ge/Si APDs. In addition to the widely discussed space charge effect in the multiplication layer, Zeng argued that an increase in incident power leads to excess holes in the absorption layer, resulting in a significant reduction in the strength of the electric field in the multiplication layer. They introduced the concept of an effective equivalent voltage to quantitatively describe this incident power effect. Zhang et al. [26] investigated the influence of surface defects and the width of the guard ring on the sidewall leakage current in Ge/Si APDs with a mesa structure. The study identified high-density surface defects and a strong electric field at the sidewall as the primary contributors to the large sidewall leakage current. Accordingly, further advancements can be made in the design, optimization, and fabrication of Ge/Si APDs, with the goal of achieving improved performance.

In this work, we focus on investigating the performance characteristics of SACM Ge-on-Si APDs, specifically their dependence on the thicknesses of the multiplication and absorption layers. Breakdown voltage, multiplication gain, bandwidth, responsivity, and quantum efficiency are among the performance parameters examined. The thickness and doping concentration of both the multiplication layer, which is responsible for internal gain generation through impact ionization, and the absorption layer, which is responsible for photon absorption and the generation of electron–hole pairs, strongly influence the APD's performance characteristics. By studying the performance characteristics and their dependence on layer thickness, this research aims to contribute to the optimization and design of Ge-on-Si APDs, with the ultimate goal of achieving improved performance in high-speed and high-sensitivity optoelectronic applications.

2. Device Structure and TCAD Physical Simulation Models

Figure 1a depicts a cross-sectional diagram of the simulated cylindrical Ge/Si APD with a diameter of 30 μm.

The structure of the device includes a p-type heavily doped (5×10^{19} cm^{-3}) Ge contact layer, a p-Ge absorption layer, a p-type doped Si charge layer, an n-Si multiplication layer, and an n-type heavily doped (5×10^{19} cm^{-3}) Si contact layer. The charge layer controls the electric field distribution in the device by making the electric field in the multiplication layer large enough to cause avalanche breakdown while also ensuring that the electric field in the absorption layer remains as low as possible without triggering an avalanche and with the carriers moving at saturation speed, thus reducing the tunneling probability caused by the narrow bandgap [16].

In order to simulate the properties of the avalanche photodetector, the threading dislocations (TDs) and their energy level distributions in the Ge layer (constituting 4.2% of the lattice mismatch between the Ge and Si) should be taken into consideration. The TDs in an epitaxial Ge layer act as the acceptor-like defects, with energy levels located in the forbidden bandgap [27]. One of the proposed methods for decreasing TDs is the two-step epitaxial growth of germanium on silicon. In this work, we suppose that a 90 nm thick low-temperature Ge seed layer has first been grown on the Si layer to confine most of the TDs in this layer (about 1×10^{10} cm^{-2}). As a result, the TD density decreases in the upper high-temperature epitaxial Ge layer to about 2×10^6 cm^{-2} [28]. These TDs are located at 0.36 eV below the conduction band [29]. In addition, the discontinuities between the valence bands and conduction bands of Ge and Si are set to 0.35 and 0.1 eV, respectively, as shown in Figure 1b.

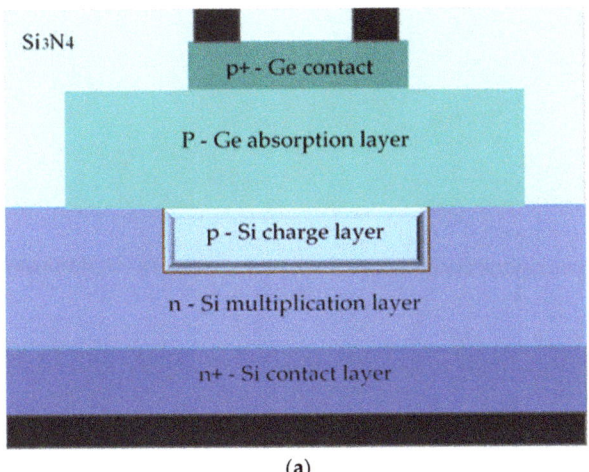
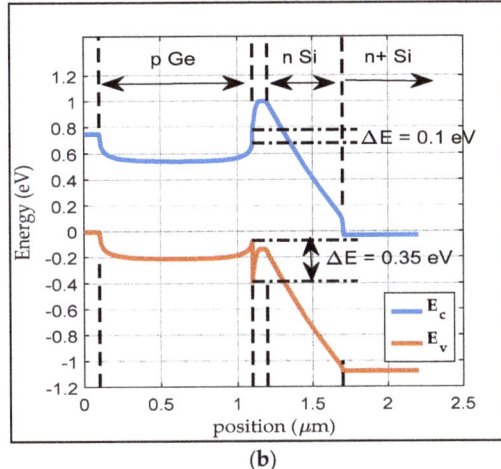

Figure 1. (**a**) The cross-sectional diagram of the Ge/Si APD; (**b**) band diagram of the center of the Ge/Si APD.

The simulation of avalanche photodiodes (APDs) involves the utilization of Synopsys Sentaurus Technology computer-aided design (TCAD) software, which is grounded in the Poisson equation, continuity equations, and carrier transport equations. The Poisson equation establishes a relationship between changes in electrostatic potential and the distribution of local charge densities, while the continuity equations and transport equations depict the mechanisms governing carrier transport, generation, and recombination processes [30].

As carriers experience acceleration in an electric field, their velocity reaches a saturation point when the electric field magnitude becomes substantial. To address this effect, the degree of effective mobility needs to be adjusted by considering that the magnitude of the drift velocity is the product of the mobility and the electric field component in the direction of current flow. To account for this behavior, the Caughey and Thomas expression [31] is utilized to incorporate field-dependent mobility. This expression enables a seamless transition between low-field and high-field behaviors:

$$\mu_n(E) = \mu_{n0} \left[\frac{1}{1 + \left(\frac{\mu_{n0} E}{\vartheta_{sat,n}} \right)^{\beta_n}} \right]^{\frac{1}{\beta_n}} \quad (1)$$

$$\mu_p(E) = \mu_{p0} \left[\frac{1}{1 + \left(\frac{\mu_{p0} E}{\vartheta_{sat,p}} \right)^{\beta_p}} \right]^{\frac{1}{\beta_p}} \quad (2)$$

where E is the parallel electric field (the electric field component in the direction of the current flow); μ_{n0} and μ_{p0} are the values of the low-field electron and hole mobility, respectively; and ($\beta_n = 2$, $\beta_p = 1$). The saturated velocities for the electrons and holes in the Si material are $\vartheta_{sat,n} = 1 \times 10^7$ cm/s, $\vartheta_{sat,p} = 7 \times 10^6$ cm/s, respectively, and those in the Ge material are $\vartheta_{sat,n} = 7 \times 10^6$ cm/s and $\vartheta_{sat,p} = 6.3 \times 10^6$ cm/s [32].

Carrier generation recombination refers to the mechanism by which a semiconductor material attempts to restore its equilibrium state after being perturbed. Phonon transitions take place when there is a trap or defect present within the forbidden energy band of the semiconductor. As previously discussed, threading dislocations (TDs) within an epitaxial Ge layer serve as acceptor-like defects, occupying energy levels within the forbidden bandgap. The defect centers enhance the level of carrier recombination due to the trap-assisted tunneling mechanism, which is described by the Shockley–Read–Hall (SRH) model [29]:

$$G_{SRH} = \frac{pn - n_{ie}^2}{\tau_p\left[n + n_{ie}exp\left(\frac{E_T-E_i}{kT_L}\right)\right] + \tau_n\left[p + n_{ie}exp\left(-\frac{E_T-E_i}{kT_L}\right)\right]} \quad (3)$$

where n and p are electron and hole concentrations, n_{ie} is the intrinsic carrier concentration, E_i is the intrinsic energy, E_T is the energy level of the TD, and τ_n and τ_p are electron and hole lifetimes, which are related to the threading dislocation density [29]:

$$\tau_n = \frac{1}{\sigma_n v_n N} \qquad \tau_p = \frac{1}{\sigma_p v_p N} \quad (4)$$

where N is the TD density in the Ge layer; v_n and v_p are the thermal velocities, whose values are $v_n = 3.1 \times 10^5$ m/s, $v_p = 1.9 \times 10^5$ m/s, respectively; and σ_n and σ_p are electron and hole capture cross sections, whose values are $\sigma_n = 3 \times 10^{-14}$ cm^2, $\sigma_p = 5 \times 10^{-14}$ cm^2 [33].

In order to simulate the surface recombination of the side walls of the cylindrical device structure, the surface recombination velocity, namely, $S_n = S_p = 1 \times 10^7$ cm/s, at the side walls is set in the surface recombination model [34].

In a strong electric field, electrons can tunnel through the bandgap via trap states. This trap-assisted tunneling (TAT) mechanism is accounted for by using the TAT model. Also, if a sufficiently high electric field exists within a device, the level of local band bending may be sufficient to allow electrons to tunnel, via internal field emission, from the valence band into the conduction band. In this study, this generation mechanism is implemented using the band-to-band tunneling (BBT) model [33].

The avalanche simulation of the Ge/Si APD Is based on Selberherr's impact-ionization model. The electron and hole ionization rates can be expressed as follows:

$$\alpha_n = A_n exp\left(-\left(\frac{B_n}{E}\right)^{C_n}\right) \quad (5)$$

$$\alpha_p = A_p exp\left(-\left(\frac{B_p}{E}\right)^{C_p}\right) \quad (6)$$

For the Si material, $A_n = 7.03 \times 10^5$ cm^{-1}, $A_p = 1.58 \times 10^6$ cm^{-1}, $B_n = 1.231 \times 10^6$ cm^{-1}, and $B_p = 2.036 \times 10^6$ cm^{-1} for $E < 4 \times 10^5$ V.cm^{-1}. $A_n = 7.03 \times 10^5$ cm^{-1}, $A_p = 6.71 \times 10^5$ cm^{-1}, $B_n = 1.231 \times 10^6$ cm^{-1}, and $B_p = 1.693 \times 10^6$ cm^{-1} for $E \geq 4 \times 10^5$ V.cm^{-1}. For the Ge material, $A_n = 1.55 \times 10^7$ cm^{-1}, $A_p = 1 \times 10^7$ cm^{-1}, $B_n = 156 \times 10^6$ cm^{-1}, and $B_p = 1.28 \times 10^6$ cm^{-1} for $E < 4 \times 10^5$ V.cm^{-1}. C_n and C_p are both set to 1 [35].

3. The Simulation Results of APD Characteristics

In order to examine the impact of different parameters in the absorption and multiplication layers on the properties of the avalanche photodiode, modifications were made to the thickness and doping concentration of both layers. Subsequently, the resulting performance was analyzed, focusing on several characteristics such as breakdown voltage, multiplication gain, bandwidth, responsivity, and quantum efficiency. The characteristics of the avalanche photodiodes were modeled and simulated under an optical input power illumination of −20 dBm at 1310 nm.

3.1. The Effect of Multiplication Layer Parameters

To explore the impact of the multiplication layer thickness and doping concentration on the performance of the avalanche photodiode, a series of simulations was conducted using ten different structures. The simulations involved varying the thickness of the multiplication layer from 0.5 µm to 1.5 µm with a step increment of 0.25 µm. Simultaneously, two distinct values were assigned to the doping concentrations of the absorption and multiplication layers: 1×10^{15} cm^{-3} and 5×10^{15} cm^{-3}. The devices' parameters used for this case study are presented in Table 1.

Table 1. Structural parameters of the Ge/Si APD related to the multiplication layer effect case study.

Layer	p$^+$-Ge	p-Ge	p-Si	n-Si	n$^+$-Si
Thickness (µm)	0.1	1	0.1	0.5; 0.75; 1; 1.25; 1.5	0.5
Doping concentration (cm^{-3})	5×10^{19}	1×10^{15} 5×10^{15}	2×10^{17}	1×10^{15} 5×10^{15}	5×10^{19}

3.1.1. The Multiplication Gain and Breakdown Voltage

Figure 2 shows the dependence of the multiplication gain (M) and the breakdown voltage on the thickness of the multiplication layer versus bias voltage for the APD structures with different doping concentrations and thicknesses of multiplication layers. To ensure a reliable comparison between the devices, the difference between the breakdown voltage (V_{bd}) and the bias voltage (V_{bias}) is used and assigned to the referenced voltage ($V_{b\text{-ref}}$):
$V_{b\text{-ref}} = V_{bd} - V_{bias}$.

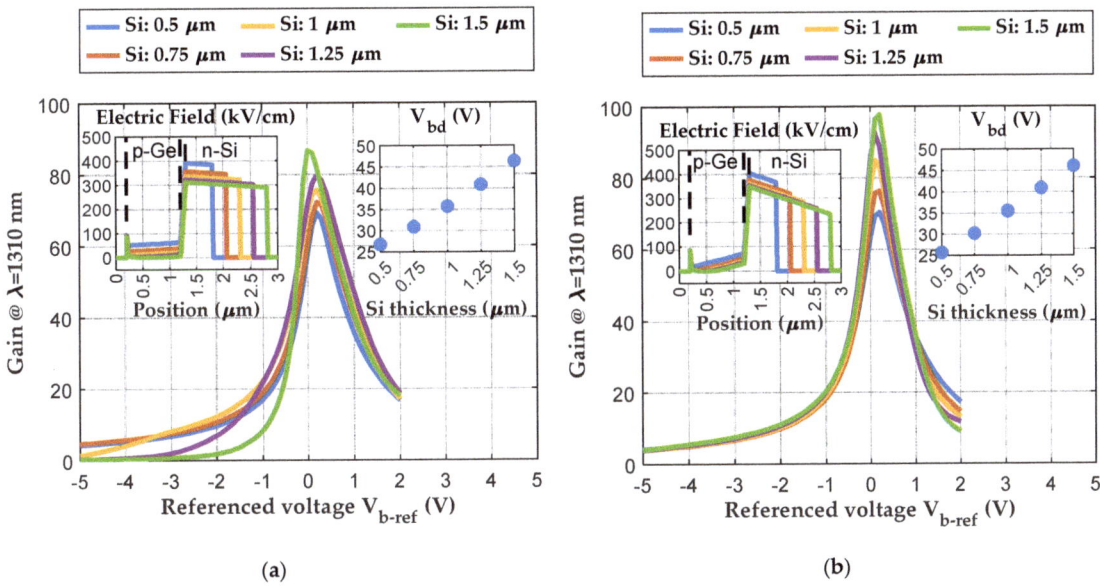

(a)   (b)

Figure 2. Dependence of the gain on the thickness of the multiplication layer versus the referenced bias voltage. The insets show the breakdown voltage and the electric field profile in the center of the devices at the respective breakdown voltages. Two different multiplication and absorption layer doping concentration densities were simulated: (a) 1×10^{15} cm^{-3}; (b) 5×10^{15} cm^{-3}.

By referring to the insets Figure 2a,b, it can be observed that an increase in the thickness of the multiplication layer leads to an increase in the breakdown voltage, and modifying the doping concentration in the absorption and multiplication layers from 1×10^{15} cm^{-3} to

5×10^{15} cm^{-3} does not significantly impact the breakdown voltage. As the thickness of the multiplication layer increases, the electric field is distributed over a larger distance, leading to a lower electric field strength across the layer. Consequently, a higher voltage is required to initiate the avalanche breakdown, leading to an increase in the breakdown voltage.

In the linear mode region where $V_{b\text{-ref}}$ is less than -0.5 V, for the low-level-doped absorption and multiplication layer (1×10^{15} cm^{-3}) structures, the multiplication gain is influenced by the bias voltage and the impact generation rate, which are determined by the thickness of the multiplication layer and the resulting electric field strength throughout the structures, as depicted in Figure 2a. On the other hand, if we consider the 5×10^{15} cm^{-3} doping concentration in the linear mode region, the higher doping concentration would cause the electric field lines to concentrate in specific regions, resulting in a non-uniform field distribution. In such a case, the electric field strength may not be effectively distributed across the entire multiplication layer, thereby limiting the potential for carrier multiplication and reducing the overall multiplication gain. Consequently, even with an increased thickness of the multiplication layer, as depicted in Figure 2b, the gain remains relatively unchanged.

Near the breakdown point, where the absolute value of $V_{b\text{-ref}}$ is less than 0.5 V, it can be seen that the electric field for all structures is higher than the breakdown field of silicon. By increasing the thickness of the multiplication layer, the avalanche region within the APD expands. This extended region allows for a larger number of carriers to participate in the multiplication process, leading to higher gain near the breakdown voltage. Therefore, an increase in the thickness of the multiplication layer leads to an increase in the multiplication gain near the breakdown voltage, as depicted in Figure 2. When the bias voltage exceeds the breakdown voltage in an avalanche photodiode, the density of electrons and holes becomes comparable to the density of donors in the multiplication region. This leads to an influence on the electric field profile due to the space charge of these electrons and holes. At the boundary between the charge and multiplication layers, there is a net excess of holes, which results in an increased magnitude of the electric field in that region. Conversely, within the multiplication region and at the edge adjacent to the n$^+$-contact layer, there is a net excess of electrons. This excess of electrons causes a decrease in the magnitude of the electric field and, consequently, a decrease in the multiplication gain.

3.1.2. The Bandwidth

Figure 3 shows the dependence of the bandwidth on the reverse bias voltage for the APD structures with different doping concentrations and thicknesses of the multiplication layer. It can be seen that the bandwidth decreases as the thickness of the multiplication layer increases. The thicker multiplication layer introduces a longer path for carriers to traverse before reaching the collection region. As carriers travel across this increased distance, their transit time increases. This prolonged transit time results in a slower response time of the device, thus limiting the bandwidth. On the other hand, if we consider the electron and hole drift velocity in the simulated structures (as depicted in the insets of Figure 3a,b), it can be seen that the electrons and holes in the n-Si multiplication layer drift at their saturation velocities due to the high electric field, but in the p-Ge absorption layer, the electron and hole drift velocities change depending on the strength of the electric field. For the low-doped absorption and multiplication layer (1×10^{15} cm^{-3}) structures, increasing the multiplication layer thickness beyond 1 μm leads to an electric field value less than 30 kV/cm in the absorption layer (as shown in Figure 2 inset); thus, the carriers do not reach their saturation speeds. As a result, the drift time increases and the bandwidth decreases (Figure 3a) compared to the higher-doped absorption and multiplication layer (5×10^{15} cm^{-3}) structures (Figure 3b). In the latter case, the electric field at the edges between the absorption layer and the charge layer exceeds 30 kV/cm, causing the carriers to drift partially at their saturation velocities. This leads to a higher bandwidth. However, as the thickness of the multiplication layer increases, the drift time increases, resulting in a reduction in bandwidth.

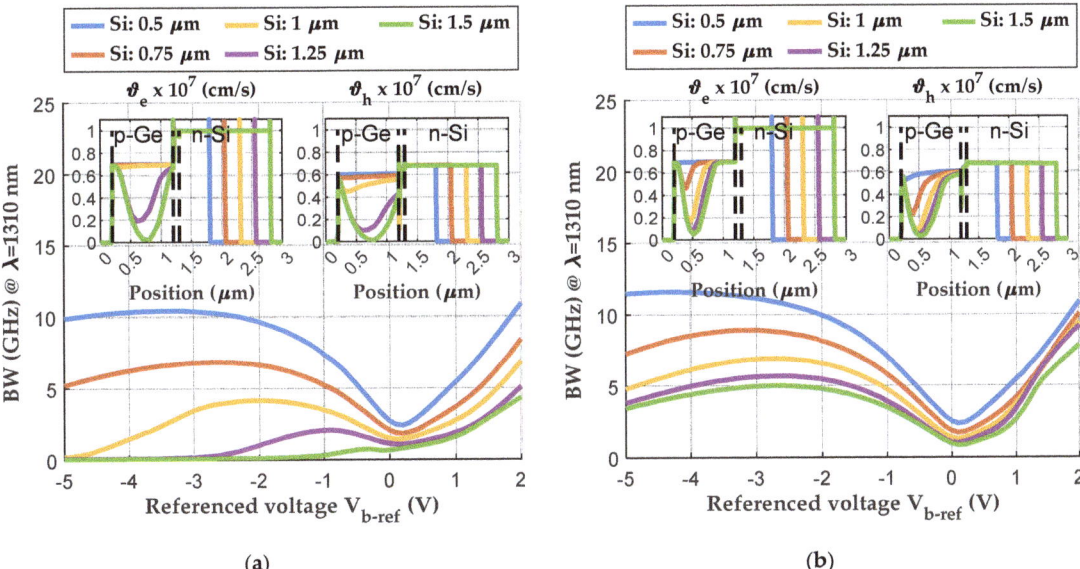

Figure 3. Dependence of the bandwidth on the thickness of the multiplication layer versus referenced bias voltage. The insets show the drift velocity for the electrons ϑ_e and for the holes ϑ_h through the center of the devices at $V_{b\text{-}ref} = -2$ V. Two different multiplication and absorption layer doping concentration densities were simulated: (**a**) 1×10^{15} cm^{-3}; (**b**) 5×10^{15} cm^{-3}.

3.1.3. The Gain Bandwidth Product, Responsivity, and Quantum Efficiency

In applications that involve pulse detection or timing, the gain bandwidth product (GBP) plays a crucial role. It determines the speed at which the APD can respond to rapid changes in the input signal. A higher gain bandwidth product allows for faster rise and fall times of pulses, thereby ensuring accurate detection and precise timing measurements. Figure 4 shows the dependence of the gain bandwidth product (GBP) on the reverse bias voltage for the APD structures with different doping concentrations and thicknesses of the multiplication layer. It can be seen that with a thinner multiplication layer, the APD typically exhibits a higher gain bandwidth product. Increasing the thickness of the multiplication layer leads to a decrease in the gain bandwidth product. This is due to the extended transit time of carriers, resulting in a limited frequency response and a reduced gain bandwidth product. Beyond the breakdown voltage, which corresponds to the gain peak, the gain bandwidth product exhibits an increase. This increase is primarily attributed to two factors: the widening of the bandwidth and a gradual decrease in gain. Now, if we consider responsivity and internal quantum efficiency, it can be seen that the structures with the higher-doped absorption and multiplication layers (5×10^{15} cm^{-3}) (shown in Figure 4b) exhibit similar levels of responsivity and quantum efficiency due to the consistent thickness of the Ge absorption layer (1 μm) across these structures. However, in the linear mode region of the structures with lower densities (1×10^{15} cm^{-3}) (shown in Figure 4a), a thinner multiplication layer tends to result in higher responsivity. This is because, as mentioned above, increasing the multiplication layer thickness beyond 1 μm leads to an electric field strength of less than 30 kV/cm in the absorption layer. Consequently, this reduction in field strength diminishes the number of drifted electron–hole pairs while augmenting the recombination rate.

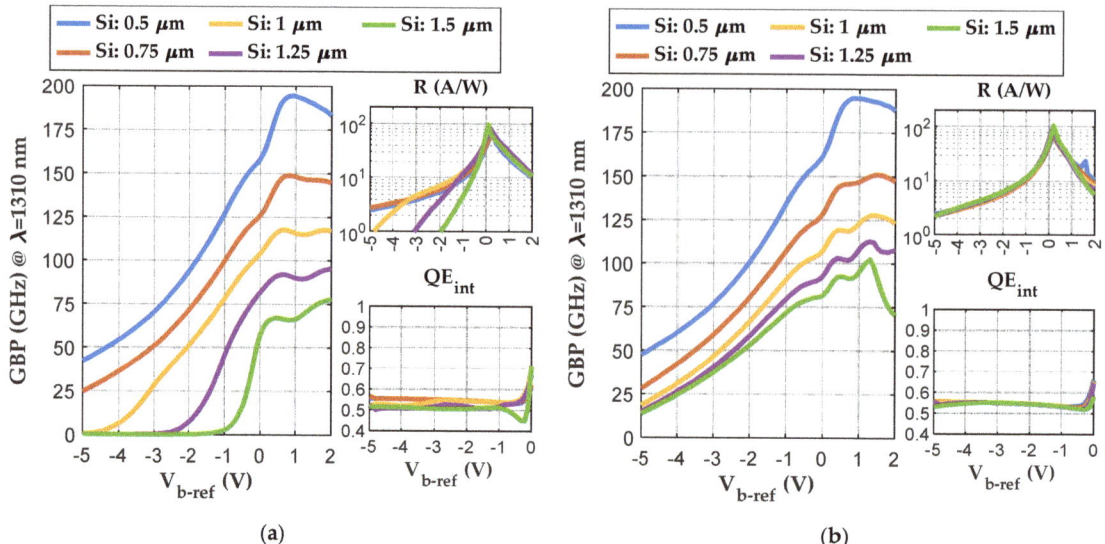

Figure 4. Dependence of the gain bandwidth product (GBP), responsivity (R), and quantum efficiency (QE) on the thickness of the multiplication layer versus referenced bias voltage. Two different multiplication and absorption layer doping densities were simulated: (**a**) 1×10^{15} cm^{-3}; (**b**) 5×10^{15} cm^{-3}.

3.1.4. The Photodiode Spectral Response

The spectral response is vital for characterizing and optimizing the performance of avalanche photodiodes, enabling their efficient use in various applications that require the precise detection and measurement of light across different wavelengths. Figure 5 shows the dependence of the internal quantum efficiency (QE) on the thickness of the multiplication layer versus the wavelength at the biasing voltage corresponding to the unity gain.

All the simulated structures share a common feature: a 1 µm thick Ge absorption layer. As a result, they are expected to exhibit a similar response to incident light with wavelengths greater than the cut-off wavelength of silicon (1.1 µm) when the thickness of the Si multiplication layer is increased. This similarity arises because only the Ge layer absorbs photons, and changing the doping concentration of the multiplication and absorption layers has a minor impact on the spectral response in this range due to the corresponding variation in the recombination rate along the structure.

On the other hand, incident light with wavelengths ranging from 0.4 µm to 1.1 µm is absorbed in the Ge layer, but some photons are also absorbed in the silicon layers. This absorption in the silicon layers enhances the quantum efficiency within this range. Altering the thickness and doping concentration of the multiplication and absorption layers affects the spectrum response in the wavelength range of 0.4 µm to 1.1 µm due to the resulting distribution of the electric field and the recombination of carriers with surface states and the bulk material. Consequently, a thinner multiplication layer tends to yield a higher spectral response.

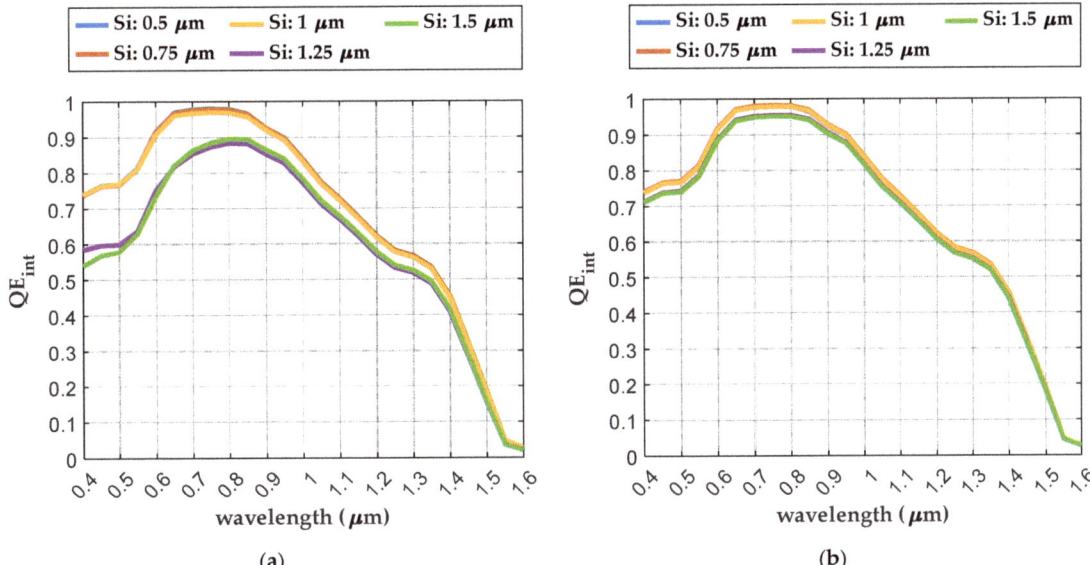

Figure 5. Dependence of the internal quantum efficiency (QE) on the thickness of the multiplication layer versus wavelength at unity gain. Two different multiplication and absorption layer doping densities were simulated: (**a**) 1×10^{15} cm^{-3}; (**b**) 5×10^{15} cm^{-3}.

In conclusion, increasing the thickness of the multiplication layer leads to an increase in the breakdown voltage and gain, as more electron–hole pairs undergo the avalanche multiplication process. However, the bandwidth of the photodiode decreases with thicker multiplication layers due to the longer carrier transit times. Considering the gain bandwidth product and breakdown voltage as a figure of merit, it can be concluded, based on the simulated results, that an optimal SACM Ge/Si APD structure should have a multiplication layer thickness of 0.5 µm and a doping concentration of 5×10^{15} cm^{-3} for both the multiplication and absorption layers; these optimal values are related to the higher GBP, better spectral response, and lower breakdown voltage.

3.2. The Effect of Absorption Layer Parameters

In order to investigate the impact of absorption layer thickness and doping concentration on the performance of the avalanche photodiode, a series of simulations was conducted on eight different structures. These simulations involved varying the thickness of the absorption layer in increments of 0.5 µm, ranging from 0.5 µm to 2 µm. Simultaneously, the multiplication layer was maintained at a constant thickness of 0.5 µm, while the absorption and multiplication layers were assigned two distinct doping concentration values: 1×10^{15} cm^{-3} and 5×10^{15} cm^{-3}. The devices' parameters used for this case study are presented in Table 2.

Table 2. Structural parameters of the Ge/Si APD related to the absorption layer effect case study.

Layer	p$^+$-Ge	p-Ge	p-Si	n-Si	n$^+$-Si
Thickness (µm)	0.1	0.5; 1; 1.5; 2	0.1	0.5	0.5
Doping Concentration (cm^{-3})	5×10^{19}	1×10^{15} 5×10^{15}	2×10^{17}	1×10^{15} 5×10^{15}	5×10^{19}

3.2.1. The Multiplication Gain and Breakdown Voltage

Figure 6 depicts the relationship between the thickness of the absorption layer, bias voltage, and the corresponding dependence of the multiplication gain (M) and breakdown voltage for different APD structures with varying doping concentrations and absorption layer thicknesses. Referring to the inset Figure 6a, it can be observed that an increase in the thickness of the absorption layer from 0.5 μm to 2 μm results in a corresponding increase in the breakdown voltage. Additionally, by changing the doping concentration in the absorption and multiplication layers from 1×10^{15} cm^{-3} to 5×10^{15} cm^{-3}, the breakdown voltage saturates beyond the 1 μm thickness of the Ge absorption layer, as indicated in the inset in Figure 6b. This saturation phenomenon can be explained by considering the electric field profiles within these structures. Upon observing the electric field profile in the inset in Figure 6b, it can be observed that the Ge absorption layer undergoes partial depletion when the thickness is extended beyond 1 μm. Consequently, this partial depletion causes the breakdown voltage to remain unchanged despite the increase in Ge thickness.

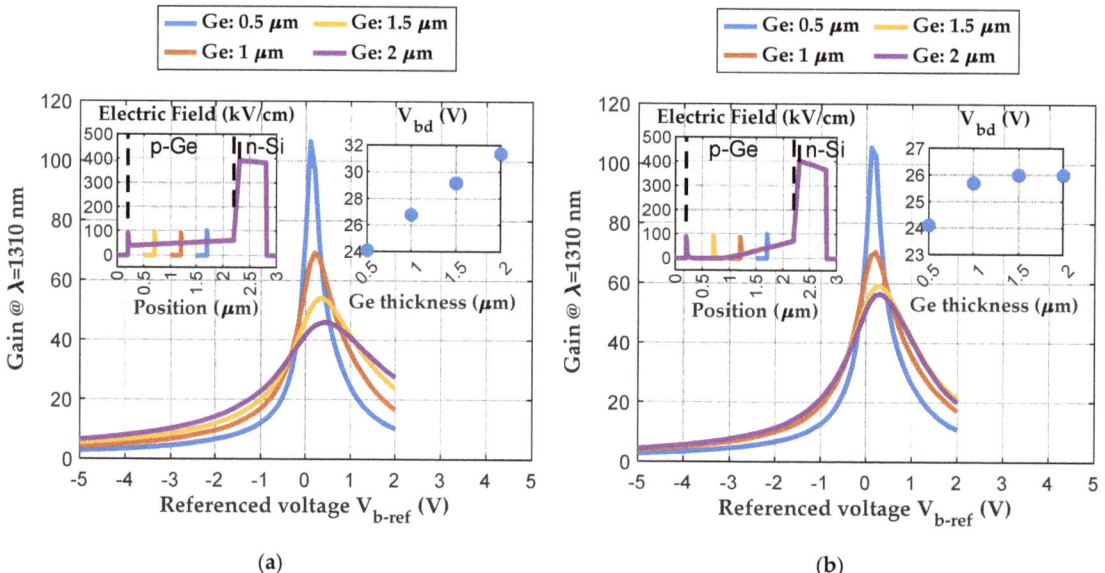

Figure 6. Dependence of gain on the thickness of the absorption layer versus referenced bias voltage. The insets show the breakdown voltage and the electric field profile through the center of the devices at the respective breakdown voltages. Two different multiplication and absorption layer doping concentration densities were simulated: (**a**) 1×10^{15} cm^{-3}; (**b**) 5×10^{15} cm^{-3}.

In the linear mode region where $V_{b\text{-ref}}$ is less than −0.5 V, as the thickness of the absorption layer increases, there is a corresponding increase in the multiplication gain. This relationship suggests that a thicker germanium absorption layer allows for a greater number of electron–hole pairs to be generated, leading to higher multiplication gain, as shown in Figure 6. On the other hand, increasing the thickness of the absorption layer results in a decrease in the multiplication gain near the breakdown voltage due to the increase in the build-up time. The build-up time refers to the time required for the avalanche multiplication process to reach its maximum gain. When the thickness of the absorption layer is increased, the build-up time tends to be prolonged. This is because a thicker absorption layer requires a longer time for the carriers to travel and accumulate energy before the avalanche multiplication process reaches its peak gain.

3.2.2. The Bandwidth

Figure 7 illustrates the relationship between the bandwidth and reverse bias voltage for the APD structures with different doping concentrations and thicknesses of the absorption layer. In the linear mode region, it can be seen that the bandwidth decreases as the thickness of the absorption layer increases. As the absorption layer becomes thicker, the carriers need more time to traverse the absorption region before reaching the multiplication layer. This increased transit time results in a slower response to high-frequency optical signals, thereby reducing the bandwidth. Conversely, for structures with a doping concentration of 1×10^{15} cm^{-3} in the absorption and multiplication layers, the insets in Figure 7a demonstrate that the electrons and holes drift at their saturation velocities. However, by altering the doping concentration to 5×10^{15} cm^{-3}, it becomes apparent, as shown in the insets of Figure 7b, that the carriers in the structures with an absorption layer thickness exceeding 1 µm do not reach their saturation velocities. Consequently, the bandwidth decreases compared to the corresponding structures with a doping concentration of 1×10^{15} cm^{-3}.

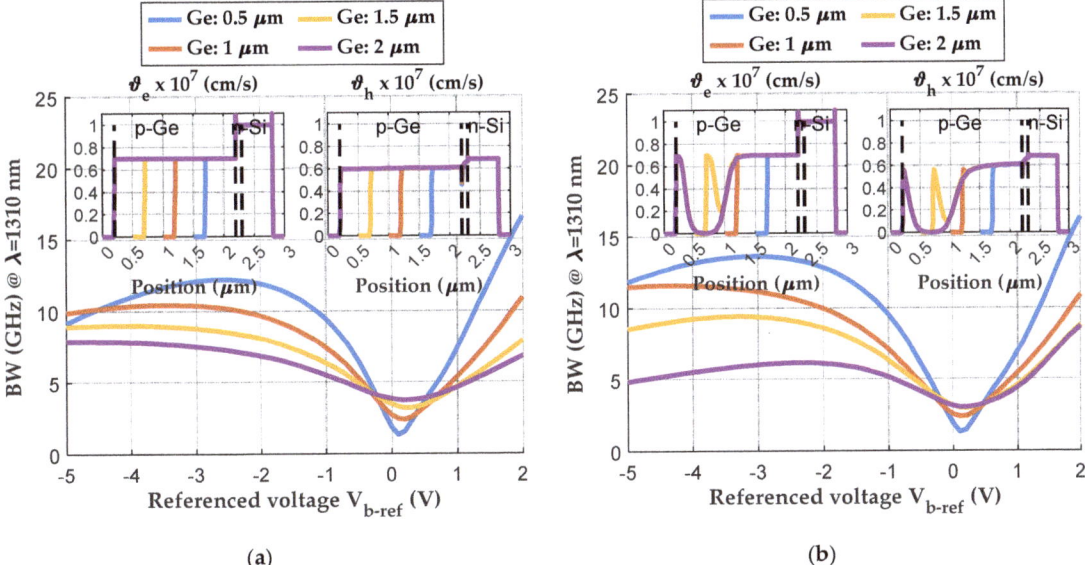

Figure 7. Dependence of the bandwidth on the thickness of the absorption layer versus referenced bias voltage. The insets show the drift velocity for the electrons ϑ_e and for the holes ϑ_h through the center of the devices at $V_{b\text{-ref}} = -2$ V. Two different multiplication and absorption layer doping concentration densities were simulated: (**a**) 1×10^{15} cm^{-3}; (**b**) 5×10^{15} cm^{-3}.

The dependence of bandwidth on the gain and build-up time near the breakdown voltage is more significant than the dependence of bandwidth on transit time. This relationship can be expressed as $BW = (2\pi M\tau)^{-1}$ [36], where M denotes multiplication gain and τ is the avalanche build-up time. Consequently, the increase in the absorption layer's thickness results in an increase in bandwidth due to the higher gain achieved with the thinner absorption layer near the breakdown voltage.

3.2.3. The Gain Bandwidth Product, Responsivity, and Quantum Efficiency

In Figure 8, the relationship between the gain bandwidth product (GBP) and reverse bias voltage is demonstrated for various APD structures with differing doping concentrations and absorption layer thicknesses. The findings indicate that the GBP is linked to the gain increase in the linear mode region due to a consistent bandwidth. Additionally, as the gain grows nearer to the breakdown voltage, there is a significant increase that outweighs the drops in bandwidth caused by the growing avalanche build-up time. If we examine the responsivity and internal quantum efficiency, thicker absorption layers demonstrate increased values for both. This outcome is due to the fact that thicker absorption layers increase the likelihood of photon absorption, resulting in a larger number of electron–hole pairs. This generates more charge carriers that are available for the multiplication process, resulting in higher values of responsivity and internal quantum efficiency.

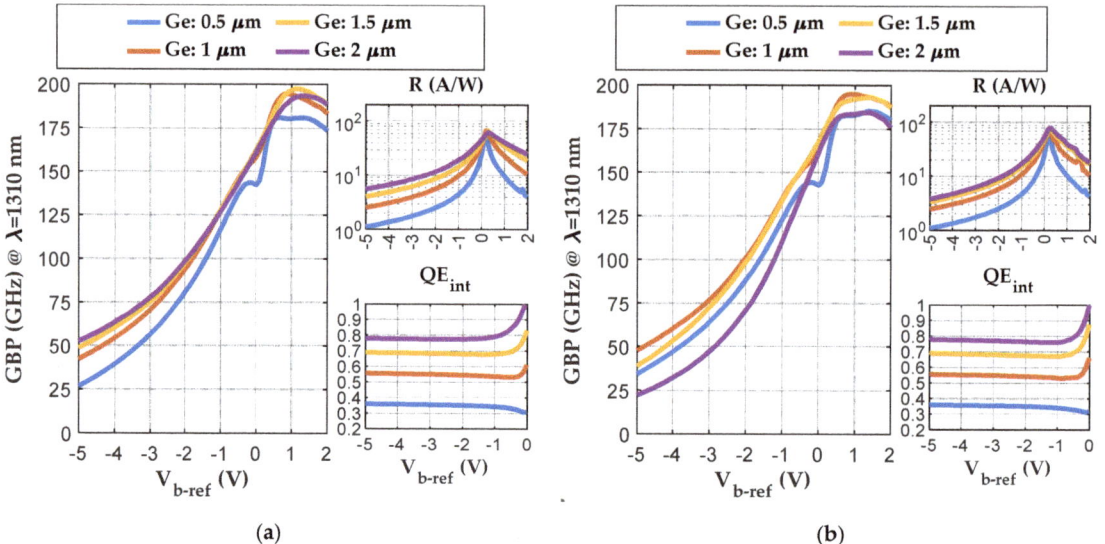

Figure 8. The dependence of the gain bandwidth product (GBP), responsivity (R), and internal quantum efficiency (QE) on the thickness of the absorption layer versus referenced bias voltage. Two different multiplication and absorption layer doping concentrations were simulated: (**a**) 1×10^{15} cm^{-3}; (**b**) 5×10^{15} cm^{-3}.

3.2.4. The Photodiode Spectral Response

The dependence of the internal quantum efficiency (QE) on the thickness of the absorption layer versus the wavelength at the biasing voltage corresponding to unity gain is illustrated in Figure 9. It is evident that all structures exhibit a comparable spectral response for incident light with wavelengths less than 0.7 µm, as the majority of photons are effectively absorbed in the silicon material. Moreover, these structures share the same silicon layer thickness.

On the other hand, the spectral response is highly influenced by the thickness of the Ge absorption layer for incident light with wavelengths ranging from 0.7 µm to 1.6 µm. As the thickness of the Ge absorption layer increases, the quantum efficiency also increases due to the increased number of absorbed photons. Figure 9a,b demonstrate that modifying the doping concentration of the multiplication and absorption layers does not impact the spectral response. This is attributed to the similar recombination rates along the structure at the relevant bias voltage corresponding to unity gain. Consequently, increasing the thickness of the absorption layer enhances the spectral response by capturing a greater number of absorbed photons.

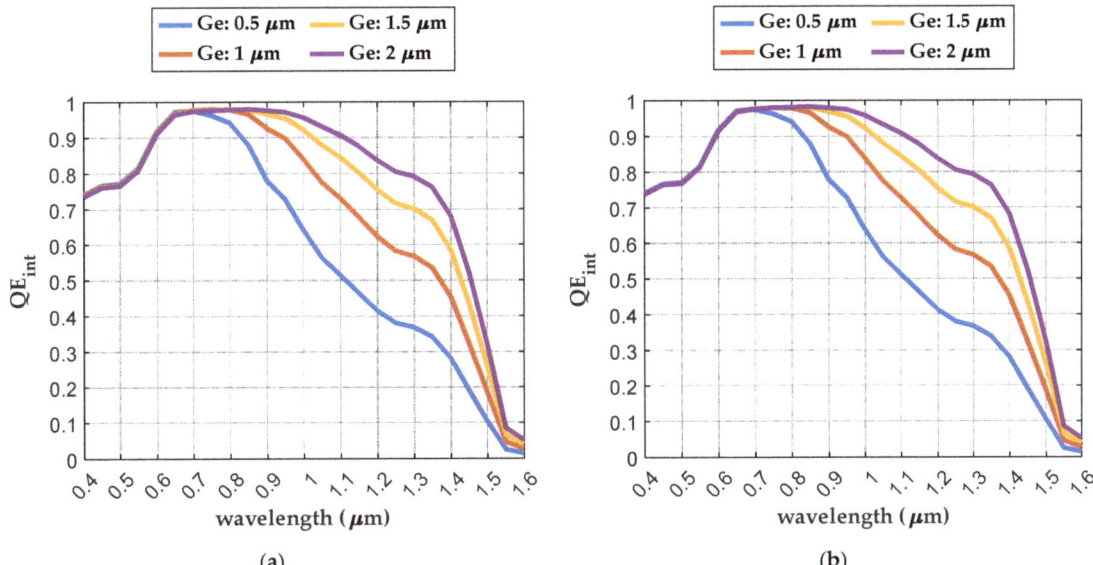

Figure 9. Dependence of the internal quantum efficiency (QE) on the thickness of the absorption layer versus wavelength at unity gain. Two different multiplication and absorption layer doping concentrations were simulated: (**a**) 1×10^{15} cm^{-3}; (**b**) 5×10^{15} cm^{-3}.

In summary, the dependence of various performance parameters on the thickness of the absorption layer in a SACM Ge/Si avalanche photodiode has been investigated. It has been observed that the gain, bandwidth, responsivity, and quantum efficiency show a clear dependence on the absorption layer thickness. A thinner absorption layer results in higher bandwidth, indicating a faster response to high-frequency optical signals. However, this also leads to a decrease in responsivity and quantum efficiency. By considering factors such as breakdown voltage, GBP, responsivity, and quantum efficiency, we can determine the optimal design for a SACM Ge/Si APD. Our simulations suggest that an optimal configuration would involve a 0.5 µm thick multiplication layer, a 1.5 µm thick absorption layer, and a doping concentration of 5×10^{15} cm^{-3} for both layers.

4. Conclusions

The separate-absorption-charge-multiplication (SACM) germanium on silicon avalanche photodiode structure has been designed, modeled, and analyzed to investigate the performance dependences on the doping concentrations and thicknesses of the absorption and multiplication layers. These performance characteristics include breakdown voltage, gain, bandwidth, responsivity, and quantum efficiency. The simulations were performed by comparing various multiplication and absorption layers' thicknesses, whereas the other structural and material parameters were left unchanged.

In general, optimizing the thickness of the multiplication and absorption layers in an APD requires a compromise between gain, bandwidth, responsivity, and quantum efficiency in order to balance the characteristics of the device. The dimensions that will ensure optimal performance depend on the specific requirements for the device and the operating conditions. Based on our analysis and modeling results, we can deduce that an optimal design for an SACM Ge/Si APD would feature a multiplication layer with a thickness of 0.5 µm, an absorption layer with a thickness of 1.5 µm, and a doping concentration of 5×10^{15} cm^{-3} for both layers.

Author Contributions: Conceptualization, H.D., K.K., A.K., R.D. and K.L.; writing—original draft preparation, H.D. and K.K.; writing—review and editing, H.D., K.K., A.K., R.D. and K.L.; supervision, A.K.; project administration A.K.; funding acquisition, A.K. All authors have read and agreed to the published version of the manuscript.

Funding: The reported study was supported by the Tomsk State University Development Programme (Priority 2030, No. 2.0.6.22).

Data Availability Statement: The authors declare that the data supporting the findings of this study are available within the article.

Conflicts of Interest: The authors declare no conflict of interest.

References

1. Benedikovic, D.; Aubin, G.; Haetmann, J.-M.; Amar, F.; Le Roux, X.; Alonso-Ramos, C.; Cassan, É.; Marris-Morini, D.; Boeuf, F.; Fédéli, J.-M.; et al. Silicon-Germanium Avalanche Receivers with fJ/bit Energy Consumption. *IEEE J. Sel. Top. Quantum Electron.* **2022**, *28*, 3802508. [CrossRef]
2. Srinivasan, S.A.; Lambrecht, J.; Guermandi, D.; Lardenois, D.; Berciano, M.; Absil, P.; Bauwelinck, J.; Yin, X.; Pantouvaki, M.; Campenhout, J.V. 56 Gb/s NRZ O-Band Hybrid BiCMOS-Silicon Photonics Receiver Using Ge/Si Avalanche Photodiode. *J. Light. Technol.* **2021**, *39*, 1409–1415. [CrossRef]
3. Zhang, J.; Kuo, B.P.-P.; Radic, S. 64 Gb/s PAM4 and 160 Gb/s 16QAM modulation reception using a low-voltage Si-Ge waveguide-integrated APD. *Opt. Express* **2020**, *28*, 23266. [CrossRef]
4. Wang, B.; Huang, Z.; Sorin, W.V.; Zeng, X.; Liang, D.; Fiorentino, M.; Beausoleil, R.G. A Low-Voltage Si-Ge Avalanche Photodiode for High-Speed and Energy Efficient Silicon Photonic Links. *J. Light. Technol.* **2020**, *38*, 3156–3163. [CrossRef]
5. Chen, H.T.; Verheyen, V.P.; De Heyn, P.; Lepage, G.; De Coster, J.; Absil, P.; Yin, X.; Bauwelinck, J.; Van Campenhout, J.; Roelkens, G. High sensitivity 10 Gb/s Si photonic receiver based on a low-voltage waveguide-coupled Ge avalanche photodetector. *Opt. Express* **2015**, *23*, 815–822. [CrossRef] [PubMed]
6. Izhnin, I.I.; Lozovoy, K.A.; Kokhanenko, A.P.; Khomyakova, K.I.; Douhan, R.M.H.; Dirko, V.V.; Voitsekhovskii, A.V.; Fitsych, O.I.; Akimenko, Y.N. Single-photon avalanche diode detectors based on group IV materials. *Appl. Nanosci.* **2022**, *12*, 253–263. [CrossRef]
7. Douhan, R.; Lozovoy, K.; Kokhanenko, A.; Deeb, H.; Dirko, V.; Khomyakova, K. Recent Advances in Si-Compatible Nanostructured Photodetectors. *Technologies* **2023**, *11*, 17. [CrossRef]
8. Hakkel, K.D.; Petruzzella, M.; Ou, F.; van Klinken, A.; Pagliano, F.; Liu, T.; van Veldhoven, R.P.J.; Fiore, A. Integrated near-infrared spectral sensing. *Nat. Commun.* **2022**, *13*, 69. [CrossRef]
9. Liu, D.; Li, T.; Tang, B.; Zhang, P.; Wang, W.; Liu, M.; Li, Z. A Near-Infrared CMOS Silicon Avalanche Photodetector with Ultra-Low Temperature Coefficient of Breakdown Voltage. *Micromachines* **2022**, *13*, 47. [CrossRef]
10. Li, Y.; Luo, X.; Liang, G.; Lo, G.-Q. Demonstration of Ge/Si Avalanche Photodetector Arrays for Lidar Application. In Proceedings of the 2019 Optical Fiber Communications Conference and Exhibition (OFC), San Diego, CA, USA, 3–7 March 2019; pp. 1–3.
11. Campbell, J.C.; Demiguel, S.; Ma, F.; Beck, A. Recent Advances in Avalanche Photodiodes. *J. Light. Technol.* **2016**, *34*, 278–285. [CrossRef]
12. Lacaita, A.; Francese, P.A.; Zappa, F.; Cova, S. Single-photon detection beyond 1 μm: Performance of commercially available germanium photodiodes. *Appl. Opt.* **1994**, *33*, 6902–6918. [CrossRef]
13. Zaoui, W.S.; Chen, H.-W.; Bowers, J.E.; Kang, Y.; Morse, M.; Paniccia, M.J.; Pauchard, A.; Campbell, J.C. Frequency response and bandwidth enhancement in Ge/Si avalanche photodiodes with over 840 GHz gain-bandwidth-product. *Opt Express* **2009**, *17*, 12641–12649. [CrossRef] [PubMed]
14. Kang, Y.; Liu, H.-D.; Morse, M.; Paniccia, M.J.; Zadka, M.; Litski, S.; Sarid, G.; Pauchard, A.; Kuo, Y.-H.; Chen, H.-W.; et al. Monolithic germanium/silicon avalanche photodiodes with 340 GHz gain-bandwidth product. *Nat. Photonics* **2009**, *3*, 59–63. [CrossRef]
15. Duan, N.; Liow, T.-Y.; Lim, A.E.-J.; Ding, L.; Lo, G.Q. 310 GHz gain-bandwidth product Ge/Si avalanche photodetector for 1550 nm light detection. *Opt. Express* **2012**, *20*, 11031–11036. [CrossRef]
16. Warburton, R.E.; Intermite, G.; Myronov, M.; Allred, P.; Leadley, D.R.; Gallacher, K.; Paul, D.J.; Pilgrim, N.J.; Lever, L.J.M.; Ikonic, Z.; et al. Ge-on-Si single-photon avalanche diode detectors: Design, modeling, fabrication, and characterization at wavelengths 1310 and 1550 nm. *IEEE Trans. Electron Devices* **2013**, *60*, 3807–3813. [CrossRef]
17. Huang, Z.; Li, C.; Liang, D.; Yu, K.; Santori, C.; Fiorentino, M.; Sorin, M.; Palermo, S.; Beausoleil, R.G. 25 Gbps low-voltage Waveguide Si-Ge Avalanche Photodiode. *Optica* **2016**, *3*, 793–798. [CrossRef]
18. Kang, Y.; Zadka, M.; Litski, S.; Sarid, G.; Morse, M.; Paniccia, M.J.; Kuo, Y.-H.; Bowers, J.; Beling, A.; Liu, H.-D.; et al. Epitaxially-grown Ge/Si avalanche photodiodes for 1.3 μm light detection. *Opt. Express* **2008**, *16*, 9365–9371. [CrossRef]
19. Zeng, X.; Huang, Z.; Wang, B.; Liang, D.; Fiorentino, M.; Beausoleil, R.G. Silicon–germanium avalanche photodiodes with direct control of electric field in charge multiplication region. *Optica* **2019**, *6*, 772–777. [CrossRef]

20. Wang, B.; Huang, Z.; Yuan, Y.; Liang, D.; Zeng, X.; Fiorentino, M.; Beausoleil, R.G. 64 Gb/s low-voltage waveguide SiGe avalanche photodiodes with distributed Bragg reflectors. *Photonics Res.* **2020**, *8*, 1118–1123. [CrossRef]
21. Samani, A.; Carpentier, O.; El-Fiky, E.; Jacques, M.; Kumar, A.; Wang, Y.; Guenin, L.; Gamache, C.; Koh, P.-C.; Plant, D.V. Highly Sensitive, 112 Gb/s O-band Waveguide Coupled Silicon-Germanium Avalanche Photodetectors. In Proceedings of the 2019 Optical Fiber Communications Conference and Exhibition (OFC), San Diego, CA, USA, 3–7 March 2019; pp. 1–3.
22. Carpentier, O.; Samani, A.; Jacques, M.; El-Fiky, E.; Alam, S.; Wang, Y.; Koh, P.-C.; Calvo, N.A.; Plant, D. High Gain-Bandwidth Waveguide Coupled Silicon Germanium Avalanche Photodiode. In Proceedings of the 2020 Conference on Lasers and Electro-Optics (CLEO), San Jose, CA, USA, 10–15 May 2020; pp. 1–2.
23. Kim, G.; Kim, S.; Kim, S.A.; Oh, J.H.; Jang, K.-S. NDR-effect vertical-illumination-type Ge-on-Si avalanche photodetector. *Opt. Lett.* **2018**, *43*, 5583–5586. [CrossRef]
24. Huang, M.; Cai, P.; Li, S.; Hou, G.; Zhang, N.; Su, T.-I.; Hong, C.; Pan, D. 56 GHz Waveguide Ge/Si Avalanche Photodiode. In Proceedings of the 2018 Optical Fiber Communications Conference and Exhibition (OFC), San Diego, CA, USA, 11–15 March 2018; pp. 1–3.
25. Zeng, Q.Y.; Pan, Z.X.; Zeng, Z.H.; Wang, J.T.; Guo, C.; Gong, Y.F.; Liu, J.C.; Gong, Z. Space charge effects on the bandwidth of Ge/Si avalanche photodetectors. *Semicond. Sci. Technol.* **2020**, *35*, 035026. [CrossRef]
26. Zhang, J.; Lin, H.; Liu, M.; Yang, Y. Research on the leakage current at sidewall of mesa Ge/Si avalanche photodiode. *AIP Adv.* **2021**, *11*, 075320. [CrossRef]
27. Masini, G.; Colace, L.; Assanto, G.; Luan, H.-C.; Kimerling, L.C. High-performance p-i-n Ge on Si photodetectors for the near infrared: From model to demonstration. *IEEE Trans. Electron Devices* **2001**, *48*, 1092–1096. [CrossRef]
28. Huang, S.; Li, C.; Zhou, Z.; Chen, C.; Zheng, Y.; Huang, W.; Lai, H.; Chen, S. Depth-dependent etch pit density in Ge epilayer on Si substrate with a self-patterned Ge coalescence island template. *Thin Solid Films* **2012**, *520*, 2307–2310. [CrossRef]
29. Wei, Y.; Cai, X.; Ran, J.; Yang, J. Analysis of dark current dependent upon threading dislocations in Ge/Si heterojunction photodetectors. *Microelectron. Int.* **2012**, *29*, 136–140. [CrossRef]
30. Synopsys and Inc. *Sentaurus Device User Guide Version H-2020.03*; Synopsys: Mountain View, CA, USA, 2017.
31. Caughey, D.M.; Thomas, R.E. Carrier mobilities in silicon empirically related to doping and field. *Proc. IEEE* **1967**, *55*, 2192–2193. [CrossRef]
32. Ke, S.; Lin, S.; Mao, D.; Ji, X.; Huang, W.; Xu, J.; Li, C.; Chen, S. Interface State Calculation of the Wafer-Bonded Ge/Si Single-Photon Avalanche Photodiode in Geiger Mode. *IEEE Trans. Electron Devices* **2017**, *64*, 2556–2563. [CrossRef]
33. Xu, Y.; Xiang, P.; Xie, X.; Huang, Y. A new modeling and simulation method for important statistical performance prediction of single photon avalanche diode detectors. *Semicond. Sci. Technol.* **2016**, *31*, 065024. [CrossRef]
34. Ke, S.; Lin, S.; Mao, D.; Ye, Y.; Ji, X.; Huang, W.; Li, C.; Chen, S. Design of wafer-bonded structures for near room temperature Geiger-mode operation of germanium on silicon single-photon avalanche photodiode. *Appl. Opt.* **2017**, *56*, 4646. [CrossRef]
35. Selberherr, S. *Analysis and Simulation of Semiconductor Devices*; Springer: New York, NY, USA, 1984. [CrossRef]
36. Ando, H.; Kanbe, H. Effect of avalanche build-up time on avalanche photodiode sensitivity. *IEEE J. Quantum Electron.* **1985**, *21*, 251–255. [CrossRef]

Disclaimer/Publisher's Note: The statements, opinions and data contained in all publications are solely those of the individual author(s) and contributor(s) and not of MDPI and/or the editor(s). MDPI and/or the editor(s) disclaim responsibility for any injury to people or property resulting from any ideas, methods, instructions or products referred to in the content.

Article

GaAs Quantum Dot Confined with a Woods–Saxon Potential: Role of Structural Parameters on Binding Energy and Optical Absorption

Hassen Dakhlaoui [1], Walid Belhadj [2,*], Haykel Elabidi [2], Fatih Ungan [3] and Bryan M. Wong [4,*]

[1] Nanomaterials Technology Unit, Basic and Applied Scientific Research Center (BASRC), Physics Department, College of Science of Dammam, Imam Abdulrahman Bin Faisal University, P.O. Box 1982, Dammam 31441, Saudi Arabia
[2] Physics Department, Faculty of Applied Science, Umm AL-Qura University, P.O. Box 715, Makkah 21955, Saudi Arabia; haelabidi@uqu.edu.sa
[3] Department of Physics, Faculty of Science, Sivas Cumhuriyet University, 58140 Sivas, Turkey
[4] Materials Science & Engineering Program, Department of Chemistry, and Department of Physics & Astronomy, University of California-Riverside, Riverside, CA 92521, USA
* Correspondence: wbbelhadj@uqu.edu.sa (W.B.); bryan.wong@ucr.edu (B.M.W.)

Citation: Dakhlaoui, H.; Belhadj, W.; Elabidi, H.; Ungan, F.; Wong, B.M. GaAs Quantum Dot Confined with a Woods–Saxon Potential: Role of Structural Parameters on Binding Energy and Optical Absorption. *Inorganics* 2023, 11, 401. https://doi.org/10.3390/inorganics11100401

Academic Editor: Sake Wang, Minglei Sun and Nguyen Tuan Hung

Received: 18 September 2023
Revised: 7 October 2023
Accepted: 10 October 2023
Published: 13 October 2023

Copyright: © 2023 by the authors. Licensee MDPI, Basel, Switzerland. This article is an open access article distributed under the terms and conditions of the Creative Commons Attribution (CC BY) license (https://creativecommons.org/licenses/by/4.0/).

Abstract: We present the first detailed study of optical absorption coefficients (OACs) in a GaAs quantum dot confined with a Woods–Saxon potential containing a hydrogenic impurity at its center. We use a finite difference method to solve the Schrödinger equation within the framework of the effective mass approximation. First, we compute energy levels and probability densities for different parameters governing the confining potential. We then calculate dipole matrix elements and energy differences, $E_{1p} - E_{1s}$, and discuss their role with respect to the OACs. Our findings demonstrate the important role of these parameters in tuning the OAC to enable blue or red shifts and alter its amplitude. Our simulations provide a guided path to fabricating new optoelectronic devices by adjusting the confining potential shape.

Keywords: optical absorption coefficient; spherical quantum dots; Schrödinger equation; hydrogenic impurity; Woods–Saxon potential

1. Introduction

The tunability of energy levels in low dimensional systems such as quantum wells (QWs), quantum wires (QWRs), and quantum dots (QDs) enable a multitude of optoelectronic devices, such as quantum cascade lasers, optical modulators, optical switches, and infrared photodetectors [1–4]. In addition, QDs are used in the creation of universal memory elements due to their spatial distribution of free carriers that are confined in three dimensions [5–8]. Generally, the position of different energy levels is determined via the geometrical shape of the confining potential of the quantum structure, such as square, parabolic, semi-parabolic, Gaussian, Razavy, Konwent, and Manning shapes [9–14]. QDs are of particular interest in optical applications due to their luminescence, potential to emit different frequencies with intense efficacies, high extinction, and prolonged lifetimes [15–17]. For these reasons, QDs are used in other technological applications such as light-emitting diodes (LEDs), electronic transistors, medical laser imaging, biosensors, quantum cascade lasers, and quantum computing architectures [18–25].

QDs generally show larger energetic separations between different levels compared to QWs and QWRs due to the three-dimensional confinement of carriers. They also give more intense density of states (DOS) than other quantum systems, which enables them to be used in amplifier applications. In addition to the geometry and shape of the confining potential, the incorporation of a hydrogenic impurity in QDs can modulate electronic and optical absorption coefficients (OACs) due to the electrostatic attraction between the free electrons and the impurity [26–32]. Previous work on OACs in QDs has focused on both theoretical and experimental studies [33–36]. For instance, Schrey et al. studied the optical absorption of quantum dots in photodetectors and analyzed the effect of QD size on their minibands [37]. Bahar et al. calculated OACs of a QD with a hydrogenic impurity in a Mathieu potential and found that the OAC and refractive index were affected by hydrostatic pressure and temperature variations [38]. Batra and coauthors examined structural parameters and the optical response of a QD with a tunable Kratzer confining potential [39]. Bassani et al. treated the effects of donor and acceptor impurities on OACs in a spherical QD [40]. The process of intraband and interband absorption in an InGaAs/GaAs QD was studied by Narvaez et al. [41], and the effects of size and distance separating QDs were evaluated by Stoleru et al. [42] The oscillator strengths between lower energy state transitions in a spherical QD with a hydrogenic impurity were calculated by Yilmaz et al. [43]. Kirak and coauthors evaluated the effect of an applied electric field on the OAC in a spherical QD with a parabolic potential under the influence of a hydrogenic impurity [44]. Fakkahi et al. studied OACs and oscillator strengths in multilayer spherical QDs under the influence of a radial electric field and hydrogenic impurities. Other works on OACs in multiple spherical QDs are also discussed in references [45–48].

Motivated by these studies, we investigate the electronic and optical properties of electrons confined in a GaAs quantum dot with a radial confinement described by the Woods–Saxon confining potential. The functional form of this potential was first proposed to describe and interpret interactive forces in the nuclear shell model [49]. Furthermore, this confining potential describes a smooth interface structure and gives an accurate description of aluminum diffusion from the AlGaAs barrier towards the GaAs quantum well. Our study commences with a calculation of the 1s and 1p energy levels and their probability densities as a function of structural parameters in the Woods–Saxon potential. We then analyze the dipole matrix elements (DMEs) and OACs as the parameters of the Woods–Saxon potential are varied in the presence of a hydrogenic impurity. Further details and approximations of our theoretical model are given in Section 2. In Section 3, our findings and the resulting physical observables are discussed. Finally, Section 4 summarizes our results.

2. Theoretical Modeling

2.1. Woods–Saxon Potential Form

We begin this section by discussing the confining potential and its structural parameters. These parameters alter the geometrical form of the potential and affect the position of different energy levels. When the confining potential is spherically symmetric, the carrier's motion is quantized and described by angular and magnetic quantum numbers, with the associated wave functions being expressed as a function of the well-known spherical harmonics. The form of the radial electronic wavefunctions is mainly determined by the geometrical shape of the confining potential. Since the energy separation, $E_{1p} - E_{1s}$, between the initial and final states, plays a major role in the OAC expression, we examine its dependence on QD size, the structural parameters of the confining potential, or both.

We first examine the radial Woods–Saxon potential, which is given by [45]

$$V_{\text{ws}}(r) = \frac{V_0}{1 + \exp[(R_0 - r)/\gamma]} + \frac{V_0}{1 + \exp[(R_0 + r)/\gamma]}. \tag{1}$$

V_0 is the height of the Woods–Saxon potential and $R_0 = R/2$, where R denotes the QD radius, and γ is a parameter characterizing the slope between the well and barrier regions.

Figure 1 depicts a schematic of the quantum dot, which consists of a GaAs core with radius $R = 25$ nm surrounded by an AlGaAs barrier. This latter has an external radius of $R_{\text{ext}} = 2R$.

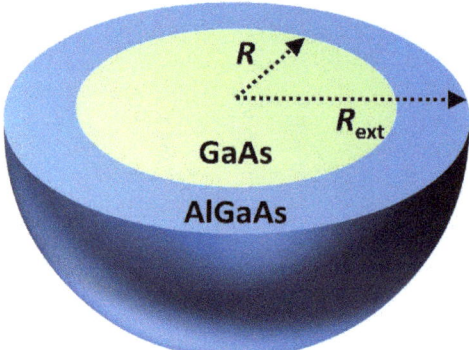

Figure 1. Schematic structure of spherical GaAs quantum dot surrounded by an AlGaAs barrier.

Before studying the optical properties of our structure, we plot the geometrical dependence of the Woods–Saxon potential on the parameter γ in Figure 2a–d. The radius of the QD is $R = 25$ nm. For $\gamma = 5$ Å, the Woods–Saxon potential resembles a square quantum well since it takes a flat form between 0 and 5 Å. However, when γ increases, the bottom of the potential becomes more parabolic. Furthermore, the top of the well becomes more curved as γ increases. For instance, the potential reaches 1500 meV at $r = 10$ nm for $\gamma = 5$ Å (Figure 2a); however, it reaches this value at $r = 15$ nm for $\gamma = 20$ Å in Figure 2d. Increasing the parameter γ influences the distribution of the confined energy levels and consequently affects the energy separation and OAC.

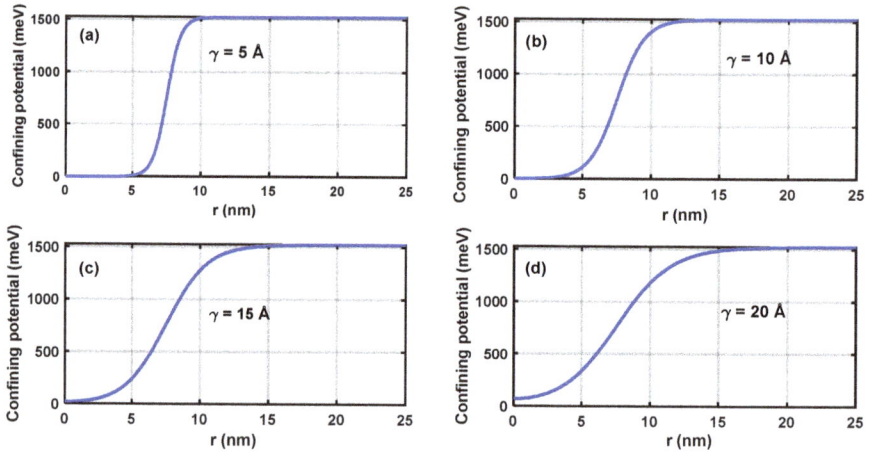

Figure 2. Woods–Saxon potential profile for (**a**) $\gamma = 5$ Å, (**b**) $\gamma = 10$ Å, (**c**) $\gamma = 15$ Å, and (**d**) $\gamma = 20$ Å. The radius of the QD is fixed at $R = 25$ nm with $R_0 = R/2$, $V_0 = 0.228$ eV, and $R_{ext} = 2R$.

2.2. Calculation of Electronic and Optical Properties

An electron in a spherical QD with a hydrogenic impurity within the effective mass approximation can be completely described by solving the radial Schrödinger equation [10,43,44]:

$$\left[-\frac{\hbar^2}{2} \vec{\nabla}_r \left(\frac{1}{m^*(r)} \vec{\nabla}_r \right) + \frac{\ell(\ell+1)\hbar^2}{2m^*(r)\, r^2} - \frac{Z\, e^2}{\varepsilon\, r} + V_{ws}(r) \right] R_{n\ell}(r) = E_{n\ell}\, R_{n\ell}(r), \qquad (2)$$

where $m^*(r)$ is the position-dependent mass of the electron, $\hbar$ represents the reduced Planck constant, ε is the dielectric constant, and ℓ is the angular quantum number. Furthermore, $R_{n\ell}(r)$ and $E_{n\ell}$ are the radial wavefunction and energy eigenvalue, respectively.

The first term in Equation (2) represents the kinetic energy, whereas the second term containing $\ell(\ell+1)$ denotes the centrifugal contribution of the potential due to the spherical symmetry of the Woods–Saxon potential. The third term represents the electron-impurity attraction. The two cases, $Z = 0$ and $Z = 1$, correspond to the absence and presence of the hydrogenic impurity, respectively. $V_{WS}(r)$ represents the Woods–Saxon potential which is a radial confinement term. To compute $E_{n\ell}$ and $R_{n\ell}(r)$, we discretized Equation (2) using the finite difference method and transformed it into a linear eigenvalue equation of the form $AX = \lambda X$, where A is a tridiagonal matrix, X represents $R_{n\ell}(r)$, and λ denotes $E_{n\ell}$. The 1D discretization of the radial Schrödinger equation was carried out with a finite difference method (FDM). Thus, Equation (2) takes the linear form:

$$R_{n\ell}(j+1)\left[-\frac{\hbar^2}{2m^* r_j(\Delta r)} - \frac{\hbar^2}{2m^*(\Delta r)^2}\right] + R_{n\ell}(j)\left[\frac{\hbar^2}{m^*(\Delta r)^2} + \frac{\ell(\ell+1)}{m^*(r_j.\Delta r)^2} + V_{WS}(j)\right]$$
$$+ R_{n\ell}(j-1)\left[\frac{\hbar^2}{2m^* r_j(\Delta r)} - \frac{\hbar^2}{2m^*(\Delta r)^2}\right] = E_{n\ell} R_{n\ell}(j), \quad (3)$$

where $r_j = j\Delta r$ $(j = 1, \ldots, N)$ and $\Delta r = \frac{R}{N}$ is the mesh discretization. Equation (3) is of the form $Hx = \lambda x$, where λ is the energy $E_{n\ell}$, x is the radial wavefunction $R_{n\ell}(j)$, and H is a tridiagonal matrix with elements given by

$$H_{ij} = \begin{cases} \frac{\hbar^2}{m^*(\Delta r)^2} + \frac{\ell(\ell+1)}{m^*(r_j.\Delta r)^2} + V_{WS}(j), & \text{if } j = i \\ \frac{\hbar^2}{2m^* r_j(\Delta r)} - \frac{\hbar^2}{2m^*(\Delta r)^2}, & \text{if } j = i - 1 \\ -\frac{\hbar^2}{2m^* r(\Delta r)} - \frac{\hbar^2}{2m^*(\Delta r)^2}, & \text{if } j = i + 1 \\ 0, & \text{otherwise} \end{cases} \quad (4)$$

In our study, we assume that the radial wavefunction at the external boundary point $(N+1)$ is zero. The dimension of matrix H is $(N \times N)$, and in all of our calculations, we set $N = 1200$ with the boundary condition $R_{n\updownarrow}(r = R_{\text{ext}}) = 0$.

Optical absorption in the QD occurs when an electron in its initial level E_i is excited to a final energy E_f after absorption of a photon with energy $\hbar\omega = (E_f - E_i)$. According to Fermi's golden rule, the OAC can be written as [45]

$$\alpha(\hbar\omega) = \frac{16\pi^2 \delta_{FS} P_{if}}{n_r V_{\text{con}}} \hbar\omega \left|M_{if}\right|^2 \delta(E_f - E_i - \hbar\omega), \quad (5)$$

where P_{if}, δ_{FS}, and V_{con} represent the electron population difference, the fine structure, and the confinement volume, respectively. n_r represents the refractive index of the GaAs semiconductor, and $\left|M_{if}\right|$ denotes the DME of the transition. Furthermore, the $\Delta\ell = \pm 1$ selection rule satisfied by the quantum number ℓ is taken into consideration.

In the present paper, we address only the transition between the $1s$ and $1p$ states. Furthermore, the δ-function in the previous equation is substituted with a Lorentzian profile:

$$\delta(E_f - E_i - \hbar\omega) = \frac{\hbar\Gamma}{\pi\left[(E_f - E_i - \hbar\omega)^2 + (\hbar\Gamma)^2\right]}, \quad (6)$$

where $\hbar\omega$ is the energy of the incident photon, and $\hbar\Gamma$ is the width at half height of the Lorentzian function. In the next section, and for simplicity of notation, we consider the initial state $(i = 1)$ to be $1s$ and the final state $(f = 2)$ to be the $1p$ state, so the term $\left|M_{if}\right|^2$

in Equation (3) is simply designated as $|M_{12}|^2$. In our study, the electromagnetic radiation is polarized along the z-axis, and $|M_{12}|^2$ is given by the following expression [50–52]:

$$|M_{12}|^2 = \frac{1}{3}\left|\int_0^\infty R_{1s}(r) r^3 R_{1p}(r) dr\right|^2, \quad (7)$$

where the $\frac{1}{3}$ pre-factor arises from integration of the spherical harmonics. In addition to the optical absorption, we have evaluated the impurity binding energy of the neutral donor, defined as $E_b = E_{n,l}^{z=0} - E_{n,l}^{z=1}$, where $E_{n,l}^{z=0}$ and $E_{n,l}^{z=1}$ denote the energy levels for QDs without and with the impurity, respectively.

3. Results and Discussion

Atomic units ($\hbar = e = m_0 = 1$) are used throughout the rest of this work, which defines the Rydberg energy ($1R_y \cong 5.6$ meV) and Bohr radius ($1a_B \cong 100$ Å). In addition, V_0 is set at 0.228 eV, which corresponds to the band offset between GaAs and $Al_xGa_{(1-x)}As$ with $x = 0.3$. Additional physical parameters used in our simulation are $\hbar\Gamma = 3$ meV, $m^* = 0.067 m_0$, and $\varepsilon = 13.11\varepsilon_0$. The radius of the QD is fixed at $R = 25$ nm.

Figure 3a–d displays the probability densities of the 1s and 1p states with the confining potential in the absence of the hydrogenic impurity (i.e., $Z = 0$) for four values of the structural parameter ($\gamma = 5, 10, 15$, and 20 nm) with $R_0 = R/2$. Increasing γ also increases the amplitudes of the probability densities of the 1s and 1p states and widens the spatial extent of the wavefunctions. For instance, when $\gamma = 5$ nm, the 1s and 1p densities decay to zero at $r = 15$ and 20 nm, respectively; however, when $\gamma = 20$ nm, both densities decrease to zero at $r = 24$ nm. This behavior is due to the slope of the Woods–Saxon potential decreasing with increasing γ (see Figure 2a–d). The spread in $V_{ws}(r)$, especially near its top, enhances the amplitudes of the densities and enlarges their geometrical distribution along the r axis. This, in turn, modifies the energy levels and DMEs between the 1s and 1p wavefunctions since their overlap is now modified. Figure 4a–d plots these densities with an on-center hydrogenic impurity. In this case, there are two confining contributions. The first one is due to the geometrical behavior of the $V_{ws}(r)$ potential due to the increase in the parameter γ, and the second one arises from the electrostatic attraction between the hydrogenic impurity and the electron in different states. This is reflected in the decrease in the amplitudes in the 1s and 1p probabilities. Note that the amplitude for the 1s density is less sensitive than that of 1p for $\gamma = 15$ and 20 nm. For these values, the impact of geometrical confinement becomes negligible compared to that of the electrostatic attraction, and no additional changes are observed for $\gamma > 20$ nm.

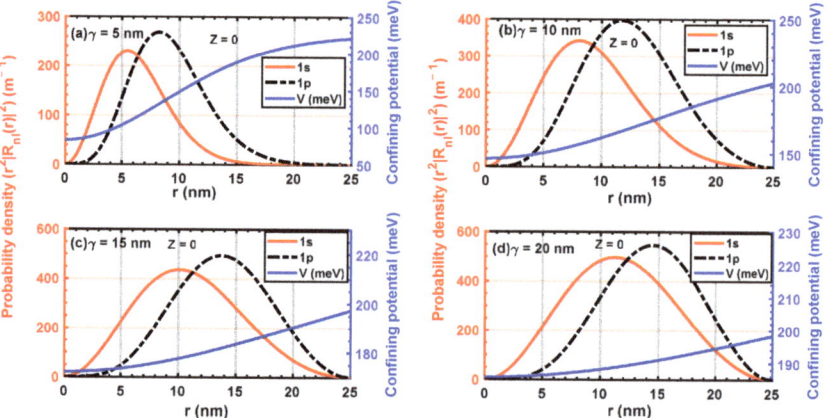

Figure 3. Confining potential and probability densities of the ground and first excited state for different values of γ: (**a**) $\gamma = 5$ nm; (**b**) $\gamma = 10$ nm; (**c**) $\gamma = 15$ nm; (**d**) $\gamma = 20$ nm. All results do not include the impurity ($Z = 0$). $R_0 = R/2$, $V_0 = 0.228$ eV, and $R_{ext} = 2R$.

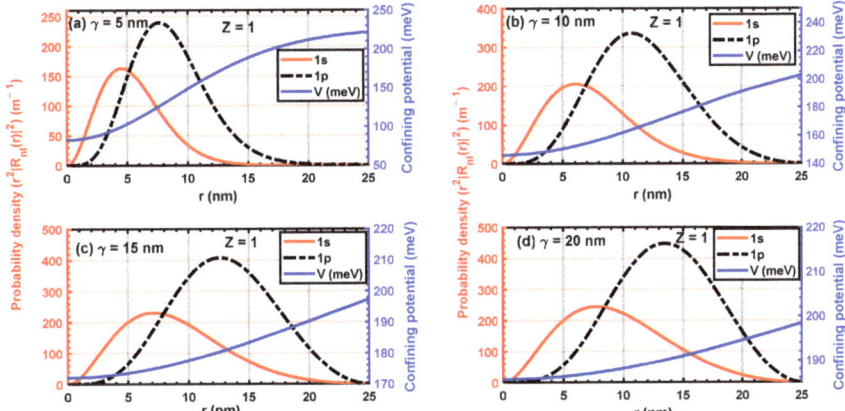

Figure 4. Confining potential and probability densities of the ground and first excited state for different values of γ: (**a**) $\gamma = 5$ nm; (**b**) $\gamma = 10$ nm; (**c**) $\gamma = 15$ nm; and (**d**) $\gamma = 20$ nm. All results include the impurity ($Z = 1$). $R_0 = R/2$, $V_0 = 0.228$ eV, and $R_{ext} = 2R$.

Figure 5 plots the energy levels E_{1p} and E_{1s}, which increase with γ. At low values of γ, the energy levels are well separated from each other; however, for higher values of γ, their separation is considerably reduced. The decrease in the energy difference between E_{1p} and E_{1s}, with and without the presence of hydrogenic impurity, is responsible for the red shift of the OAC, which we discuss later. Furthermore, for all values of γ, the energy levels in the presence of the hydrogenic impurity are less than those without the hydrogenic impurity. This is due to the attraction between the electron and the impurity, which causes the electron to be near the impurity at the center of the QD.

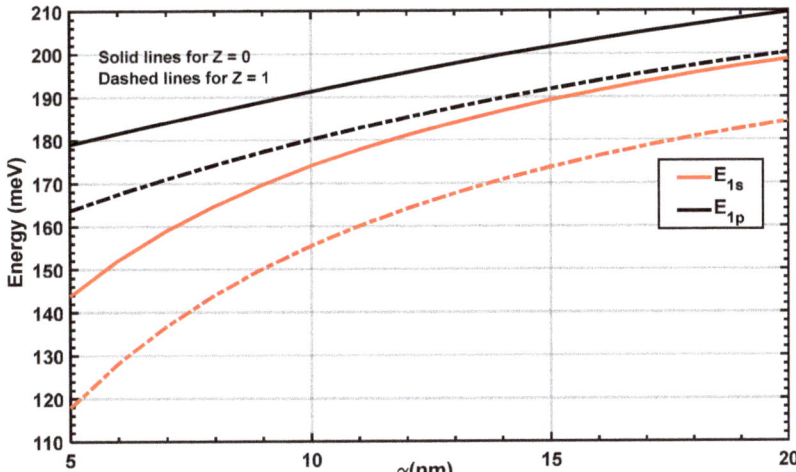

Figure 5. Variation of E_{1s} and E_{1p} for different values of γ with ($Z = 1$) and without ($Z = 0$) impurities. $R_0 = R/2$, $V_0 = 0.228$ eV, and $R_{ext} = 2R$.

From Equation (5), the OAC is proportional to $|M_{12}|^2$, which controls the amplitude of the OAC and explains the overlap between the $1s$ and $1p$ wavefunctions. Figure 6 plots its variation with the energy separation $(E_{1p} - E_{1s})$ as a function of γ. For $\gamma = 5$ nm, the values of $|M_{12}|^2$ with ($Z = 1$) and without impurity ($Z = 0$) are similar. However, when γ is increased, $|M_{12}|^2$ increases and takes higher values for $Z = 0$ than for $Z = 1$. This result is due to the change in the overlap between the 1s and 1p wavefunctions. In addition, Figure 5 shows that the energy separation $(E_{1p} - E_{1s})$ decreases for both cases (with and without impurity), resulting in a red shift in the OAC.

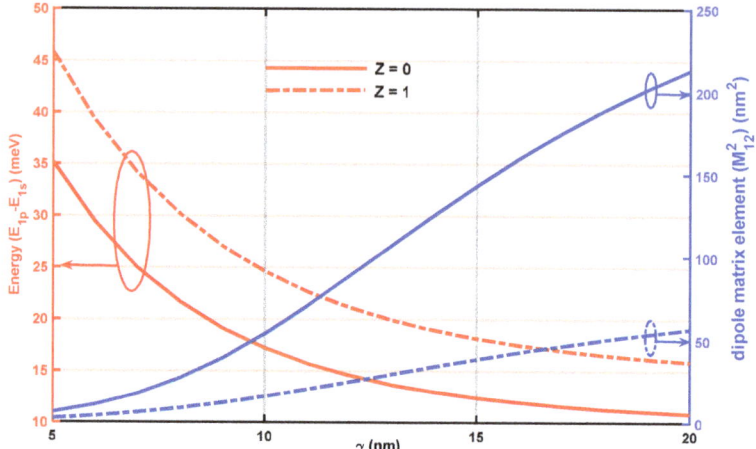

Figure 6. Variation of the energy separation $(E_{1p} - E_{1s})$ and dipole matrix element $|M_{12}|^2$ as a function of the parameter γ for $Z = 0$ (solid line) and $Z = 1$ (dashed line). $R_0 = R/2$, $V_0 = 0.228$ eV, and $R_{ext} = 2R$.

Figure 7 plots the OAC as a function of photon energy for $\gamma = 5$, 10, and 20 nm. We report results for two cases: with $(Z = 1)$ and without $(Z = 0)$ the hydrogenic impurity. The OAC amplitudes move towards lower energies (red shift) with increasing γ. This variation is in accordance with the variation of $(E_{1p} - E_{1s})$, previously shown in Figure 5. In addition, we note that the OAC amplitudes in the presence of the hydrogenic impurity are always smaller than those without the hydrogenic impurity. This is due to the difference in the DME with and without the presence of the impurity, as shown in Figure 5.

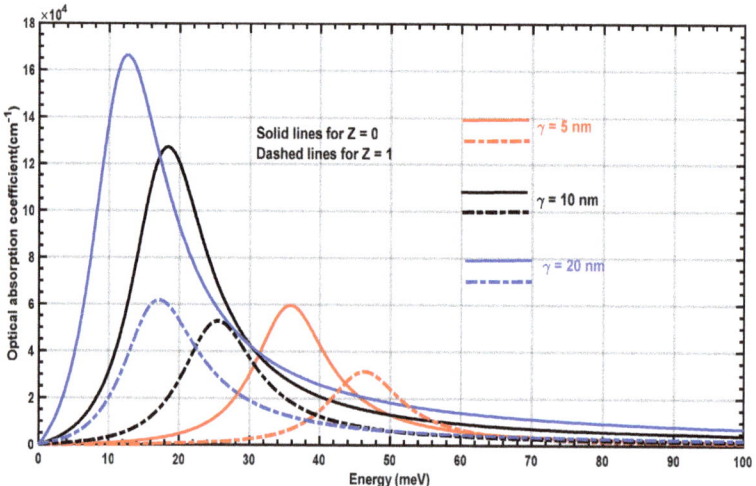

Figure 7. OAC as a function of incident photon energy for different γ values with $(Z = 1)$ and without $(Z = 0)$ impurities. $R_0 = R/2$, $V_0 = 0.228$ eV, and $R_{ext} = 2R$.

Figure 8 plots the binding energies of the 1s and 1p states as a function of γ. Both states gradually decrease with γ. For lower values, they decrease rapidly; however, for higher values ($\gamma > 15$ nm), the binding energies show a small variation. This behavior in binding energy for the $1s$ and $1p$ states is explained by the strong attraction near the center of the QD; however, for higher values of r, this attraction is reduced compared to the geometrical confinement, and consequently, the binding energy remains constant for all higher values of r.

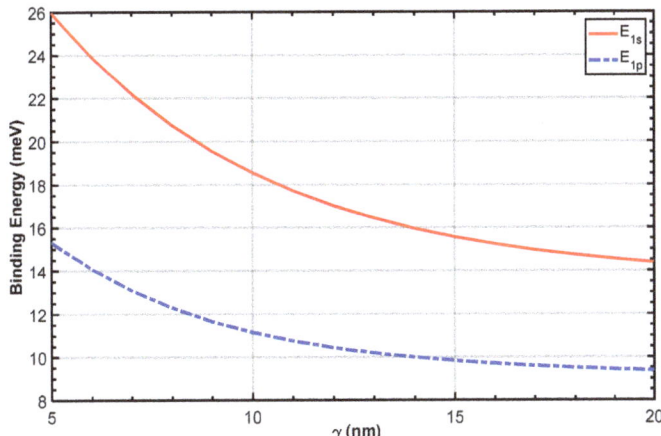

Figure 8. Binding energy for $1s$ and $1p$ states as a function of γ. $R_0 = R/2$, $V_0 = 0.228$ eV, and $R_{\text{ext}} = 2R$.

We now turn our attention to the effect of R_0. Figure 9a–d plots the probability densities of the lowest electronic states $1s$ and $1p$ with the confining potential in the absence of the hydrogenic impurity (i.e., $Z = 0$) for $R_0 = 7, 12, 17$, and 22 nm, with $\gamma = 10$ nm. Increasing R_0 enlarges the potential and minimizes its values at the center and surface of the quantum dot. Consequently, the two probability densities maintain the same spread; however, their amplitudes increase with R_0. The amplitude of the $1p$ density is more sensitive than that of $1s$ when R_0 increases. The influence of the hydrogenic impurity on these densities is shown in Figure 10a–d. The densities have the same spread along the radius of the quantum dot, but their amplitudes are reduced due to the electrostatic attraction introduced by the hydrogenic impurity. To evaluate the effect of the on-center impurity on the OAC, Figure 10 plots its variation as a function of the incident energy for three values of R_0. The OAC peak moves towards higher energy (blue shift) when R_0 increases from 8 to 18 nm. Subsequently, it moves in the direction of low energies, exhibiting a red shift. This double behavior can be interpreted via the variation in the energy separation between the $1s$ and $1p$ energy levels.

Figure 11 plots the variation of the $1s$ and $1p$ energy levels as a function of R_0, which shows a gradual decrease for the two cases (with and without impurity). This decrease is due to the enlargement of the confining potential with R_0 as shown in Figures 9 and 10. However, the slope of this decrease is slightly different. Figure 12 plots the energy separation $E_{1p} - E_{1s}$ as a function of the parameter R_0, which shows that this separation increases up to $R_0 = 16$ nm but subsequently decreases. This behavior confirms the red and blue shift shown in the OAC variation in Figure 13. In addition, Figure 12 shows the variation of the dipole matrix element $|M_{12}|^2$ as a function of R_0. This physical quantity decreases up to $R_0 = 16$ nm and then subsequently increases. This arises from the variation of the overlap between the R_{1p} and R_{1s} wave functions, which agree with the OAC trends shown in Figure 12. For $R_0 < 16$ nm, the OAC amplitude diminishes; however, for $R_0 > 16$ nm, the amplitude subsequently increases.

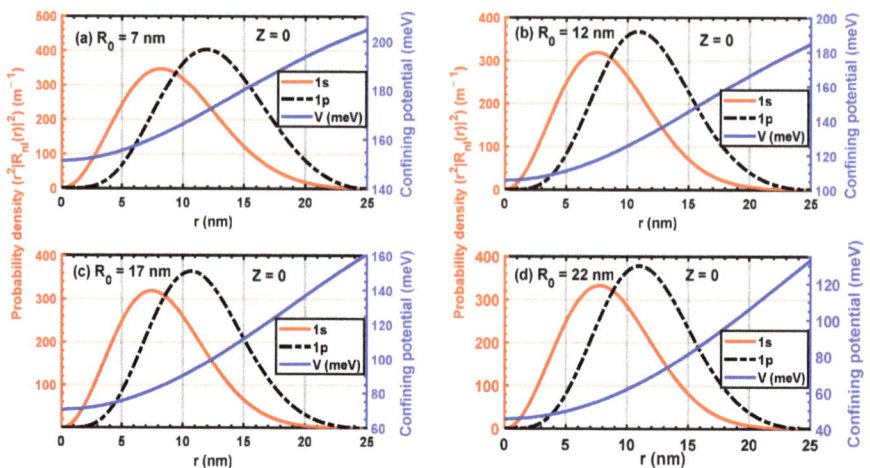

Figure 9. Confining potential and probability densities of the ground and first excited states for different values of R_0: (**a**) $R_0 = 7$ nm; (**b**) $R_0 = 12$ nm; (**c**) $R_0 = 17$ nm; and (**d**) $R_0 = 22$ nm. All results do not include the impurity ($Z = 0$). $\gamma = 10$ nm, $V_0 = 0.228$ eV, and $R = 25$ nm.

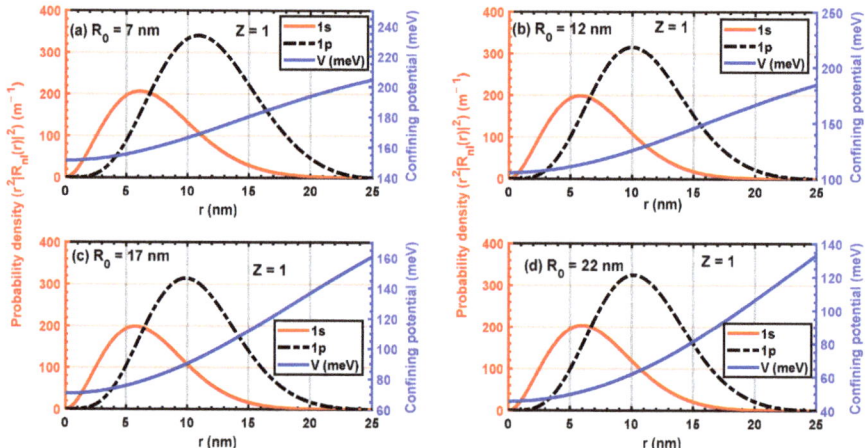

Figure 10. Confining potential and probability densities of the ground and first excited states for different values of R_0: (**a**) $R_0 = 7$ nm; (**b**) $R_0 = 12$ nm; (**c**) $R_0 = 17$ nm; and (**d**) $R_0 = 22$ nm. All results include the impurity ($Z = 1$). $\gamma = 10$ nm, $V_0 = 0.228$ eV, and $R = 25$ nm.

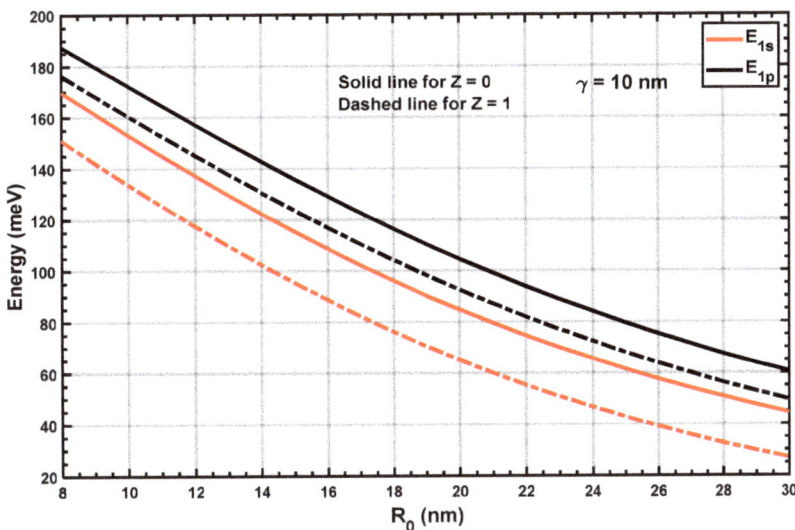

Figure 11. Variation of energy levels E_{1s} and E_{1p} for different values of R_0, with ($Z = 1$) and without ($Z = 0$) impurities. $V_0 = 0.228$ eV and $R = 25$ nm.

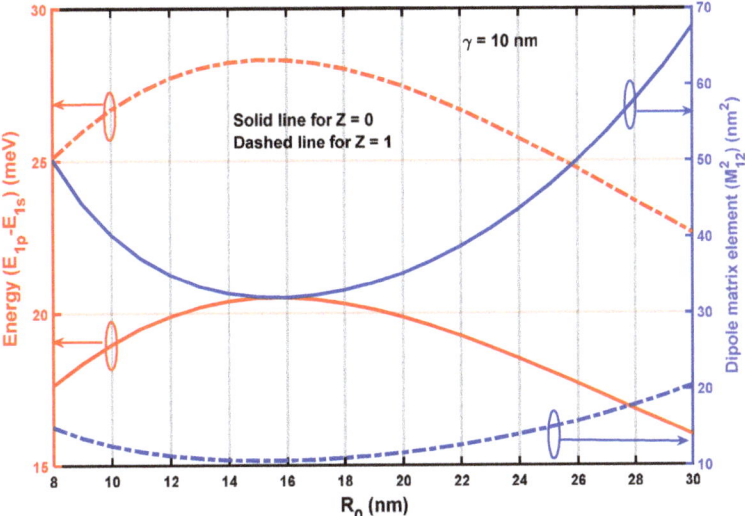

Figure 12. Variation of $E_{1p} - E_{1s}$ and $|M_{12}|^2$ with (dashed curve) and without (solid curve) the on-center impurity as a function of R_0. $V_0 = 0.228$ eV and $R = 25$ nm.

Finally, Figure 14 displays the binding energy as a function of R_0. The binding energy increases up to $R_0 = 16$ nm and subsequently diminishes gradually. Consequently, this critical value of R_0 can play an important role in shifting the OAC from red to blue as well as controlling the variation of the binding energy toward high or low values.

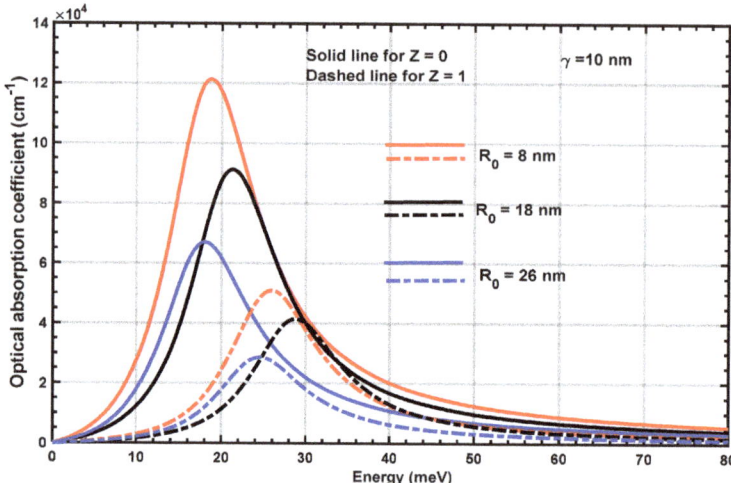

Figure 13. OAC as a function of incident photon energy for different values of R_0. Results are with ($Z = 1$) and without ($Z = 0$) impurities. $V_0 = 0.228$ eV and $R = 25$ nm.

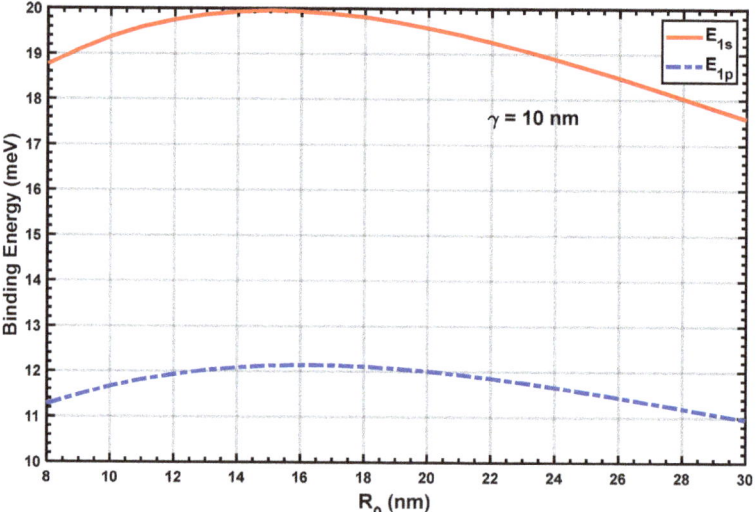

Figure 14. Binding energy for the $1s$ and $1p$ states as a function of R_0. $\gamma = 10$ nm, $V_0 = 0.228$ eV, and $R = 25$ nm.

4. Conclusions

In summary, we have presented the first study of the optical and electronic properties of a GaAs spherical QD with a Woods–Saxon potential in the presence of a hydrogenic impurity. By solving the radial part of the Schrödinger equation using the finite difference method, we obtain energy levels of $1s$ and $1p$ states and their probability densities. These quantities allow us to calculate dipole matrix elements, energy separations, OACs, and binding energies as a function of the parameters R_0 and γ. Our results indicate that increasing γ leads to a red shift of the OAC; however, an increase in R_0 initially gives rise to a blue shift and, subsequently, a red shift. We also demonstrated that the variation of the OAC amplitude is determined via the dipole matrix element, which effectively captures the overlap between R_{1p} and R_{1s}. Moreover, our findings indicate that the insertion of a hydrogenic impurity at the center of the QD considerably decreases the energy levels due to the strong attraction between the free electrons and the hydrogenic impurity. Our numerical calculations provide mechanistic insight into the electronic transport and optical properties of spherical QDs.

Author Contributions: H.D. and W.B. were responsible for the analytical and numerical calculations; H.E. was responsible for formal analysis and writing of the manuscript; F.U. and B.M.W. were responsible for the formal analysis and writing of the manuscript. All authors have read and agreed to the published version of the manuscript.

Funding: H.D., W.B. and H.E. acknowledge the Deputyship for Research and Innovation, Ministry of Education in Saudi Arabia for funding this research work through project number IFP22UQU4331235DSR206. B.M.W. acknowledges support from the UC Riverside Committee on Research grant.

Data Availability Statement: Not applicable.

Conflicts of Interest: The authors declare no conflict of interest.

References

1. Dakhlaoui, H.; Belhadj, W.; Musa, M.O.; Ungan, F. Binding energy, electronic states, and optical absorption in a staircase-like spherical quantum dot with hydrogenic impurity. *Eur. Phys. J. Plus* **2023**, *138*, 519. [CrossRef]
2. Durante, F.; Alves, P.; Karunasiri, G.; Hanson, N.; Byloos, M.; Liu, H.C.; Bezinger, A.; Buchanan, M. NIR, MWIR and LWIR quantum well infrared photodetector using interband and intersubband transitions. *Infrared Phys. Technol.* **2007**, *50*, 182.
3. Imamura, K.; Sugiyama, Y.; Nakata, Y.; Muto, S.; Yokoyama, N. New optical memory structure using self-assembled InAs quantum dots. *Jpn. J. Appl. Phys.* **1995**, *34*, L1445. [CrossRef]
4. Fickenscher, M.; Shi, T.; Jackson, H.E.; Smith, L.M.; Yarrison-Rice, J.M.; Zheng, C.; Miller, P.; Etheridge, J.; Wong, B.M.; Gao, Q.; et al. Optical, structural, and numerical investigations of GaAs/AlGaAs core–multishell nanowire quantum well tubes. *Nano Lett.* **2013**, *13*, 1016–1022. [CrossRef]
5. Marent, A.; Nowozin, T.; Geller, M.; Bimberg, D. The QD-Flash: A quantum dot-based memory device. *Semicond. Sci. Technol.* **2011**, *26*, 014026. [CrossRef]
6. Bonato, L.; Arikan, I.F.; Desplanque, L.; Coinon, C.; Wallart, X.; Wang, Y.; Ruterana, P.; Bimberg, D. Hole localization energy of 1.18 eV in GaSb quantum dots embedded in GaP. *Phys. Status Solidi B* **2016**, *253*, 1869. [CrossRef]
7. Abramkin, D.S.; Atuchin, V.V. Novel InGaSb/AlP Quantum dots for non-volatile memories. *Nanomaterials* **2022**, *12*, 3794. [CrossRef]
8. Abramkin, D.S.; Petrushkov, M.O.; Bogomolov, D.B.; Emelyanov, E.A.; Yesin, M.Y.; Vasev, A.V.; Bloshkin, A.A.; Koptev, E.S.; Putyato, M.A.; Atuchin, V.V.; et al. Structural properties and energy spectrum of novel GaSb/AlP self-assembled quantum dots. *Nanomaterials* **2023**, *13*, 910. [CrossRef]
9. AL-Naghmaisha, A.; Dakhlaoui, H.; Ghrib, T.; Wong, B.M. Effects of magnetic, electric, and intense laser fields on the optical properties of AlGaAs/GaAs quantum wells for terahertz photodetectors. *Phys. B Condens. Matter.* **2022**, *635*, 413838. [CrossRef]
10. Almansour, S.; Dakhlaoui, H.; Algrafy, E. The effect of hydrostatic pressure, temperature and magnetic field on the nonlinear optical properties of asymmetrical Gaussian potential quantum wells. *Chin. Phys. Lett.* **2016**, *33*, 027301.
11. Turkoglu, A.; Dakhlaoui, H.; Mora-Ramos, M.E.; Ungan, F. Optical properties of a quantum well with Razavy confinement potential: Role of applied external field. *Phys. E* **2021**, *134*, 114919. [CrossRef]
12. Dakhlaoui, H.; Altuntas, I.; Mora-Ramos, M.E.; Ungan, F. Numerical simulation of linear and nonlinear optical properties in heterostructure based on triple Gaussian quantum wells: Effects of applied external fields and structural parameters. *Eur. Phys. J. Plus* **2021**, *136*, 894. [CrossRef]
13. Dakhlaoui, H.; Ungan, F.; Martínez-Orozco, J.C.; Mora-Ramos, M.E. Theoretical investigation of linear and nonlinear optical properties in an heterostructure based on triple parabolic barriers: Effects of external fields. *Phys. B* **2021**, *607*, 412782. [CrossRef]
14. Dakhlaoui, H.; Belhadj, W.; Durmuslar, A.S.; Ungan, F.; Abdelkader, A. Numerical study of optical absorption coefficients in Manning-like AlGaAs/GaAs double quantum wells: Effects of doped impurities. *Phys. E Low Dimens. Syst. Nanostruct.* **2023**, *147*, 115623. [CrossRef]
15. Dakhlaoui, H.; Belhadj, W.; Musa, M.O.; Ungan, F. Electronic states and optical characteristics of GaAs Spherical quantum dot based on Konwent-like confining potential: Role of the hydrogenic impurity and structure parameters. *Optik* **2023**, *277*, 170684. [CrossRef]
16. Yoffe, A.D. Semiconductor quantum dots and related systems: Electronic, optical, luminescence and related properties of low dimensional systems. *Adv. Phys.* **2001**, *50*, 1. [CrossRef]
17. Nirmal, M.; Brus, L. Luminescence Photophysics in semiconductor nanocrystals. *Acc. Chem. Res.* **1999**, *32*, 407. [CrossRef]
18. Sargent, E.H. Colloidal quantum dot solar cells. *Nat. Photon.* **2012**, *6*, 133. [CrossRef]
19. Bouzaiene, L.; Alamri, H.; Sfaxi, L.; Maaref, H. Simultaneous effects of hydrostatic pressure, temperature and electric field on optical absorption in InAs/GaAs lens shape quantum dot. *J. Alloys Compd.* **2016**, *655*, 172. [CrossRef]

20. Ben Mahrsia, R.; Choubani, M.; Bouzaiene, L.; Maaref, H. Second-harmonic generation in vertically coupled InAs/GaAs quantum dots with a Gaussian potential distribution: Combined effects of electromagnetic fields, pressure, and temperature. *Electron. Mater.* **2015**, *44*, 2792. [CrossRef]
21. Al-Marhaby, F.A.; Al-Ghamdi, M.S. Experimental investigation of stripe cavity length effect on threshold current density for InP/AlGaInP QD laser diode. *Opt. Mater.* **2022**, *127*, 112191. [CrossRef]
22. Chuang, C.H.M.; Brown, P.R.; Bulović, V.; Bawendi, M.G. Improved performance and stability in quantum dot solar cells through band alignment engineering. *Nat. Mater.* **2014**, *13*, 796. [CrossRef]
23. De Franceschi, S.; Kouwenhoven, L.; Schönenberger, C.; Wernsdorfer, W. Hybrid superconductor–quantum dot devices. *Nat. Nanotechnol.* **2010**, *5*, 703. [CrossRef] [PubMed]
24. Gao, X.; Yang, L.; Petros, J.A.; Marshall, F.F.; Simons, J.W.; Nie, S. In vivo molecular and cellular imaging with quantum dots. *Cur. Opin. Biotechnol.* **2005**, *16*, 63. [CrossRef]
25. Medintz, I.L.; Clapp, A.R.; Mattoussi, H.; Goldman, E.R.; Fisher, B.; Mauro, J.M. Self-assembled nanoscale biosensors based on quantum dot FRET donors. *Nat. Mater.* **2003**, *2*, 630. [CrossRef] [PubMed]
26. Loss, D.; DiVincenzo, D.P. Quantum computation with quantum dots. *Phys. Rev. A* **1998**, *57*, 120. [CrossRef]
27. Al-Sheikhi, A.; Al-Abedi, N.A.A. The luminescent emission and quantum optical efficiency of $Cd_{1-x}Sr_xSe$ QDs developed via ions exchange approach for multicolor-lasing materials and LED applications. *Optik* **2021**, *227*, 166035. [CrossRef]
28. Al-Ahmadi, N.A. The anti-crossing and dipping spectral behavior of coupled nanocrystal system under the influence of the magnetic field. *Results Phys.* **2021**, *22*, 103835.
29. Galiautdinov, A. Ground state of an exciton in a three-dimensional parabolic quantum dot: Convergent perturbative calculation. *Phys. Lett. A* **2018**, *382*, 72. [CrossRef]
30. Vahdani, M.R.K.; Rezaei, G. Intersubband optical absorption coefficients and refractive index changes in a parabolic cylinder quantum dot. *Phys. Lett. A* **2010**, *374*, 637. [CrossRef]
31. Khordad, R. Use of modified Gaussian potential to study an exciton in a spherical quantum dot. *Superlattices Microstruct.* **2013**, *54*, 7. [CrossRef]
32. Sakiroglu, S.; Kasapoglu, S.; Restrepo, R.L.; Duque, C.A.; Sökmen, I. Intense laser field-induced nonlinear optical properties of Morse quantum well. *Phys. Stat. Solidi B* **2017**, *254*, 1600457. [CrossRef]
33. Prasad, V.; Silotia, P. Effect of laser radiation on optical properties of disk-shaped quantum dot in magnetic fields. *Phys. Lett. A* **2011**, *375*, 3910. [CrossRef]
34. Lee, S.W.; Hirakava, K.; Shimada, Y. Bound-to-continuum intersubband photoconductivity of self-assembled InAs quantum dots in modulation-doped heterostructures. *Appl. Phys. Lett.* **1999**, *75*, 1428. [CrossRef]
35. Klimov, V.I.; McBranch, D.W.; Leatherdale, C.A.; Bawendi, M.G. Electron and hole relaxation pathways in semiconductor quantum dots. *Phys. Rev. B* **1999**, *60*, 13740. [CrossRef]
36. Mackowski, S.; Kyrychenko, F.; Karczewski, G.; Kossut, J.; Heiss, W.; Prechtl, G. Thermal carrier escape and capture in CdTe quantum dots. *Phys. Stat. Solidi B* **2001**, *224*, 465. [CrossRef]
37. Sauvage, S.; Boucaud, P.; Brunhes, T.; Immer, V.; Finkman, E.; Gerard, J.M. Midinfrared absorption and photocurrent spectroscopy of InAs/GaAs self-assembled quantum dot. *Appl. Phys. Lett.* **2001**, *78*, 2327. [CrossRef]
38. Schrey, F.F.; Rebohle, L.; Muller, T.; Strasser, G.; Unterrainer, K.; Nguyen, D.P.; Regnault, N.; Ferreira, R.; Bastard, G. Intraband transitions in quantum dot–superlattice heterostructures. *Phys. Rev. B* **2005**, *72*, 155310. [CrossRef]
39. Bahar, M.K.; Baser, P. Nonlinear optical characteristics of thermodynamic effects-and electric field-triggered Mathieu quantum dot. *Micro Nanostruct.* **2022**, *170*, 207371. [CrossRef]
40. Batra, K.; Prasad, V. Spherical quantum dot in Kratzer confining potential: Study of linear and nonlinear optical absorption coefficients and refractive index changes. *Eur. Phys. J. B* **2018**, *91*, 298. [CrossRef]
41. Buczko, R.; Bassani, F. Bound and resonant electron states in quantum dots: The optical spectrum. *Phys. Rev. B* **1996**, *54*, 2667. [CrossRef]
42. Narvaez, G.A.; Zunger, A. Calculation of conduction-to-conduction and valence-to-valence transitions between bound states in (In, Ga)As/GaAs quantum dots. *Phys. Rev. B* **2007**, *75*, 085306. [CrossRef]
43. Stoleru, V.G.; Towe, E. Oscillator strength for intraband transitions in (In, Ga)As/GaAs. *Appl. Phys. Lett.* **2003**, *83*, 5026. [CrossRef]
44. Yilmaz, S.; Safak, H. Oscillator strengths for the intersubband transitions in a $CdS–SiO_2$ quantum dot with hydrogenic impurity. *Phys. E* **2007**, *36*, 40. [CrossRef]
45. Costa, L.S.D.; Prudente, F.V.; Acioli, P.H.; Neto, J.S.; Vianna, J.D.M. A study of confined quantum systems using the Woods-Saxon potential. *J. Phys. B At. Mol. Opt. Phys.* **1999**, *32*, 2461. [CrossRef]
46. Fakkahi, A.; Sali, A.; Jaouane, M.; Arraoui, R.; Ed-Dahmouny, A. Study of photoionization cross section and binding energy of shallow donor impurity in multilayered spherical quantum dot. *Phys. E* **2022**, *143*, 115351. [CrossRef]
47. Sali, A.; Satori, H.; Fliyou, M.; Loumrhari, H. The photoionization cross-section of impurities in quantum dots. *Phys. Status Solidi* **2002**, *232*, 209. [CrossRef]
48. Holovatsky, V.; Chubrei, M.; Yurchenko, O. Impurity photoionization cross-section and intersubband optical absorption coefficient in multilayer spherical quantum dots. *Phys. Chem. Solid. State* **2021**, *22*, 630. [CrossRef]
49. Arraoui, R.; Sali, A.; Ed-Dahmouny, A.; Jaouane, M.; Fakkahi, A. Polaronic mass and non-parabolicity effects on the photoionization cross section of an impurity in a double quantum dot. *Superlattice Microst.* **2021**, *159*, 107049. [CrossRef]

50. Sahin, M.; Koksal, K. The linear optical properties of a multi-shell spherical quantum dot of a parabolic confinement for cases with and without a hydrogenic impurity. *Semicond. Sci. Technol.* **2012**, *27*, 125011. [CrossRef]
51. Ed-Dahmouny, A.; Arraoui, R.; Jaouane, M.; Fakkahi, A.; Sali, A.; Es-Sbai, N.; El-Bakkari, K.; Zeiri, N.; Duque, C.A. The influence of the electric and magnetic fields on donor impurity electronic states and optical absorption coefficients in a core/shell $GaAs/Al_{0.33}Ga_{0.67}As$ ellipsoidal quantum dot. *Eur. Phys. J. Plus* **2023**, *138*, 774. [CrossRef]
52. Ed-Dahmouny, A.; Zeiri, N.; Arraoui, R.; Es-Sbai, N.; Jaouane, M.; Fakkahi, A.; Sali, A.; El-Bakkari, K.; Duque, C.A. The third-order nonlinear optical susceptibility in an ellipsoidal core-shell quantum dot embedded in various dielectric surrounding matrices. *Phys. E* **2023**, *153*, 115784. [CrossRef]

Disclaimer/Publisher's Note: The statements, opinions and data contained in all publications are solely those of the individual author(s) and contributor(s) and not of MDPI and/or the editor(s). MDPI and/or the editor(s) disclaim responsibility for any injury to people or property resulting from any ideas, methods, instructions or products referred to in the content.

Article

The Effect of Cation Incorporation on the Elastic and Vibrational Properties of Mixed Lead Chloride Perovskite Single Crystals

Syed Bilal Junaid, Furqanul Hassan Naqvi and Jae-Hyeon Ko *

School of Nano Convergence Technology, Nano Convergence Technology Center, Hallym University, Chuncheon 24252, Republic of Korea; 43497@hallym.ac.kr (S.B.J.); furqanhassan@hallym.ac.kr (F.H.N.)
* Correspondence: hwangko@hallym.ac.kr

Abstract: In recent years, there have been intense studies on hybrid organic–inorganic compounds (HOIPs) due to their tunable and adaptable features. This present study reports the vibrational, structural, and elastic properties of mixed halide single crystals of $MA_xFA_{1-x}PbCl_3$ at room temperature by introducing the FA cation at the A-site of the perovskite crystal structure. Powder X-ray diffraction analysis confirmed that its cubic crystal symmetry is similar to that of $MAPbCl_3$ and $FAPbCl_3$ with no secondary phases, indicating a successful synthesis of the $MA_xFA_{1-x}PbCl_3$ mixed halide single crystals. Structural analysis confirmed that the FA substitution increases the lattice constant with increasing FA concentration. Raman spectroscopy provided insight into the vibrational modes, revealing the successful incorporation of the FA cation into the system. Brillouin spectroscopy was used to investigate the changes in the elastic properties induced via the FA substitution. A monotonic decrease in the sound velocity and the elastic constant suggests that the incorporation of large FA cations causes distortion within the inorganic framework, altering bond lengths and angles and ultimately resulting in decreased elastic constants. An analysis of the absorption coefficient revealed lower attenuation coefficients as the FA content increased, indicating reduced damping effects and internal friction. The current findings can facilitate the fundamental understanding of mixed lead chloride perovskite materials and pave the way for future investigations to exploit the unique properties of mixed halide perovskites for advanced optoelectronic applications.

Keywords: lead halide perovskites; $MA_{1-x}FA_xPbCl_3$; Raman spectroscopy; Brillouin spectroscopy

Citation: Junaid, S.B.; Naqvi, F.H.; Ko, J.-H. The Effect of Cation Incorporation on the Elastic and Vibrational Properties of Mixed Lead Chloride Perovskite Single Crystals. *Inorganics* 2023, 11, 416. https://doi.org/10.3390/inorganics11100416

Academic Editors: Sake Wang, Minglei Sun and Nguyen Tuan Hung

Received: 28 September 2023
Revised: 13 October 2023
Accepted: 17 October 2023
Published: 22 October 2023

Copyright: © 2023 by the authors. Licensee MDPI, Basel, Switzerland. This article is an open access article distributed under the terms and conditions of the Creative Commons Attribution (CC BY) license (https://creativecommons.org/licenses/by/4.0/).

1. Introduction

Hybrid organic–inorganic perovskites (HOIPs), particularly 3D perovskites denoted as the formula ABX_3, have recently attracted considerable attention due to their versatile properties [1–6]. Here, X denotes halide anions (I^-, Br^-, Cl^-) located inside corner-sharing BX_6 octahedra, where B denotes divalent metal cations (Pb^{2+}, Sn^{2+}), and A represents monovalent cations (FA^+, MA^+, Cs^+). In particular, perovskites, such as $MAPbX_3$ (MA = $CH_3NH_3^+$, methylammonium) and $FAPbX_3$ (FA = $CH(NH_2)_2^+$, formamidinium), have attracted significant attention due to their remarkable photovoltaic performance and exceptional optical and photoluminescent properties. Additionally, their low-cost solution processability enhances their potential, making them promising candidates for implementation in solar cells, light-emitting diodes, and photodetectors [1–12].

$APbX_3$ compounds have been extensively investigated as highly effective materials for photovoltaics [7,8,12]. Their power conversion efficiencies have exceeded 25% throughout the last ten years, largely due to their high absorption coefficients, modifiable band gaps, and other contributing factors [13–17]. Nonetheless, the features of HOIPs can be significantly altered via the cation and anion compositions and structural phase transitions, leading to promising opportunities for further optimization [18–20]. In particular, $FAPbX_3$ compounds (X = I, Br, Cl) within the HOIPs have attracted considerable attention due to their narrower and more preferable band gap, improved thermal stability at higher temperatures, and a more symmetrical crystal structure in contrast to their counterparts with MA [21–24]. Within the MAPbX$_3$ family, MAPbCl$_3$ has wide-ranging applications in ultraviolet (UV) photodetection due to its wide band gap of 2.88 eV [25]. This compound undergoes two distinct phase transitions with decreasing temperature from cubic to tetragonal at 163 K and then to orthorhombic phases at 159 K [26,27]. In contrast, FAPbCl$_3$ is a distinctive compound among the FAPbX$_3$ compounds due to its high stability, which has proven to be critical in the development of highly sensitive ammonia gas sensors [28] and polymer solar cells with anode interfacial layers (AIL) that have achieved power conversion efficiency (PCE) of 8.75% [29]. Moreover, the ambipolar transport properties of FAPbCl$_3$ also play a significant role in the fields of UV detection and optoelectronics [29,30]. Differential scanning calorimetry (DSC) and quasi-elastic neutron scattering (QENS) experiments indicate that it undergoes phase transitions from cubic to tetragonal at 271 K and then from tetragonal to orthorhombic phase at 258 K [31,32]. Furthermore, only a few density functional theory (DFT) studies thus far [33–35] have focused on their electronic, optical, and elastic properties. However, no research studies have yet integrated mixed A site cation composition for lead chloride. Therefore, our aim is to fill this research gap by analyzing the mixed MA$_x$FA$_{1-x}$PbCl$_3$ system.

The integration of mixed A-site cation compositions, such as hybrid FA/MA single crystals, has been used to improve device performances [36–41]. These compositional variations provide clear benefits, including the ability to precisely adjust band gaps and exhibit remarkable stability under different stress conditions, resulting in the successful demonstration of the most successful solar cells using these mixed A-site cation perovskites to date [36–41]. In regard to band gap tuning, studies have revealed a significant redshift in the fundamental band gap when FA replaces MA in MAPbI$_3$ [42]. Researchers have employed various techniques, such as photoluminescence (PL) measurements [43], absorption [44], diffuse reflectance [45], and ellipsometry spectra [46], to investigate this phenomenon. Furthermore, hybrid FA$_x$MA$_{1-x}$PbI$_3$ perovskites were investigated in Raman and PL studies. The resulting temperature–composition phase diagram confirms a slight redshift in the fundamental band gap when replacing MA cations with FA cations [47].

In regard to improved photovoltaic applications, recent studies show that MAPbI$_3$ achieves impressive power conversion efficiencies (PCEs) exceeding 17% [6,9]. Nevertheless, they face challenges like thermal and moisture instability issues [7,10–12]. Similarly, FAPbI$_3$ perovskite also shows potential as a photovoltaic material due to its narrower bandgap (E$_g$) of 1.48 eV, aligning more closely with the Shockley–Queisser ideal bandgap of approximately 1.40 eV for single-junction solar cells. This alignment indicates the possibility of achieving even higher PCEs [48,49]. However, there are challenges, particularly the undesirable creation of a yellow non-perovskite δ-phase during the annealing process, which hinders the efficiency of resulting photovoltaic solar cells (PVSCs) due to their insulating properties. To overcome these obstructions inherent within initial perovskite materials, researchers have explored the concept of mixing A-site cations as a strategy to achieve improved PCEs. Grätzel et al. successfully demonstrated an approach to achieving higher PCEs in perovskite solar cells [23]. In particular, they synthesized the perovskite MA$_{0.6}$FA$_{0.4}$PbI$_3$ via the partial substitution of MA$^+$ ions with FA$^+$ ions in MAPbI$_3$, resulting in an enhanced efficiency compared to parent perovskites [23].

The use of cation–halide mixtures is a widely used method to stabilize the cubic perovskite structure at operational temperatures [37,50–53]. In particular, FA-dependent perovskites, such as FAPbI$_3$, typically undergo a phase transition from the cubic α-phase to the undesired hexagonal δ-phase at room temperature, affecting its photovoltaic behavior. However, the introduction of cation–halide mixtures helps to stabilize the cubic phase [37,50–53]. Furthermore, studies on mixed MA$_{1-x}$FA$_x$PbBr$_3$ hybrids have used methods such as specific heat and differential scanning calorimetry to investigate the phase transition behavior. These studies have revealed a significant reduction in phase transition temperatures and increased the stability of the cubic phase caused by the A-site mixing [54]. These findings emphasize the considerable influence of mixed A-site cations on the characteristics of lead halide perovskites.

Our research aims to investigate mixed $MA_xFA_{1-x}PbCl_3$ single crystals, which, to the best of our knowledge, have not been studied before. We successfully analyzed the changes in the optical and acoustic phonon modes induced via the FA substitution at room temperature. We introduced 30% and 40% FA at the A-site cation of $MAPbCl_3$ and studied their properties using Raman and Brillouin spectroscopic techniques. The cation stoichiometries (i.e., x values) were precisely determined using nuclear magnetic resonance (NMR) spectroscopy. Raman spectroscopy was used to investigate optical phonon modes in the low-frequency range, corresponding to lattice vibrations. The observed decrease in wavenumber with increasing FA content indicates changes in the Pb-Cl bond strength within the lattice. Furthermore, Brillouin spectroscopy was used to obtain acoustic phonon velocities and associated elastic constants. Our results indicate that these parameters decrease with increasing FA content, suggesting that the lattice rigidity is reduced due to the larger cation radius of the FA cation compared to the MA cation, resulting in the lattice distortion and the softer lattice structure.

2. Results and Discussion

The synthesized samples were initially analyzed using solid-state 1H NMR spectroscopy to determine the exact FA/MA ratios. Figure 1 illustrates the 1H magic-angle spinning (MAS) spectra for $MA_xFA_{1-x}PbCl_3$ (x = 1, 0.7, 0.6) single crystals. The spectrum of the $MAPbCl_3$ composition indicates two distinguishable peaks at 6.52 and 3.29 ppm, with a ratio of 1:1. Interestingly, the signals of the $-NH^{3+}$ and $-CH^3$ protons, which have previously been linked to the $MAPbX_3$ series (X = I, Br, Cl) [55], were correlated with these peaks. The NMR data analysis indicated a discrepancy between the real compositions of the samples and the nominal compositions employed during synthesis. The $MA_xFA_{1-x}PbCl_3$ had values of x at 1, 0.87, and 0.77 with percentage errors of 28% and 24% for the mixed systems, respectively. It is worth noting that the MA content decreased significantly in the mixed crystal compared to the starting ratio. This emphasizes the importance of monitoring the actual MA/FA ratio for this and similar systems after the materials are synthesized. The discrepancy between the nominal and measured stoichiometries may be due to the different solubilities and reactivities of the MA and FA precursors in the solution.

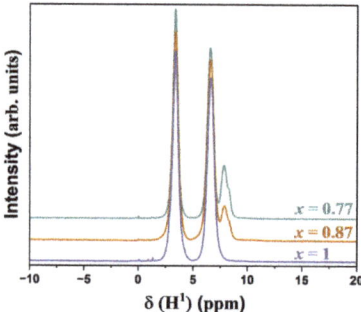

Figure 1. Solid-state (H^1) NMR spectra of the $MA_xFA_{1-x}PbCl_3$ (x = 1, 0.87, 0.77) single crystals.

The powder X-ray diffraction (PXRD) pattern of $MA_xFA_{1-x}PbCl_3$ with varying compositions is shown in Figure 2. The diffraction pattern consistent with the cubic XRD patterns of $MAPbCl_3$ [56] and $FAPbCl_3$ [57] emphasizes the crystallographic cubic structure of the synthesized compounds. Notably, the absence of additional peaks and peak splitting in the PXRD pattern of the mixed crystals at room temperature confirms the high-quality cubic crystal symmetry ($Pm\bar{3}m$) shared with $MAPbCl_3$ and $FAPbCl_3$ [56,57].

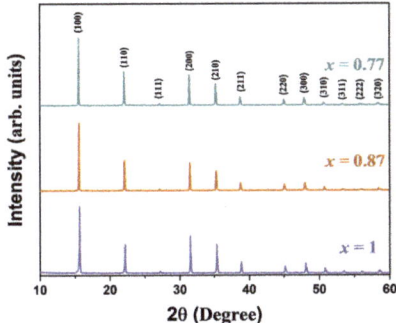

Figure 2. PXRD patterns of synthesized MA$_x$FA$_{1-x}$PbCl$_3$ (x = 1, 0.87, 0.77) single crystals.

Figure 3 shows the changes in the lattice constant, the unit cell volume, and the density as a function of FA content within the MA$_x$FA$_{1-x}$PbCl$_3$ system. Figure 3a,b show that the lattice constant and the unit cell volume exhibit a linearly increasing trend with increasing FA content. This observation supports a strong link between structural changes induced via the FA substitution and their influence on material properties. There is an increase in the lattice constant from 5.67 to 5.69 Å with increasing FA content. This increase in the lattice constant and the unit cell volume can be attributed to the weakening of electrostatic forces resulting from the change in the atomic packing. In particular, the substitution of smaller MA cations by larger FA cations within the perovskite structure causes structural distortion and a decrease in the atomic packing density, as shown in Figure 3c. The absolute values of all these parameters are reported in Table 1.

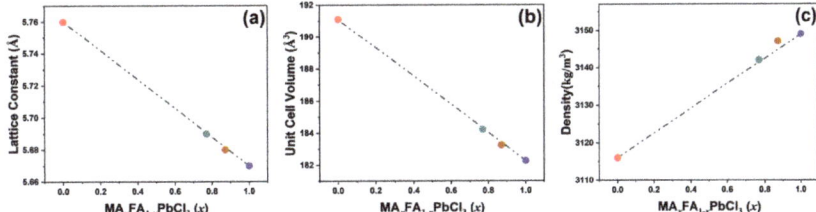

Figure 3. (a) Lattice constant, (b) unit-cell volume, and (c) density of MA$_x$FA$_{1-x}$PbCl$_3$ (x = 1, 0.87, 0.77) single crystals. The lattice constant of FAPbCl$_3$ was taken from ref. [57].

Table 1. Summary of crystallographic and physical properties for MA$_x$FA$_{1-x}$PbCl$_3$ (x = 1, 0.87, 0.77) single crystals. Refractive indices for MAPbCl$_3$ [58] and FAPbCl$_3$ [33] were interpolated to get values for all mixed compositions. The lattice constant of FAPbCl$_3$ was taken from ref. [57].

Composition	Lattice Parameter (Å)	Unit-Cell Volume (Å^3)	Refractive Index (n)	Density (kg/m^3)
MAPbCl$_3$	5.67	182.28	1.90 [58]	3149
MA$_{0.87}$FA$_{0.13}$PbCl$_3$	5.68	183.25	1.93	3147
MA$_{0.77}$FA$_{0.23}$PbCl$_3$	5.69	184.22	1.95	3142
FAPbCl$_3$	5.76 [57]	191.10	2.10 [33]	3116

Raman spectroscopy has been employed for exploring the vibrational dynamics of the mixed lead halide perovskite material, as it gives us insights into the lattice dynamics, phonon modes, and other material properties. Figure 4a–c displays the Raman spectra of MA$_x$FA$_{1-x}$PbCl$_3$ single crystals at room temperature, covering a broad range of wavenumbers from 10 to 3500 cm^{-1}. For the sake of clarity, this study categorizes the data into low, medium, and high wavenumber regions. The Raman spectra exhibit significant peaks that correspond to lattice vibrational modes and internal modes of the constituent cations. In particular, these cations are the MA cation (for $x = 1$) and the MA/FA cations (for $x = 0.87$, 0.77). It is worth noting that $MA_xFA_{1-x}PbCl_3$ ($x = 0.87, 0.77$) produces additional peaks in the frequency domain at 522 cm^{-1}, 1117 cm^{-1}, 1396 cm^{-1}, 1713 cm^{-1}, 3228 cm^{-1}, and 3336 cm^{-1}. These peaks correspond to the bending mode of the FA$^+$ cation [11,59], the symmetric stretching of CN [11,59], the rocking mode of NH$_2$ [11,59,60], the asymmetric stretching of NH$_2$ [60,61], and the stretching of NH$_2$ (3228 cm^{-1}, 3336 cm^{-1}) [60–62], respectively. This provides strong evidence for the successful substitution of the FA cation within the crystal lattice.

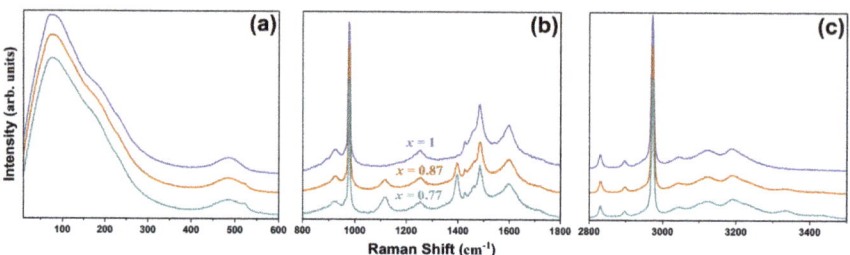

Figure 4. Room temperature Raman spectra of $MA_xFA_{1-x}PbCl_3$ (x = 1, 0.87, 0.77) single crystals in (**a**) 10–600 cm^{-1}, (**b**) 800–1800 cm^{-1}, and (**c**) 2800–3500 cm^{-1} wavenumber ranges.

The peak positions for all the observed vibrational modes were determined in terms of the curve fitting analysis using the Lorentzian function. A graphical representation of this fitting process is provided in the Supplementary Materials, with the resulting individual fitting lines for the Raman spectra recorded at room temperature (Figures S1–S3). The vibrational mode wavenumbers obtained from the fitting analyses are listed, as shown in Table 2, with their respective mode assignments.

Table 2. Mode assignments of Raman peaks of $MA_xFA_{1-x}PbCl_3$ (x = 1, 0.87, 0.77) single crystals at room temperature.

MAPbCl$_3$ (cm^{-1})	MA$_{0.87}$FA$_{0.13}$PbCl$_3$ (cm^{-1})	MA$_{0.77}$FA$_{0.23}$PbCl$_3$ (cm^{-1})	Mode Assignment
36	36	36	PbCl$_6$ motion [62,63]
63	62	62	PbCl$_6$ motion [63]
90	89	88	δ_s (Cl–Pb–Cl) [59]
122	123	123	δ_{as} (Cl–Pb–Cl) [59]
182	177	174	ν_{as} (Pb–Cl) [59,61]
239	232	228	R of MA$^+$ cation [63]
484	486	485	τ (MA) [64]
	522	522	δ (FA) [11,61]
922	924	923	ρ (MA) [61]
962	962	961	
976	978	978	ν (C–N) [59]
	1117	1116	ν_s (CN) [11,61]
1246	1247	1247	ρ (MA) [59,62]
	1396	1395	ρ (NH$_2^+$) [11,60,61]
1427	1428	1428	δ_s (CH$_3$) [59]

Table 2. Cont.

MAPbCl$_3$ (cm^{-1})	MA$_{0.87}$FA$_{0.13}$PbCl$_3$ (cm^{-1})	MA$_{0.77}$FA$_{0.23}$PbCl$_3$ (cm^{-1})	Mode Assignment
1455	1458	1458	δ_{as} (CH$_3$) [59]
1485	1486	1486	δ_s (NH$_3$) [59]
1597	1600	1600	δ_{as} (NH$_3$) [59]
	1715	1716	ν_{as} (CN) [60,61]
2830	2831	2831	Combination modes [65]
2897	2898	2898	ν_s (C–H) [62]
2947	2949	2948	sym. CH$_3$ stretch [65]
2972	2973	2973	ν_{as} (C–H) [59,62,65]
3042	3042	3041	ν_s (CH$_3$) [59,62,65]
3120	3120	3118	ν_s (NH$_3$) [62]
3189	3190	3190	ν_s (NH$_3$) [62]
3221	3229	3228	ν (NH$_2$) [60–62]
	3336	3336	ν (NH$_2$) [60,61]
	3438	3436	sym. NH$_3^+$ stretch [65]

δ: bending; ρ: rocking; ν: stretching; s/as: symmetric/asymmetric; τ: torsion; R: rotation.

Figure 4a shows the low-frequency lattice modes below 600 cm^{-1}, including the characteristic torsional (τ) mode of the MA cation and the characteristic bending (δ) mode of the FA cation. In particular, the bending (δ) mode of the FA cation is more pronounced, indicating a significant effect of the FA inclusion. It is also worth noting that there is a slight reduction in the mode frequencies of the low-frequency lattice modes with increasing FA content. The small reduction in mode frequencies is probably due to the complex interplay between the FA and MA cations, which has a significant effect on the bonding characteristics of the PbCl$_6$ octahedra. When larger FA cations occupy the MA site, they generate local stress distortions, leading to changes in the lattice behavior and intermolecular angles.

The results for the mid-frequency modes from 800 to 1800 cm^{-1} are shown in Figure 4b. These vibrational data show that FA has been integrated into the mixed crystal structure, as evidenced in the clear observation of the symmetric stretching mode of CN [11,59] (~1117 cm^{-1}) and the rocking mode of NH$_2$ [11,59,60] (~1396 cm^{-1}) of FA. The mode intensity tends to increase with increasing FA content as seen in the bending mode of the FA cation. These vibrations demonstrate how cations interact within the crystal lattice and provide valuable insights into the complex nature of localized structural changes resulting from the coexistence of different cationic entities.

The internal modes observed between 2800 and 3500 cm^{-1} are shown in Figure 4c. Although there are only minor shifts in mode frequencies, new internal modes associated with the stretching of the NH$_2$ mode [60,61] (~3228 cm^{-1}, 3336 cm^{-1}) appear, revealing the successful integration of the FA cation in the crystal structure. The distinct properties of the FA and MA cations induce a rearrangement of atomic interactions, resulting in local perturbations as evidenced by the spectra.

Overall, the Raman spectroscopy results suggest that small changes in certain Raman modes indicate the presence of local structural distortions within the crystal. These deviations can be attributed to various factors, such as the different ionic radii of the FA and MA cations and their different interactions with the PbCl$_6$ octahedra. The random substitution of the larger FA cation into the MA site generates local heterogeneous stresses throughout the lattice, which, in turn, affects the lattice dynamics and interactions with neighboring PbCl$_6$ octahedra and can alter the Pb-Cl bond length and the Pb-Cl-Pb bond angle.

Figure 5 shows the Brillouin spectra at room temperature for the mixed single crystals of $MA_xFA_{1-x}PbCl_3$. Two distinct doublets can be seen in these spectra, corresponding to the longitudinal acoustic (LA) and transverse acoustic (TA) modes. The mode frequencies of both the LA and TA modes shift to lower values as the FA content increases. This phenomenon indicates changes in the elastic properties of the perovskite lattice due to the introduction of FA cations. The larger FA^+ cation leads to extended Pb-Cl bond lengths and weaker bond strengths within the inorganic framework of halide perovskites. Consequently, a softer lattice structure and lower elastic moduli are expected. The Brillouin spectra were fitted using the Voigt function, a convolution of the Gaussian instrumental function, and the Lorentzian phonon response function. The sound velocity (V) can be calculated from the acoustic mode frequency (v_B) using the following equation:

$$V = \frac{v_B \lambda}{2n} \quad (1)$$

where λ is the wavelength of the excitation source (532 nm), and n is the refractive index of the crystal. However, the experimental refractive indices for $MA_xFA_{1-x}PbCl_3$ (x = 0.87, 0.77) are not reported. Therefore, approximate refractive indices were obtained via linear interpolation between the experimental value of $MAPbCl_3$ (n = 1.90) [58] and the theoretical value of $FAPbCl_3$ (n = 2.10) [33] obtained from DFT calculations, as shown in Figure S4 in the Supplementary Information. From these data, the longitudinal and transverse sound velocities were obtained and plotted in Figure 6a,b. Over the investigated composition range, the longitudinal and transverse sound velocities show a monotonic decrease with increasing FA content. In particular, for the composition change from x = 1 to x = 0.77, the longitudinal sound velocity decreases from 3574 m/s to 3471 m/s, and the transverse sound velocity decreases from 1087 m/s to 961 m/s. The increase in the lattice constant, caused via the incorporation of the larger FA cation, seems to be the primary factor contributing to the gradual decrease in the sound velocity. In addition, the reduced dipole moment of the FA cation could lead to lower acoustic phonon velocities. Similar behavior was observed by Ma et al. [66] in the mixed system of $FA_xMA_{1-x}PbBr_3$.

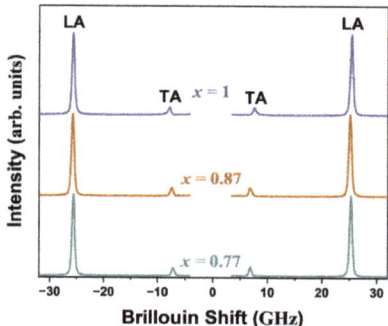

Figure 5. Room temperature Brillouin spectra of $MA_xFA_{1-x}PbCl_3$ (x = 1, 0.87, 0.77) single crystals. The LA and the TA denote the longitudinal and the transverse acoustic modes, respectively.

The LA and TA modes observed in this scattering geometry correspond to the elastic constants C_{11} and C_{44}, respectively. These elastic constants can be determined from the sound velocity (V) and the crystal density (ρ) using the following relation:

$$C_{ij} = \rho V^2 \quad (2)$$

The calculated densities for the $MA_xFA_{1-x}PbCl_3$ (x = 1, 0.87, 0.77) single crystals are 3149 kg/m³, 3147 kg/m³, and 3142 kg/m³, respectively. These values were determined using the lattice parameter obtained from PXRD and the chemical formula. The elastic constants C_{11} and C_{44} are shown in Figure 7a,b, respectively. Both C_{11} and C_{44} show a gradual decrease from 40.22 GPa to 37.85 GPa and from 3.72 GPa to 2.90 GPa, respectively, with an increase in FA cations. The results are presented in Table 3, together with the reported values for FAPbCl$_3$ obtained from DFT calculations.

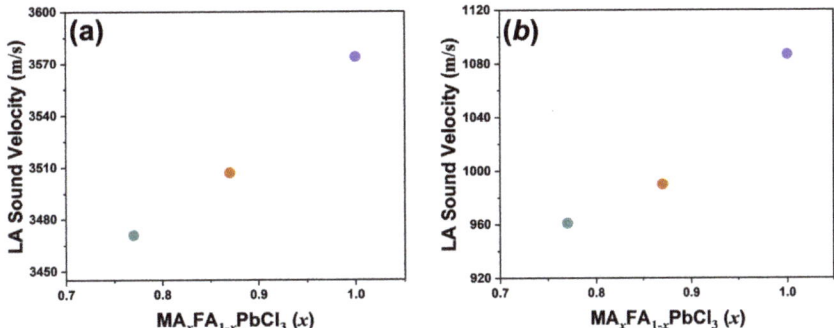

Figure 6. (**a**) Longitudinal acoustic (LA) and (**b**) transverse acoustic (TA) sound velocities as a function of composition x in MA$_x$FA$_{1-x}$PbCl$_3$ (x = 1, 0.87, 0.77) single crystals.

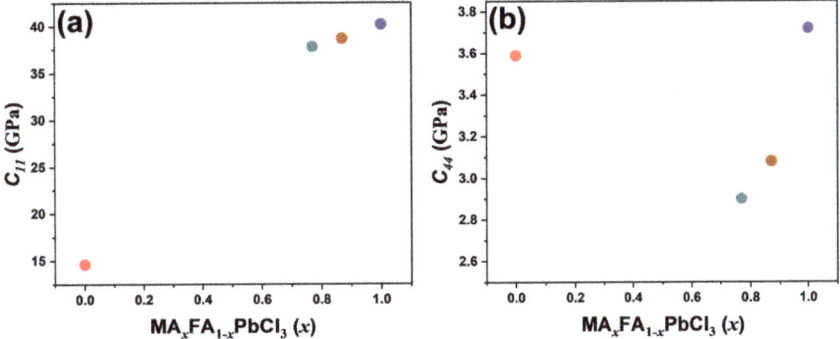

Figure 7. (**a**) C_{11} and (**b**) C_{44} elastic constants as a function of composition x in MA$_x$FA$_{1-x}$PbCl$_3$ (x = 1, 0.87, 0.77) single crystals. (DFT values of FAPbCl$_3$ are from ref. [34]).

Table 3. Summary of calculated sound velocities and elastic constants for MA$_x$FA$_{1-x}$PbCl$_3$ (x = 1, 0.87, 0.77) single crystals. Values of elastic constants for pure FAPbCl$_3$ were from ref [34].

Composition	V_{LA} (m/s)	V_{TA} (m/s)	C_{11} (GPa)	C_{44} (Gpa)
MAPbCl$_3$ (Brillouin)	3574	1087	40.22	3.72
MA$_{0.87}$FA$_{0.13}$PbCl$_3$ (Brillouin)	3507	990	38.71	3.08
MA$_{0.77}$FA$_{0.23}$PbCl$_3$ (Brillouin)	3471	961	37.85	2.90
FAPbCl$_3$ (DFT calculation)	-	-	14.64 [34]	3.59 [34]

The incorporation of larger FA$^+$ cations distorts the lattice, resulting in a monotonic decrease in both elastic constants and sound velocities in the mixed system. A recent study has suggested that the FA cations in FAPbBr$_3$ have weaker hydrogen bonding between H atoms and halide ions compared to that between MA cations and halide ions in MAPbBr$_3$ [67]. This suggests that the larger cation size of FA may distort the crystal structure and weaken the bonding between FA cations and halide ions as the FA concentration increases. This results in increased Pb-Cl bond length and weaker bond strength in the inorganic framework of halide perovskites, leading to greater flexibility and lower elastic moduli. Conversely, the smaller MA cations are likely to contribute to stronger elastic interactions, hence, pure MAPbCl$_3$ shows the highest value for the sound velocity and the associated elastic constants. Figure 7 shows that the DFT results are deviating significantly from the changing trend of the experimental elastic constants, indicating that the DFT approach needs to be refined to get more reliable results of elastic properties.

The absorption coefficient (α) was calculated from the full width at half maximum (FWHM, Γ_B) of the LA and TA modes using the following formula:

$$\alpha = \frac{\pi \Gamma_B}{V} \quad (3)$$

The calculated absorption coefficients for both acoustic modes, as shown in Figure 8, show a consistent decrease with increasing FA content. This trend indicates that materials with higher FA content have lower absorption or attenuation coefficients, implying a weaker presence of damping effects or internal friction. This could be due to the larger ionic radii and lower polarizability of the FA cation, which could potentially lead to a weaker interaction with the inorganic framework and a smaller bandgap. In addition, the heterogeneity introduced via the MA/FA random substitution is much smaller than the wavelength of LA and TA mode (the order of 100 nm) and, thus, does not affect the acoustic absorption process. These results highlight the importance of investigating how changes in composition not only affect the mechanical properties but also shape the mechanisms of energy dissipation within the MA$_x$FA$_{1-x}$PbCl$_3$ mixed halide perovskite system.

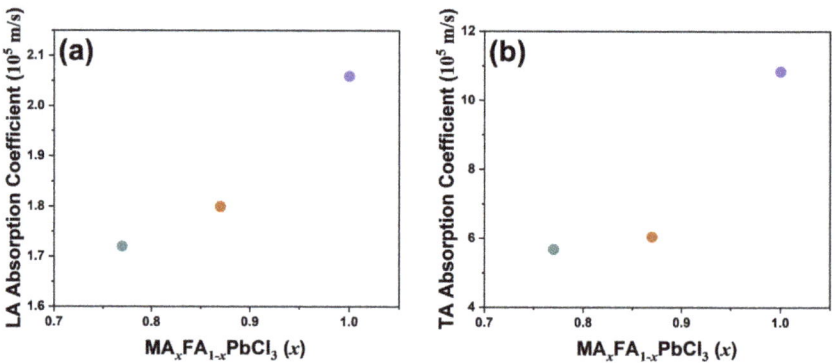

Figure 8. (a) Longitudinal acoustic (LA) and (b) transverse acoustic (TA) absorption coefficients as a function of composition x in MA$_x$FA$_{1-x}$PbCl$_3$ (x = 1, 0.87, 0.77) single crystals.

3. Materials and Methods

3.1. Chemicals

Lead chloride (PbCl$_2$, 99.999%), methylamine (CH$_3$NH$_2$, 40% in water), hydrochloric acid (HCl, 37%, ACS reagent), formamidinium chloride (CH(NH$_2$)$_2$Cl, 99.9%), dimethyl sulfoxide (DMSO, anhydrous $\geq$ 99.9%), diethyl ether (HPLC grade, $\geq$99.9%), and ethanol (anhydrous 99.5%) were purchased from Sigma Aldrich (St. Louis, MO, USA). Methylammonium chloride (MACl) was synthesized according to previously reported methods [26].

3.2. Single Crystals Synthesis

The MA$_x$FA$_{1-x}$PbCl$_3$ (x = 1, 0.7, 0.6) single crystals were synthesized using the solvent evaporation method. The MACl, PbCl$_2$, and FACl powders were dissolved in DMSO (10 mL) in equimolar ratios at 40 °C with continuous stirring in a glove box under nitrogen gas environment. The resulting dissolved solution was filtered into a crystallization dish using a 0.22 μm syringe filter. Slow evaporation was promoted by covering the dish with aluminum foil and leaving it undisturbed at a constant temperature of 85 °C for 1–2 days. Transparent MA$_x$FA$_{1-x}$PbCl$_3$ (x = 1, 0.7, 0.6) single crystals were formed, which were then washed with dichloromethane and dried in a vacuum oven at 60 °C for 12 h to obtain the final product.

3.3. Characterization Techniques

In this study, the powder XRD pattern was obtained using a high-resolution XRD spectrometer (PANalytical; X'pert PRO MPD, Malvern, UK) at room temperature with Cu-K radiation (λ = 1.5406 Å) in the 2θ angular range from 10 to 60°. Prior to measurement, the single crystals were crushed into crystalline powders. The obtained XRD patterns were analyzed using the PANalytical software (X'pert highscore v1.1) for further analysis.

Moreover, 1H solid-state room-temperature NMR spectra were acquired on a (1H = 400.13 MHz) Bruker Avance II+ spectrometer using TopSpin 2.1 software (at KBSI Seoul Western Center, Seoul, Republic of Korea); spectra were collected with a 4 mm magic-angle spinning (MAS) probe under 12 kHz spinning conditions. Quantitative 1H single pulse experiments were performed with a pulse length of 1.2 µs, a recycle delay of 100, and 8 scans. The pulse length and recycle delay were carefully calibrated prior to acquisition of the final spectra to ensure full relaxation of the magnetization and to meet the conditions for quantitative data acquisition. Tetramethylsilane (TMS) was used as a calibration sample. The analysis of the obtained data was carried out using the MestReNova program.

Raman measurements were conducted using a standard Raman spectrometer (LabRam HR800, Horiba Co., Kyoto, Japan) in the frequency range of 10 to 3500 cm^{-1}. A diode-pumped solid-state green laser with a wavelength of 532 nm was utilized to probe the samples. Backscattering geometry was used in terms of an optical microscope (BX41, Olympus, Tokyo, Japan) with a 50× magnification objective lens. The phonon propagation direction for the measurement was [100] in the cubic phase. Prior to measurement, the Raman spectrometer was calibrated using a silicon substrate as a reference sample with a single peak at 520 cm^{-1}. The measured Raman spectra were corrected using the Bose-Einstein correction factor.

Brillouin spectra were acquired using a standard tandem multi-pass Fabry–Perot interferometer (TFP-2, JRS Co., Zurich, Switzerland) with a 532 nm excitation source. A modified microscope with a backscattering geometry (BH-2, Olympus, Tokyo, Japan) was used for the measurement. The signal was identified and averaged over 1024 channels using a conventional photon-counting instrument linked with a multichannel analyzer. The free spectral range was set to 33 GHz to include both the LA and the TA modes in the Brillouin spectrum.

4. Conclusions

The mixed halide perovskite single crystals MA$_x$FA$_{1-x}$PbCl$_3$ have been grown successfully, providing valuable insights into their structural, vibrational, and elastic properties. Solid-state 1H NMR spectroscopy revealed that the actual composition of the grown crystals differed significantly from the nominal values. This highlights the importance of controlling the composition after synthesis. Powder X-ray diffraction analysis confirmed the uniform crystallographic structure of the synthesized compounds, which maintained the cubic symmetry common to MAPbCl$_3$ and FAPbCl$_3$. The inclusion of FA cations led to an increase in the lattice constant, the unit cell volume, and a decrease in the density, indicating structural changes due to FA substitution. Raman spectroscopy revealed wavenumber shifts in the vibrational modes and the appearance of new modes, particularly associated with the FA cations, indicating the successful incorporation of the FA cations into the lattice structure and local structural distortions caused by them. These distortions may be attributed to the different ionic radii and interactions of the FA and MA cations with the PbCl$_6$ octahedra.

Brillouin scattering results showed that the incorporation of the FA cation induced a monotonic decrease in both the longitudinal and transverse sound velocities, along with the corresponding elastic constants, ultimately affecting the elasticity of the perovskite framework. It suggests that incorporating different cations in the perovskite lattice is one way to control the elastic properties of this system. The obtained absorption coefficients show that the mixed crystals with higher FA content have lower absorption coefficients, indicating reduced damping effects and internal friction. This may be due to weaker interactions between FA cations and the inorganic framework. In summary, our study highlights the complex relationship between the composition and physical properties of $MA_xFA_{1-x}PbCl_3$ mixed halide perovskites. These findings are not only relevant to their potential use in optoelectronic devices but also improve our fundamental understanding of these materials. Further research is needed to fully explore their range of properties and improve their functionality.

Supplementary Materials: The following supporting information can be downloaded at: https://www.mdpi.com/article/10.3390/inorganics11100416/s1, Figure S1: Raman spectra and best-fitted curves of $MAPbCl_3$ single crystal at room temperature. Figure S2: Raman spectra and best-fitted curves of $MA_{0.87}FA_{0.13}PbCl_3$ single crystal at room temperature. Figure S3: Raman spectra and best-fitted curves of $MA_{0.77}FA_{0.23}PbCl_3$ single crystal at room temperature. Figure S4: Interpolated refractive indices for $MA_xFA_{1-x}PbCl_3$ single crystals.

Author Contributions: Methodology, S.B.J. and J.-H.K.; Validation, S.B.J. and F.H.N.; Formal analysis, S.B.J. and F.H.N.; Investigation, S.B.J. and F.H.N.; Resources, J.-H.K.; Data curation, S.B.J. and F.H.N.; Writing—original draft, S.B.J.; Writing—review and editing, J.-H.K.; Visualization, S.B.J.; Supervision, J.-H.K.; Project administration, J.-H.K.; Funding acquisition, J.-H.K. All authors have read and agreed to the published version of the manuscript.

Funding: This work was supported by the National Research Foundation of Korea (NRF) grant funded by the Korea government (MSIT) (No. RS-2023-00219703).

Data Availability Statement: The data in this study are available upon request to the corresponding author.

Conflicts of Interest: The authors declare no conflict of interest.

References

1. Quan, L.N.; Rand, B.P.; Friend, R.H.; Mhaisalkar, S.G.; Lee, T.W.; Sargent, E.H. Perovskites for Next-Generation Optical Sources. *Chem. Rev.* **2019**, *119*, 7444–7477. [CrossRef] [PubMed]
2. Mączka, M.; Ptak, M.; Gągor, A.; Stefańska, D.; Sieradzki, A. Layered Lead Iodide of [Methylhydrazinium]$_2$PbI$_4$ with a Reduced Band Gap: Thermochromic Luminescence and Switchable Dielectric Properties Triggered by Structural Phase Transitions. *Chem. Mater.* **2019**, *31*, 8563–8575. [CrossRef]
3. Li, X.; Hoffman, J.M.; Kanatzidis, M.G. The 2D halide perovskite rulebook: How the spacer influences everything from the structure to optoelectronic device efficiency. *Chem. Rev.* **2021**, *121*, 2230–2291. [CrossRef] [PubMed]
4. Mączka, M.; Zaręba, J.K.; Gągor, A.; Stefańska, D.; Ptak, M.; Roleder, K.; Kajewski, D.; Soszyński, A.; Fedoruk, K.; Sieradzki, A. [Methylhydrazinium]$_2$PbBr$_4$, a Ferroelectric Hybrid Organic-Inorganic Perovskite with Multiple Nonlinear Optical Outputs. *Chem. Mater.* **2021**, *33*, 2331–2342. [CrossRef]
5. Mączka, M.; Gągor, A.; Stroppa, A.; Gonçalves, J.N.; Zaręba, J.K.; Stefańska, D.; Pikul, A.; Drozd, M.; Sieradzki, A. Two-dimensional metal dicyanamide frameworks of BeTriMe[M(dca)$_3$(H$_2$O)] (BeTriMe = benzyltrimethylammonium; dca = dicyanamide; M = Mn^{2+}, Co^{2+}, Ni^{2+}): Coexistence of polar and magnetic orders and nonlinear optical threshold temperature sensing. *J. Mater. Chem. C* **2020**, *8*, 11735–11747. [CrossRef]
6. Mączka, M.; Zienkiewicz, J.A.; Ptak, M. Comparative Studies of Phonon Properties of Three-Dimensional Hybrid Organic-Inorganic Perovskites Comprising Methylhydrazinium, Methylammonium, and Formamidinium Cations. *J. Phys. Chem. C* **2022**, *126*, 4048–4056. [CrossRef]
7. Grätzel, M. The light and shade of perovskite solar cells. *Nat. Mater.* **2014**, *13*, 838–842. [CrossRef]
8. Šimėnas, M.; Balčiū Nas, S.; Gągor, A.; Pieniążek, A.; Tolborg, K.; Kinka, M.; Klimavicius, V.; Svirskas, Š.N.; Kalendra, V.; Ptak, M.; et al. Mixology of MA1-xEAxPbI3Hybrid Perovskites: Phase Transitions, Cation Dynamics, and Photoluminescence. *Chem. Mater.* **2022**, *34*, 10104–10112. [CrossRef]
9. Glazer, A.M. Simple ways of determining perovskite structures. *Acta Crystallogr. Sect. A* **1975**, *31*, 756–762. [CrossRef]
10. Petrov, A.A.; Goodilin, E.A.; Tarasov, A.B.; Lazarenko, V.A.; Dorovatovskii, P.V.; Khrustalev, V.N. Formamidinium iodide: Crystal structure and phase transitions. *Acta Crystallogr. Sect. E Crystallogr. Commun.* **2017**, *73*, 569–572. [CrossRef]
11. Ruan, S.; McMeekin, D.P.; Fan, R.; Webster, N.A.S.; Ebendorff-Heidepriem, H.; Cheng, Y.B.; Lu, J.; Ruan, Y.; McNeill, C.R. Raman Spectroscopy of Formamidinium-Based Lead Halide Perovskite Single Crystals. *J. Phys. Chem. C* **2020**, *124*, 2265–2272. [CrossRef]

12. Snaith, H.J. Perovskites: The emergence of a new era for low-cost, high-efficiency solar cells. *J. Phys. Chem. Lett.* **2013**, *4*, 3623–3630. [CrossRef]
13. Kojima, A.; Teshima, K.; Shirai, Y.; Miyasaka, T. Organometal halide perovskites as visible-light sensitizers for photovoltaic cells. *J. Am. Chem. Soc.* **2009**, *131*, 6050–6051. [CrossRef] [PubMed]
14. Park, N.G. Organometal perovskite light absorbers toward a 20% efficiency low-cost solid-state mesoscopic solar cell. *J. Phys. Chem. Lett.* **2013**, *4*, 2423–2429. [CrossRef]
15. Li, Z.; Klein, T.R.; Kim, D.H.; Yang, M.; Berry, J.J.; Van Hest, M.F.A.M.; Zhu, K. Scalable fabrication of perovskite solar cells. *Nat. Rev. Mater.* **2018**, *3*, 18017. [CrossRef]
16. Jeong, J.; Kim, M.; Seo, J.; Lu, H.; Ahlawat, P.; Mishra, A.; Yang, Y.; Hope, M.A.; Eickemeyer, F.T.; Kim, M.; et al. Pseudo-halide anion engineering for α-FAPbI3 perovskite solar cells. *Nature* **2021**, *592*, 381–385. [CrossRef] [PubMed]
17. Yoo, J.J.; Seo, G.; Chua, M.R.; Park, T.G.; Lu, Y.; Rotermund, F.; Kim, Y.K.; Moon, C.S.; Jeon, N.J.; Correa-Baena, J.P.; et al. Efficient perovskite solar cells via improved carrier management. *Nature* **2021**, *590*, 587–593. [CrossRef]
18. Frost, J.M.; Butler, K.T.; Brivio, F.; Hendon, C.H.; Van Schilfgaarde, M.; Walsh, A. Atomistic origins of high-performance in hybrid halide perovskite solar cells. *Nano Lett.* **2014**, *14*, 2584–2590. [CrossRef]
19. Frost, J.M.; Walsh, A. What Is Moving in Hybrid Halide Perovskite Solar Cells? *Acc. Chem. Res.* **2016**, *49*, 528–535. [CrossRef]
20. Egger, D.A.; Rappe, A.M.; Kronik, L. Hybrid Organic-Inorganic Perovskites on the Move. *Acc. Chem. Res.* **2016**, *49*, 573–581. [CrossRef] [PubMed]
21. Pang, S.; Hu, H.; Zhang, J.; Lv, S.; Yu, Y.; Wei, F.; Qin, T.; Xu, H.; Liu, Z.; Cui, G. NH2CH=NH2PbI3: An alternative organolead iodide perovskite sensitizer for mesoscopic solar cells. *Chem. Mater.* **2014**, *26*, 1485–1491. [CrossRef]
22. Eperon, G.E.; Stranks, S.D.; Menelaou, C.; Johnston, M.B.; Herz, L.M.; Snaith, H.J. Formamidinium lead trihalide: A broadly tunable perovskite for efficient planar heterojunction solar cells. *Energy Environ. Sci.* **2014**, *7*, 982–988. [CrossRef]
23. Pellet, N.; Gao, P.; Gregori, G.; Yang, T.Y.; Nazeeruddin, M.K.; Maier, J.; Grätzel, M. Mixed-organic-cation perovskite photovoltaics for enhanced solar-light harvesting. *Angew. Chem. Int. Ed.* **2014**, *53*, 3151–3157. [CrossRef]
24. Levchuk, I.; Osvet, A.; Tang, X.; Brandl, M.; Perea, J.D.; Hoegl, F.; Matt, G.J.; Hock, R.; Batentschuk, M.; Brabec, C.J. Brightly Luminescent and Color-Tunable Formamidinium Lead Halide Perovskite FAPbX3 (X = Cl, Br, I) Colloidal Nanocrystals. *Nano Lett.* **2017**, *17*, 2765–2770. [CrossRef]
25. Cheng, Z.; Liu, K.; Yang, J.; Chen, X.; Xie, X.; Li, B.; Zhang, Z.; Liu, L.; Shan, C.; Shen, D. High-Performance Planar-Type Ultraviolet Photodetector Based on High-Quality CH3NH3PbCl3 Perovskite Single Crystals. *ACS Appl. Mater. Interfaces* **2019**, *11*, 34144–34150. [CrossRef] [PubMed]
26. Naqvi, F.H.; Ko, J.H. Structural Phase Transitions and Thermal Degradation Process of MAPbCl3 Single Crystals Studied by Raman and Brillouin Scattering. *Materials* **2022**, *15*, 8151. [CrossRef] [PubMed]
27. Lee, J.; Naqvi, F.; Ko, J.H.; Kim, T.; Ahn, C. Acoustic Anomalies and the Critical Slowing-Down Behavior of MAPbCl3 Single Crystals Studied by Brillouin Light Scattering. *Materials* **2022**, *15*, 3692. [CrossRef] [PubMed]
28. Parfenov, A.A.; Yamilova, O.R.; Gutsev, L.G.; Sagdullina, D.K.; Novikov, A.V.; Ramachandran, B.R.; Stevenson, K.J.; Aldoshin, S.M.; Troshin, P.A. Highly sensitive and selective ammonia gas sensor based on FAPbCl3 lead halide perovskites. *J. Mater. Chem. C* **2021**, *9*, 2561–2568. [CrossRef]
29. Wang, J.; Peng, J.; Sun, Y.; Liu, X.; Chen, Y.; Liang, Z. FAPbCl3 Perovskite as Alternative Interfacial Layer for Highly Efficient and Stable Polymer Solar Cells. *Adv. Electron. Mater.* **2016**, *2*, 1600329. [CrossRef]
30. Gong, J.; Li, X.; Guo, P.; Zhang, I.; Huang, W.; Lu, K.; Cheng, Y.; Schaller, R.D.; Marks, T.J.; Xu, T. Energy-distinguishable bipolar UV photoelectron injection from LiCl-promoted FAPbCl3 perovskite nanorods. *J. Mater. Chem. A* **2019**, *7*, 13043–13049. [CrossRef]
31. Govinda, S.; Kore, B.P.; Swain, D.; Hossain, A.; De, C.; Guru Row, T.N.; Sarma, D.D. Critical Comparison of FAPbX3 and MAPbX3 (X = Br and Cl): How Do They Differ? *J. Phys. Chem. C* **2018**, *122*, 13758–13766. [CrossRef]
32. Sharma, V.K.; Mukhopadhyay, R.; Mohanty, A.; García Sakai, V.; Tyagi, M.; Sarma, D.D. Influence of the Halide Ion on the A-Site Dynamics in FAPbX3 (X = Br and Cl). *J. Phys. Chem. C* **2022**, *126*, 7158–7168. [CrossRef]
33. Nations, S.; Gutsev, L.; Ramachandran, B.; Aldoshin, S.; Duan, Y.; Wang, S. First-principles study of the defect-activity and optical properties of FAPbCl3. *Mater. Adv.* **2022**, *3*, 3897–3905. [CrossRef]
34. Roknuzzaman, M.; Alarco, J.A.; Wang, H.; Du, A.; Tesfamichael, T.; Ostrikov, K.K. Ab initio atomistic insights into lead-free formamidinium based hybrid perovskites for photovoltaics and optoelectronics. *Comput. Mater. Sci.* **2019**, *169*, 109118. [CrossRef]
35. Pachori, S.; Agarwal, R.; Shukla, A.; Rani, U.; Verma, A.S. Mechanically stable with highly absorptive formamidinium lead halide perovskites [(HC(NH2)2PbX3; X = Br, Cl]: Recent advances and perspectives. *Int. J. Quantum Chem.* **2021**, *121*, e26671. [CrossRef]
36. Yang, W.S.; Noh, J.H.; Jeon, N.J.; Kim, Y.C.; Ryu, S.; Seo, J.; Seok, S. Il High-performance photovoltaic perovskite layers fabricated through intramolecular exchange. *Science* **2015**, *348*, 1234–1237. [CrossRef]
37. Charles, B.; Dillon, J.; Weber, O.J.; Islam, M.S.; Weller, M.T. Understanding the stability of mixed A-cation lead iodide perovskites. *J. Mater. Chem. A* **2017**, *5*, 22495–22499. [CrossRef]
38. Lee, J.W.; Kim, D.H.; Kim, H.S.; Seo, S.W.; Cho, S.M.; Park, N.G. Formamidinium and cesium hybridization for photo- and moisture-stable perovskite solar cell. *Adv. Energy Mater.* **2015**, *5*, 1501310. [CrossRef]
39. Yi, C.; Luo, J.; Meloni, S.; Boziki, A.; Ashari-Astani, N.; Grätzel, C.; Zakeeruddin, S.M.; Röthlisberger, U.; Grätzel, M. Entropic stabilization of mixed A-cation ABX3 metal halide perovskites for high performance perovskite solar cells. *Energy Environ. Sci.* **2016**, *9*, 656–662. [CrossRef]

40. Saliba, M.; Matsui, T.; Domanski, K.; Seo, J.Y.; Ummadisingu, A.; Zakeeruddin, S.M.; Correa-Baena, J.P.; Tress, W.R.; Abate, A.; Hagfeldt, A.; et al. Incorporation of rubidium cationsinto perovskite solar cells improvesphotovoltaic performance. *Science* **2016**, *354*, 203–206. [CrossRef]
41. Saliba, M.; Matsui, T.; Seo, J.Y.; Domanski, K.; Correa-Baena, J.P.; Nazeeruddin, M.K.; Zakeeruddin, S.M.; Tress, W.; Abate, A.; Hagfeldt, A.; et al. Cesium-containing triple cation perovskite solar cells: Improved stability, reproducibility and high efficiency. *Energy Environ. Sci.* **2016**, *9*, 1989–1997. [CrossRef]
42. Weber, O.J.; Charles, B.; Weller, M.T. Phase behaviour and composition in the formamidinium-methylammonium hybrid lead iodide perovskite solid solution. *J. Mater. Chem. A* **2016**, *4*, 15375–15382. [CrossRef]
43. Jesper Jacobsson, T.; Correa-Baena, J.P.; Pazoki, M.; Saliba, M.; Schenk, K.; Grätzel, M.; Hagfeldt, A. Exploration of the compositional space for mixed lead halogen perovskites for high efficiency solar cells. *Energy Environ. Sci.* **2016**, *9*, 1706–1724. [CrossRef]
44. Yang, Z.; Chueh, C.C.; Liang, P.W.; Crump, M.; Lin, F.; Zhu, Z.; Jen, A.K.Y. Effects of formamidinium and bromide ion substitution in methylammonium lead triiodide toward high-performance perovskite solar cells. *Nano Energy* **2016**, *22*, 328–337. [CrossRef]
45. Pisanu, A.; Ferrara, C.; Quadrelli, P.; Guizzetti, G.; Patrini, M.; Milanese, C.; Tealdi, C.; Malavasi, L. The $FA_{1-x}MA_xPbI_3$ System: Correlations among Stoichiometry Control, Crystal Structure, Optical Properties, and Phase Stability. *J. Phys. Chem. C* **2017**, *121*, 8746–8751. [CrossRef]
46. Alonso, M.I.; Charles, B.; Francisco-López, A.; Garriga, M.; Weller, M.T.; Goñi, A.R. Spectroscopic ellipsometry study of $FA_xMA_{1-x}PbI_3$ hybrid perovskite single crystals. *J. Vac. Sci. Technol. B* **2019**, *37*, 062901. [CrossRef]
47. Francisco-López, A.; Charles, B.; Alonso, M.I.; Garriga, M.; Campoy-Quiles, M.; Weller, M.T.; Goñi, A.R. Phase Diagram of Methylammonium/Formamidinium Lead Iodide Perovskite Solid Solutions from Temperature-Dependent Photoluminescence and Raman Spectroscopies. *J. Phys. Chem. C* **2020**, *124*, 3448–3458. [CrossRef]
48. Chen, T.; Chen, W.L.; Foley, B.J.; Lee, J.; Ruff, J.P.C.; Ko, J.Y.P.; Brown, C.M.; Harriger, L.W.; Zhang, D.; Park, C.; et al. Origin of long lifetime of band-edge charge carriers in organic–inorganic lead iodide perovskites. *Proc. Natl. Acad. Sci. USA* **2017**, *114*, 7519–7524. [CrossRef]
49. Miyata, A.; Mitioglu, A.; Plochocka, P.; Portugall, O.; Wang, J.T.W.; Stranks, S.D.; Snaith, H.J.; Nicholas, R.J. Direct measurement of the exciton binding energy and effective masses for charge carriers in organic-inorganic tri-halide perovskites. *Nat. Phys.* **2015**, *11*, 582–587. [CrossRef]
50. Chen, T.; Foley, B.J.; Park, C.; Brown, C.M.; Harriger, L.W.; Lee, J.; Ruff, J.; Yoon, M.; Choi, J.J.; Lee, S.H. Entropy-driven structural transition and kinetic trapping in formamidinium lead iodide perovskite. *Sci. Adv.* **2016**, *2*, e1601650. [CrossRef]
51. Stoumpos, C.C.; Malliakas, C.D.; Kanatzidis, M.G. Semiconducting tin and lead iodide perovskites with organic cations: Phase transitions, high mobilities, and near-infrared photoluminescent properties. *Inorg. Chem.* **2013**, *52*, 9019–9038. [CrossRef] [PubMed]
52. Jeon, N.J.; Noh, J.H.; Yang, W.S.; Kim, Y.C.; Ryu, S.; Seo, J.; Seok, S. Il Compositional engineering of perovskite materials for high-performance solar cells. *Nature* **2015**, *517*, 476–480. [CrossRef] [PubMed]
53. Binek, A.; Hanusch, F.C.; Docampo, P.; Bein, T. Stabilization of the trigonal high-temperature phase of formamidinium lead iodide. *J. Phys. Chem. Lett.* **2015**, *6*, 1249–1253. [CrossRef] [PubMed]
54. Šimėnas, M.; Balčiū Nas, S.; Svirskas, Š.N.; Kinka, M.; Ptak, M.; Kalendra, V.; Gągor, A.; Szewczyk, D.; Sieradzki, A.; Grigalaitis, R.; et al. Phase Diagram and Cation Dynamics of Mixed MA1- xFA xPbBr3Hybrid Perovskites. *Chem. Mater.* **2021**, *33*, 5926–5934. [CrossRef]
55. Baikie, T.; Barrow, N.S.; Fang, Y.; Keenan, P.J.; Slater, P.R.; Piltz, R.O.; Gutmann, M.; Mhaisalkar, S.G.; White, T.J. A combined single crystal neutron/X-ray diffraction and solid-state nuclear magnetic resonance study of the hybrid perovskites $CH_3NH_3PbX_3$ (X = I, Br and Cl). *J. Mater. Chem. A* **2015**, *3*, 9298–9307. [CrossRef]
56. Maculan, G.; Sheikh, A.D.; Abdelhady, A.L.; Saidaminov, M.I.; Haque, M.A.; Murali, B.; Alarousu, E.; Mohammed, O.F.; Wu, T.; Bakr, O.M. $CH_3NH_3PbCl_3$ Single Crystals: Inverse Temperature Crystallization and Visible-Blind UV-Photodetector. *J. Phys. Chem. Lett.* **2015**, *6*, 3781–3786. [CrossRef]
57. Askar, A.M.; Karmakar, A.; Bernard, G.M.; Ha, M.; Terskikh, V.V.; Wiltshire, B.D.; Patel, S.; Fleet, J.; Shankar, K.; Michaelis, V.K. Composition-Tunable Formamidinium Lead Mixed Halide Perovskites via Solvent-Free Mechanochemical Synthesis: Decoding the Pb Environments Using Solid-State NMR Spectroscopy. *J. Phys. Chem. Lett.* **2018**, *9*, 2671–2677. [CrossRef] [PubMed]
58. He, C.; Zha, G.; Deng, C.; An, Y.; Mao, R.; Liu, Y.; Lu, Y.; Chen, Z. Refractive Index Dispersion of Organic–Inorganic Hybrid Halide Perovskite $CH_3NH_3PbX_3$ (X = Cl, Br, I) Single Crystals. *Cryst. Res. Technol.* **2019**, *54*, 1900011. [CrossRef]
59. Niemann, R.G.; Kontos, A.G.; Palles, D.; Kamitsos, E.I.; Kaltzoglou, A.; Brivio, F.; Falaras, P.; Cameron, P.J. Halogen Effects on Ordering and Bonding of $CH_3NH_3^+$ in $CH_3NH_3PbX_3$ (X = Cl, Br, I) Hybrid Perovskites: A Vibrational Spectroscopic Study. *J. Phys. Chem. C* **2016**, *120*, 2509–2519. [CrossRef]
60. Kucharska, E.; Hanuza, J.; Ciupa, A.; Mączka, M.; Macalik, L. Vibrational properties and DFT calculations of formamidine-templated Co and Fe formates. *Vib. Spectrosc.* **2014**, *75*, 45–50. [CrossRef]
61. Kontos, A.G.; Manolis, G.K.; Kaltzoglou, A.; Palles, D.; Kamitsos, E.I.; Kanatzidis, M.G.; Falaras, P. Halogen-NH_2^+ Interaction, Temperature-Induced Phase Transition, and Ordering in $(NH_2CHNH_2)PbX_3$ (X = Cl, Br, I) Hybrid Perovskites. *J. Phys. Chem. C* **2020**, *124*, 8479–8487. [CrossRef]

62. Leguy, A.M.A.; Goñi, A.R.; Frost, J.M.; Skelton, J.; Brivio, F.; Rodríguez-Martínez, X.; Weber, O.J.; Pallipurath, A.; Alonso, M.I.; Campoy-Quiles, M.; et al. Dynamic disorder, phonon lifetimes, and the assignment of modes to the vibrational spectra of methylammonium lead halide perovskites. *Phys. Chem. Chem. Phys.* **2016**, *18*, 27051–27066. [CrossRef]
63. Maalej, A.; Abid, Y.; Kallel, A.; Daoud, A.; Lautié, A.; Romain, F. Phase transitions and crystal dynamics in the cubic perovskite-$CH_3NH_3PbCl_3$. *Solid State Commun.* **1997**, *103*, 279–284. [CrossRef]
64. MącZka, M.; Ptak, M.; Vasconcelos, D.L.M.; Giriunas, L.; Freire, P.T.C.; Bertmer, M.; Banys, J.; Simenas, M. NMR and Raman Scattering Studies of Temperature-and Pressure-Driven Phase Transitions in $CH_3NH_2NH_2PbCl_3$ Perovskite. *J. Phys. Chem. C* **2020**, *124*, 26999–27008. [CrossRef]
65. Glaser, T.; Müller, C.; Sendner, M.; Krekeler, C.; Semonin, O.E.; Hull, T.D.; Yaffe, O.; Owen, J.S.; Kowalsky, W.; Pucci, A.; et al. Infrared Spectroscopic Study of Vibrational Modes in Methylammonium Lead Halide Perovskites. *J. Phys. Chem. Lett.* **2015**, *6*, 2913–2918. [CrossRef] [PubMed]
66. Ma, L.; Li, W.; Yang, K.; Bi, J.; Feng, J.; Zhang, J.; Yan, Z.; Zhou, X.; Liu, C.; Ji, Y.; et al. A- or X-site mixture on mechanical properties of $APbX_3$ perovskite single crystals. *APL Mater.* **2021**, *9*, 041112. [CrossRef]
67. Sun, S.; Isikgor, F.H.; Deng, Z.; Wei, F.; Kieslich, G.; Bristowe, P.D.; Ouyang, J.; Cheetham, A.K. Factors Influencing the Mechanical Properties of Formamidinium Lead Halides and Related Hybrid Perovskites. *ChemSusChem* **2017**, *10*, 3740–3745. [CrossRef] [PubMed]

Disclaimer/Publisher's Note: The statements, opinions and data contained in all publications are solely those of the individual author(s) and contributor(s) and not of MDPI and/or the editor(s). MDPI and/or the editor(s) disclaim responsibility for any injury to people or property resulting from any ideas, methods, instructions or products referred to in the content.

Article

Engineering Band Gap of Ternary Ag₂Te_xS_{1−x} Quantum Dots for Solution-Processed Near-Infrared Photodetectors

Zan Wang [1,2,3,4], Yunjiao Gu [1,2,3], Daniil Aleksandrov [1,2,3], Fenghua Liu [1,2,3], Hongbo He [1,2,3,*] and Weiping Wu [1,2,3,*]

1. Laboratory of Thin Film Optics, Shanghai Institute of Optics and Fine Mechanics, Chinese Academy of Sciences, Shanghai 201800, China
2. State Key Laboratory of High Field Laser Physics, Shanghai Institute of Optics and Fine Mechanics, Chinese Academy of Sciences, Shanghai 201800, China
3. University of Chinese Academy of Sciences, Beijing 100049, China
4. Department of Optics and Optical Engineering, University of Science and Technology of China, Hefei 230026, China
* Correspondence: hbhe@siom.ac.cn (H.H.); wuwp@siom.ac.cn (W.W.)

Citation: Wang, Z.; Gu, Y.; Aleksandrov, D.; Liu, F.; He, H.; Wu, W. Engineering Band Gap of Ternary Ag₂Te_xS_{1−x} Quantum Dots for Solution-Processed Near-Infrared Photodetectors. *Inorganics* 2024, 12, 1. https://doi.org/10.3390/inorganics12010001

Academic Editor: Richard Dronskowski

Received: 8 October 2023
Revised: 8 December 2023
Accepted: 11 December 2023
Published: 19 December 2023

Copyright: © 2023 by the authors. Licensee MDPI, Basel, Switzerland. This article is an open access article distributed under the terms and conditions of the Creative Commons Attribution (CC BY) license (https://creativecommons.org/licenses/by/4.0/).

Abstract: Silver-based chalcogenide semiconductors exhibit low toxicity and near-infrared optical properties and are therefore extensively employed in the field of solar cells, photodetectors, and biological probes. Here, we report a facile mixture precursor hot-injection colloidal route to prepare Ag₂Te_xS_{1−x} ternary quantum dots (QDs) with tunable photoluminescence (PL) emissions from 950 nm to 1600 nm via alloying band gap engineering. As a proof-of-concept application, the Ag₂Te_xS_{1−x} QDs-based near-infrared photodetector (PD) was fabricated via solution-processes to explore their photoelectric properties. The ICP-OES results reveal the relationship between the compositions of the precursor and the samples, which is consistent with Vegard's equation. Alloying broadened the absorption spectrum and narrowed the band gap of the Ag₂S QDs. The UPS results demonstrate the energy band alignment of the Ag₂Te_{0.53}S_{0.47} QDs. The solution-processed Ag₂Te_xS_{1−x} QD-based PD exhibited a photoresponse to 1350 nm illumination. With an applied voltage of 0.5 V, the specific detectivity is 0.91×10^{10} Jones and the responsivity is 0.48 mA/W. The PD maintained a stable response under multiple optical switching cycles, with a rise time of 2.11 s and a fall time of 1.04 s, which indicate excellent optoelectronic performance.

Keywords: Ag₂Te_xS_{1−x} QDs; ternary alloying; band gap engineering; photodetectors

1. Introduction

Near-infrared (NIR) light has several advantages, including great penetration depth, strong confidentiality, and excellent anti-electromagnetic interference properties [1–4]. Thus, NIR materials have found extensive applications across diverse domains such as bioimaging, optical communication, photodetectors, and solar cells [5–11]. Currently, NIR materials are mainly concentrated on PbS, PbSe, HgTe QDs, rare-earth nanocrystals, and single-walled carbon nanotubes (SWNTs) [12–16]. Among them, PbS, PbSe, and HgTe QDs contain the toxic heavy metals lead and mercury and rare-earth nanocrystals with non-tunable band gaps and narrow absorption windows. For SWNTs, their application is limited by their broad length distribution, spanning hundreds of nanometers. Therefore, it becomes imperative to foster the creation of novel materials with continuously tunable band gap and high biocompatibility. Silver-based chalcogenide (Ag₂S, Ag₂Se, and Ag₂Te) QDs, as classic NIR materials, possess excellent properties such as low toxicity, wide absorption windows, and good biocompatibility [17–19]. Among these, Ag₂S QDs are one of the most extensively studied semiconductor materials and have ultralow solubility product constants (Ksp(Ag₂S) = $6.69 \times 10^{−50}$), which enable the minimal release of Ag ions in biosystems, thus ensuring their potential application in biomedical, ReRAMs, and optoelectronic devices [20–22]. However, due to the large band gap of Ag$_2$S (1.1 eV), its photoluminescence (PL) emission peak is limited to less than 1200 nm, and photodetectors based on this material cannot respond to light in longer wavelength bands and have poor power conversion efficiencies [23]. Therefore, expanding the band gap of Ag$_2$S and improving its photoelectric performance is very important for the application of Ag$_2$S materials in the field of optoelectronic devices.

Compared to traditional binary QDs, ternary alloy QDs not only retain the quantum size effect of nanomaterials but also enable effective control of the band gap by adjusting the composition [24,25]. Smith et al. [26] studied the influence of different reactive anion precursors on the size and composition of PbSe$_x$Te$_{1-x}$, PbS$_x$Te$_{1-x}$, and PbS$_x$Se$_{1-x}$ QDs. The results demonstrated that the highly reactive Chalcogenide precursors bis(trimethylsilyl) (TMS$_2$) help to achieve uniform anion incorporation. In addition, it was possible to modulate the excitonic absorption and fluorescence peak of alloy QDs by adjusting the anions ratios. The next generation of electroluminescent displays based on quantum dots requires the development of efficient and stable Cd-free blue emission devices, which remains a challenge due to the poor photophysical properties of blue emission materials. Jang et al. [27] proposed a method by which to synthesize efficient blue-emitting ZnTeSe QDs. They found that the hydrofluoric acid and zinc chloride additives effectively improved the luminescence efficiency by eliminating the layer faults in the ZnSe crystal structure. Moreover, chloride passivation via liquid or solid ligand exchange results in slow radiative recombination, high thermal stability, and efficient charge transport characteristics, and the fluorescence peak was adjusted to 457 nm, while the photoluminescence quantum yield (PLQY) was elevated to a remarkable 100% by controlling the Te doping level. In addition, Ren et al. [28] proposed a water-phase synthesis method for the preparation of NIR CdHgTe alloy quantum dots. CdHgTe QDs are obtained by heating a mixture of CdCl$_2$, Hg(ClO$_4$)$_2$ and NaHTe in the presence of a thiol stabilizer and exhibit PL emission peaks in the range from 600 to 830 nm, which can be adjusted according to size and composition. The PLQY of CdHgTe QDs is about 20–50%, which depends on its emission wavelength and composition. Compared to other reported NIR quantum dots (such as CdTe/CdHgTe and InAs), the prepared CdHgTe alloy quantum dots have a much narrower emission spectrum, with a full width at half-maximum (FWHM) of only 60–80 nm. HRTEM and XRD characterization show that CdHgTe QDs have a good crystal structure and monodispersity. In order to improve the photostability of CdHgTe QDs and reduce their cytotoxicity, CdS nanocrystal shells were added to the surface of CdHgTe QDs. Kim et al. [29] investigated the synthesis of eco-friendly materials AgBiS$_2$ QDs and the effect of heat treatment on their properties. Increasing the heat treatment temperature reduces the number of surface functional groups, including N (amine) and S (thiol) groups, and there are fewer defects on the particle surface. However, heat treatment at 300 °C reduces PL intensity even when the ligands are fully removed. By measuring the photocurrent response of the AgBiS$_2$ photodetector to near-infrared light, the photocurrent of the AgBiS$_2$ photodetector is the highest after heat treatment at 200 °C. Heat treatment removes excessive protective agents and ligands in the inks and improves the photocurrent response of AgBiS$_2$.

Herein, for the first time, we develop a facile mixture precursor hot-injection method for the synthesis of Ag$_2$Te$_x$S$_{1-x}$ alloy QDs. The composition of the alloy QDs was regulated by adjusting the ratio of precursor S to precursor Te. Through augmenting the amount of the precursor Te, the fluorescence peak of the alloy QDs redshifted and the absorption spectrum broadened, thereby achieving controllable tuning of the band gap. Furthermore, a photodetector based on Ag$_2$Te$_{0.53}$S$_{0.47}$ QDs as the photosensitive layer has been constructed. The narrow band gap photosensitive layer enables the device to exhibit photoresponse to 1350 nm illumination. The device maintained a stable response and exhibited excellent optoelectronic performance under multiple light switching cycles. This is also the first report on the preparation of Ag$_2$Te$_x$S$_{1-x}$ QDs and the investigation of their optoelectronic properties.

2. Experimental Section

2.1. Materials

Silver acetate (AgAc, 99.99%, Alfa Aesar, Haverhill, MA, USA), sulfur powder (S, 99.95%, Aladdin, Shanghai, China), oleylamine (OAm, 80%, Acros, Geel, Belgium), tellurium powder (Te, 99.999%, Alfa Aesar), tri-n-butylphosphine (TBP, >95.0%, TCI, Tokyo, Japan), and 1,2-ethanedithiol (EDT, 98.0%, Alfa Aesar); all these reagents were used without purification.

2.2. Preparation of TBP-Te and TBP-S Precursors

The precursors of TBP-Te and TBP-S were prepared in a glove box. TBP-Te: 2.5 mmol (0.32 g) Te powder was dissolved in 5 mL TBP, employing the use of ultrasonic treatment for a duration of 24 h. TBP-S: 2.5 mmol (0.08 g) S powder was dissolved in 5 mL TBP with vigorous magnetic stirring and ultrasonic treatment for 1 h. The total number of moles of Te and S was kept at 0.10 mmol and the two precursors with different atomic ratios.

2.3. Synthesis of $Ag_2Te_xS_{1-x}$ QDs

In the conventional hot-injection method, a blend comprising AgAc (0.067 g, 0.4 mmol) and Oam (16 mL) was introduced into a 100 mL three-neck flask at room temperature. Subsequently, the solution underwent a remove oxygen process through vigorous magnetic stirring while maintaining a vacuum for 30 min. Afterwards, the temperature of the reaction was subsequently elevated to 120 °C in the presence of a nitrogen atmosphere, resulting in the attainment of a transparent solution. Then, 0.2 mL mixture precursor (TBP-S and TBP-Te) was injected into the solution while vigorously stirring at a speed of 1000 rpm. This temperature was maintained for a duration of 5 min to ensure the consistent growth of the QDs. The products were precipitated with ethanol and recovered by centrifugation at a speed of 12,000 rpm for 10 min. The products were dispersed in chloroform (15 mg/mL). A series of $Ag_2Te_xS_{1-x}$ samples were used, referred to as S/Te-A (S:Te = 1:0), S/Te-B (S:Te = 2:1), S/Te-C (S:Te = 1:1), and S/Te-D (S:Te = 0:1).

2.4. Photodetector Fabrication

The photodetector device fabrication and ligand exchange protocol is based on previous research [30]. At first, this process begins with cleaning the glass substrates. Then the $Ag_2Te_xS_{1-x}$ QDs solution was deposited onto the glass substrate by employing the spin-coating technique under a speed of 1500 rpm and kept for 30 s. Subsequently, a volume of 500 µL of ligand solution (consisting of 5 mM EDT dissolved in acetonitrile) was gently applied onto the film for a duration of 30 s, followed by removal of the ligand solution using a spinning method. Afterwards, the QD film was immersed in acetonitrile 3 times to eliminate any remaining unbound ligands. The ligand exchange and spin-coating were iterated 6 times. Finally, 5 nm/100 nm-thick Ti/Au metal electrodes were deposited through an interdigitated shadow mask (the channel width is 50 µm and the size is 18 mm × 18 mm) through a thermal evaporator.

2.5. Characterizations

Talos F200X (ThermoFisher, Waltham, MA, USA) was used to test transmission electron microscope (TEM) and high-resolution TEM (HRTEM) images of $Ag_2Te_xS_{1-x}$ QDs under an acceleration voltage of 200 kV. X-ray diffraction (XRD) patterns were measured by a PANalytical Empyrean diffractometer (Malven PANalytical, Almelo, Holland) (Cu-Kα λ = 1.54056 Å) in the range of 20–80°. An inductively coupled plasma optical emission spectrometer (ICP-OES) was performed on PerkinElmer Avio 500 (PerkinElmer, Waltham, MA, USA). The absorption spectra were recorded with a PerkinElmer Lambda 750 spectrometer (PerkinElmer, Waltham, MA, USA). The NIR fluorescence spectra were executed on an Applied NanoFluorescence Spectrometer (Houston, TX, USA), applying an excitation laser source of 785 nm. The X-ray photoelectron spectroscopy (XPS) were collected on a PHI-5000 VersaProbe instrument (ULVAC-PHI, Chigasaki, Japan) using Al/Kα radiation as excitation source, the electron emission angle is 34°, the diameter of the X-ray spot is 250 µm, the base pressure is 4.17×10^{-7} mBar during the analyses, the samples are etched for 60 s with 1 keV Ar$^+$ ions before the XPS scan to remove the surface layer which adsorb water and oxygen in the air. All the XPS spectra recorded for Ag 3d, Te 3d, and S 2p were referenced to the C 1s peak, which was calibrated by positioning it to 289.58 eV−Φ. The 289.58 eV is the binding energy position of the C 1s peak with respect to the vacuum level, and Φ is the work function of $Ag_2Te_{0.53}S_{0.47}$ (~5.16 eV) [31,32]. Thus, the set value of the extraneous polluted carbon is 284.42 eV. The ultraviolet photoelectron spectroscopy (UPS) was measured by using a Thermo ESCALAB 250Xi instrument (ThermoFisher, Waltham, MA, USA) with He I radiation. The Keithley 4200-SCS semiconductor analyzer equipped with a Signatone S-1160 Probe Station (Signatone, Gilroy, CA, USA) is employed to characterize the optoelectronic performance of devices.

3. Results and Discussion

Figure 1 displays the low-magnification TEM and HRTEM images of $Ag_2Te_xS_{1-x}$ QDs, which exhibit excellent dispersibility, clear lattice patterns, and good crystallinity. Figure 1a shows TEM image of Ag_2S QDs (denoted as S/Te-A), which appear as uniformly spherical particles. With the increase in the Te content, the morphology of the QDs gradually transformed from uniform spherical morphology to a branch-like structure, as shown in Figure 1b,c (denoted as S/Te-B and S/Te-C, respectively). When all the S elements are replaced by Te, the morphology changes to a branch-like structure composed of interconnected spherical particles (S/Te-D), as depicted in Figure 1d.

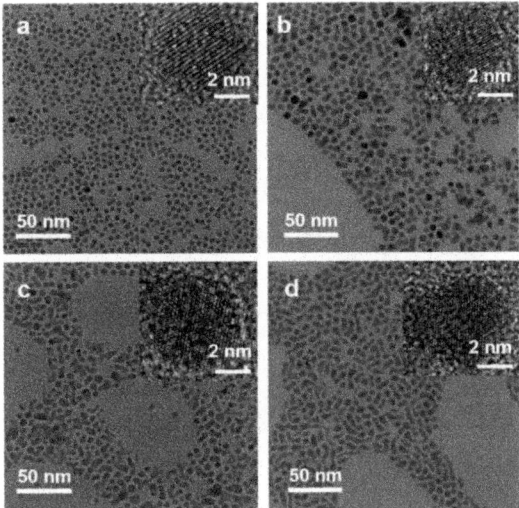

Figure 1. TEM and HRTEM images of $Ag_2Te_xS_{1-x}$ QDs: (**a**) S/Te-A, (**b**) S/Te-B, (**c**) S/Te-C, and (**d**) S/Te-D.

Similar results have been reported in other studies on Ag_2Te QDs, which elucidate that this morphology is mainly attributed to the driving force generated by strong dipole-dipole interactions, leading to the formation of a branch-like structure [33]. Additionally, we performed a statistical analysis of the size distributions of the different $Ag_2Te_xS_{1-x}$ QDs (100 particles counted in each image, Figure S1). The diameters of S/Te-A, S/Te-B, S/Te-C, and S/Te-D were measured as 3.92 ± 0.19 nm, 5.45 ± 0.31 nm, 5.32 ± 0.51 nm, and 5.23 ± 0.32 nm, respectively.

To investigate the crystal structure of the $Ag_2Te_xS_{1-x}$ QDs, we measured the XRD patterns of the samples with different S/Te ratios. Figure 2 shows the S/Te-A sample consisting of Ag_2S QDs, with diffraction peaks matching those of Ag_2S crystals (JCPDS:14-0072) and exhibiting a relatively broad peak width, which is attributed to the small size of the QDs.

The characteristic diffraction peaks of Ag_2S in the $Ag_2Te_xS_{1-x}$ QDs gradually weaken with increasing Te content, indicating a phase transition from monoclinic Ag_2S crystals (JCPDS:14-0072) to monoclinic Ag_2Te crystals (JCPDS:81-1820). Finally, the characteristic diffraction peaks of the S/Te-D sample coincided with those of pure Ag_2Te crystals (JCPDS:81-1820), exhibiting broadened diffraction peaks. ICP-OES was used to analyze the actual atomic ratios of the S and Te anions in the samples. Figure S2 illustrates the relationship between Te/(Te + S) for the precursors and the products of the four samples. When the molar ratio of the Te source to the S source is 1:2, the Te-to-S ratio in the product is 0.54:1 ($Ag_2Te_{0.35}S_{0.65}$). When the molar ratio of the Te source to the S source turns to 1:1, the Te-to-S ratio in the product becomes 1.13:1 ($Ag_2Te_{0.53}S_{0.47}$) owing to the slightly higher reactivity of the Te source compared to that of the S source. These results demonstrate that the manipulation of the anion ratio in the precursor mixture is an effective means of controlling the anion content within the alloy QDs.

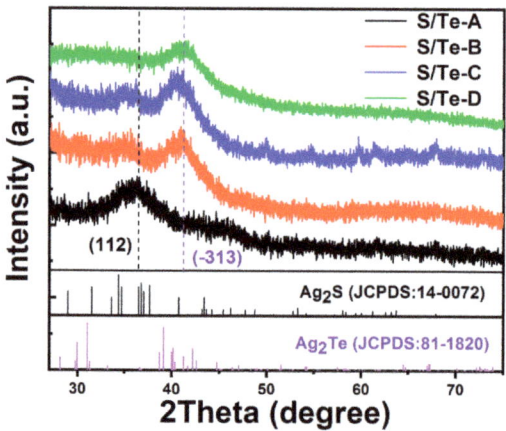

Figure 2. XRD patterns of four different $Ag_2Te_xS_{1-x}$ QDs.

XPS was used to characterize the elements and their valence states in the alloyed QDs. Figure 3a shows the low-resolution scan spectrum of the $Ag_2Te_{0.53}S_{0.47}$ QDs, which demonstrates the sample contains Ag, Te, O, S, and C elements. Figure 3b–d show the high-resolution scan spectra of Ag 3d, Te 3d, and S 2p. As illustrated in Figure 3b, the high-resolution Ag 3d region spectra of $Ag_2Te_{0.53}S_{0.47}$ QDs show two symmetric peaks, Ag $3d_{5/2}$ and Ag $3d_{3/2}$, which correspond to binding energies of 367.7 eV and 373.7 eV, respectively. This confirms that the oxidation state of silver is univalent (Ag^+) [34]. In Figure 3c, the high-resolution Te 3d region spectra show two peaks, Te $3d_{5/2}$ and Te $3d_{3/2}$, which correspond to binding energies of 572.0 eV and 582.4 eV, respectively, suggesting that the valence of telluride is negative bivalence (Te^{2-}) [35,36]. Furthermore, the spectrum of the S 2p region was characterized and analysed (Figure 3d). The spectrum contains multiple peaks, which must be separated. Fitting the low-binding-energy peaks yields two distinct peaks with binding energies measuring 162.6 eV and 161.3 eV. These peaks can be attributed to the S $2p_{1/2}$ and S $2p_{3/2}$, respectively, in agreement with the Ag-S binding energies [37]. An additional peak appears in the high-energy region (approximately 168.3 eV), indicating that a small fraction of S may have undergone oxidation to form sulfate ions upon exposure to air [38].

Figure 4 shows the PL emission spectra of $Ag_2Te_xS_{1-x}$ QDs with four different S/Te ratios under 785 nm excitation. The dashed lines represent the spectral regions that could not be detected because of the response range of the spectrometer (900 nm–1700 nm). The fluorescence peak of the $Ag_2Te_xS_{1-x}$ QDs exhibits a gradual redshift, transitioning from 955 nm to 1255 nm, 1470 nm, and ultimately reaching 1605 nm. When the size of the direct band gap semiconductor decreases, the PL emission should blue-shift. However, the size of $Ag_2Te_xS_{1-x}$ QDs decreases, and the PL peak redshift contradicts the above theory (S/Te-B to S/Te-D). This anomalous phenomenon can be explained by Vegard's law, which describes the relationship between the band gap of ternary alloy materials and their elemental composition of elements [39]. The band gap of $Ag_2Te_xS_{1-x}$ alloy can be described as follows:

$$E_{alloy} = xE_t + (1-x)E_s \tag{1}$$

where x is the proportion of component t; and E_{alloy}, E_t, and E_s are the band gap energies of the alloy material, pure t, and pure s, respectively. Because the band gap of Ag_2Te (0.06 eV) [40] is much smaller than that of Ag_2S (1.1 eV) [41], as the Te component increases, the value of x increases, leading to a decrease in the band gap of the alloy QDs and a redshift in the PL emission peaks.

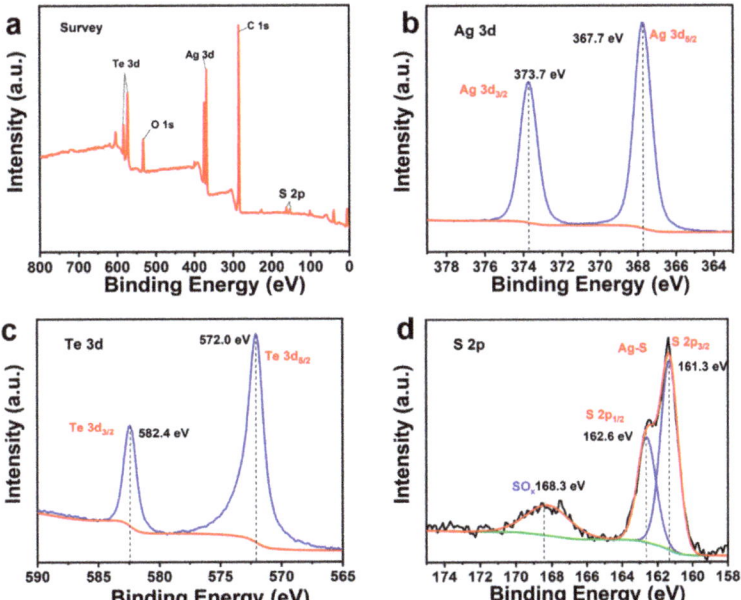

Figure 3. (a) Survey XPS spectra of $Ag_2Te_xS_{1-x}$ QDs. High-resolution XPS spectra for (b) Ag 3d, (c) Te 3d, and (d) S 2p. The overlapped spectra of S were fitted with the Gaussian function.

Figure 5 shows the absorption spectra of $Ag_2Te_xS_{1-x}$ QDs with four different S/Te ratios. The S/Te-A to S/Te-D samples do not exhibit distinct excitonic absorption peaks, which is similar to previous reports of Ag_2S QDs and Ag_2Te QDs [42,43]. The optical band gap values of $Ag_2Te_xS_{1-x}$ QDs are calculated by the extrapolated energy intercept of the Tauc plot. For direct bandgap materials, the Tauc equation can be described as follows [44]:

$$\alpha h\nu = A(h\nu - E_g)^{1/2} \tag{2}$$

where α is the absorption coefficient; $h\nu$ is the incident photon energy; A is a proportionality constant; and E_g is the optical band gap, respectively. The band gap was calculated to be 1.45 eV, 1.07 eV, 0.89 eV, and 0.84 eV for S/Te-A, S/Te-B, S/Te-C, and S/Te-D, respectively. As the Te content increased, the absorption wavelength ranges of the alloy QDs broadened, and the band gap decreased. This result is consistent with the alterations observed in the PL emission spectra of $Ag_2Te_xS_{1-x}$ QDs. Furthermore, the result reveals that the adjustment of Te content effectively controls the band gap of $Ag_2Te_xS_{1-x}$ QDs.

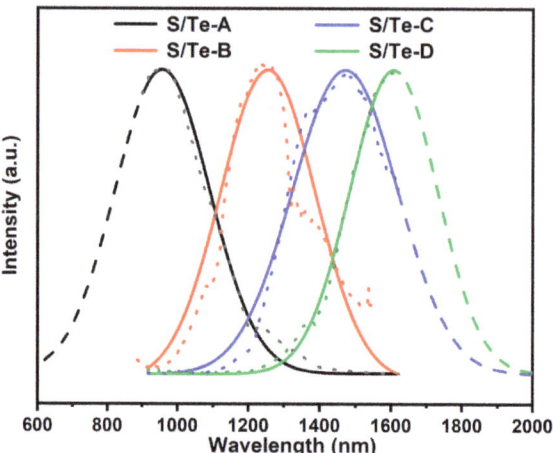

Figure 4. PL emission spectra of the obtained four different $Ag_2Te_xS_{1-x}$ QDs under an excitation of 785 nm (the dotted lines are the real raw experimental data, and the solid lines are the corresponding Gaussian fitting data).

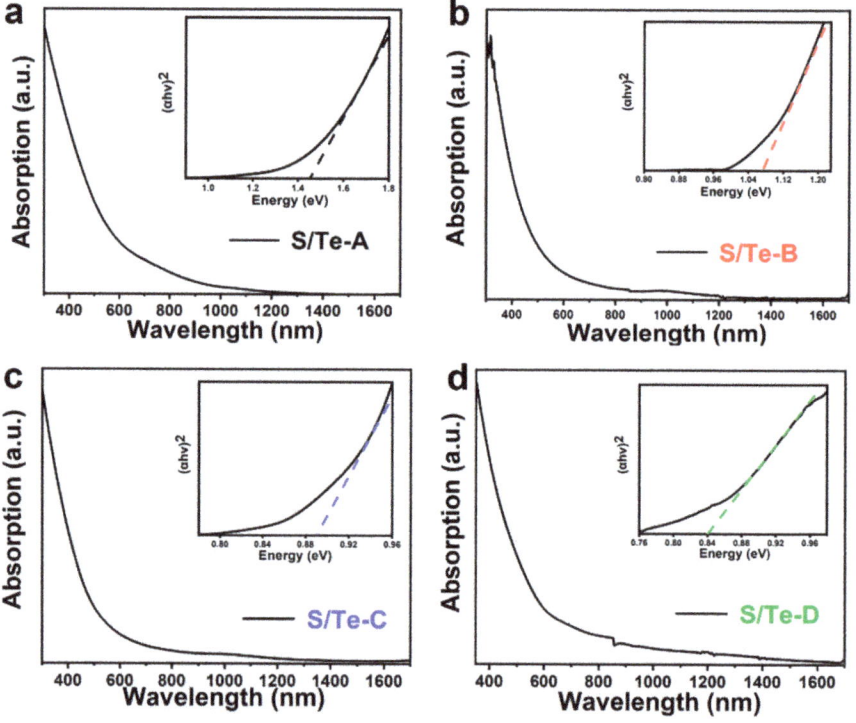

Figure 5. Absorption spectrum and Tauc plot for estimating E_g (insert) of four different $Ag_2Te_xS_{1-x}$ samples: (**a**) S/Te-A, (**b**) S/Te-B, (**c**) S/Te-C, and (**d**) S/Te-D.

UPS was used to measure the kinetic energy of the $Ag_2Te_{0.53}S_{0.47}$ QDs, as shown in Figure 6. The test sample was prepared by spin-coating it into a thin film on a glass substrate. Figure 6a shows the survey of UPS spectrum of the $Ag_2Te_{0.53}S_{0.47}$ QDs. Figure 6b portrays the region of high binding energy cutoff. The tangent value in the cutoff region is approximately 16.6 eV, indicating that the Fermi level positioning of the alloy QDs is

4.61 eV. Combined with the low binding energy cutoff region shown in Figure 6c (with a tangent value of approximately 0.54 eV), the energy level of the valence band maximum (E_{VB}) was calculated to be 5.15 eV (relative to the vacuum level E_{Vac}). Finally, combined with the band gap of $Ag_2Te_{0.53}S_{0.47}$ QDs (E_g = 0.89 eV), the energy level of the conduction band minimum (E_{CB}) can be calculated to be 4.26 eV. Therefore, the accurate depiction of the energy band alignment for the $Ag_2Te_{0.53}S_{0.47}$ QDs is depicted in Figure S3.

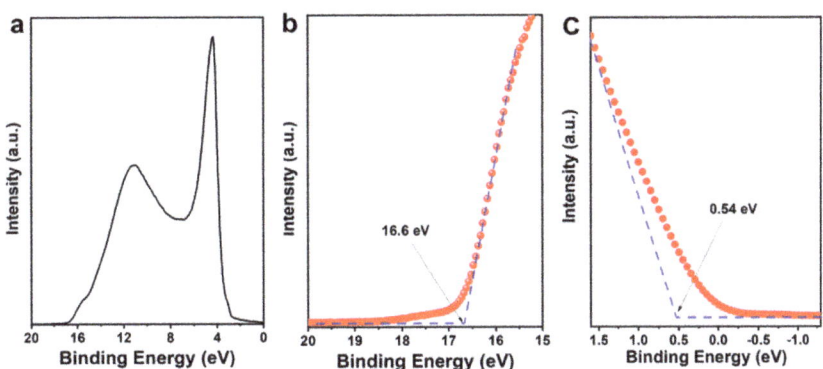

Figure 6. (a) Survey UPS spectrum of the $Ag_2Te_{0.53}S_{0.47}$ QDs. Amplified areas of (b) the high binding energy regions and (c) the low binding energy regions.

As a proof-of-concept application, $Ag_2Te_xS_{1-x}$ QDs were used as the active layer for the photodetector (PD) device. The photoelectric properties of $Ag_2Te_xS_{1-x}$ QD-based PD were investigated (Figure S4). Figure S4a,b show that neither the Ag_2S QD-based or $Ag_2Te_{0.53}S_{0.47}$ QD-based PD have any response to 1350 nm illumination. This can be attributed to the large band gap of these two QDs, which cannot be excited by 1350 nm light to generate photocurrent. With the increase in Te component, both $Ag_2Te_{0.53}S_{0.47}$ QD-based and Ag_2Te QD-based PD generate photocurrent under 1350 nm illumination, and the light response of $Ag_2Te_{0.53}S_{0.47}$ QDs is more obvious than that of Ag_2Te QDs. Thus, composition adjustment can not only expand the band gap of QDs, but also help to improve the performance of photodetectors.

In order to evaluate the performance of $Ag_2Te_xS_{1-x}$ QD-based PDs, the corresponding figures of merit were calculated The two representative parameters (responsivity R and special detectivity D^*) were determined as follows [45]:

$$R = \frac{I_{ph} - I_d}{P} \quad (3)$$

$$D^* = \frac{A^{1/2} \cdot R}{(2q \cdot I_d)^{1/2}} \quad (4)$$

where I_{ph} is the photocurrent; I_d is the dark current; P is the radiated power (product of active area and incident light density); A is the active area of the photodetector; and q is the electron charge. At an applied voltage of 0.5 V and an illumination intensity of 2 mW/cm^2, the D^* of the $Ag_2Te_{0.53}S_{0.47}$ QD-based PD reaches 0.91×10^{10} Jones, and the R is 0.48 mA/W. In addition, the response speed of the $Ag_2Te_{0.53}S_{0.47}$ QD-based PD was determined using the rise time (τ_r) and fall time (τ_f). Figure 7 shows the response speed curve of the detector at a bias voltage of 0.5 V and an illumination intensity of 2 mW/cm^2. Figure 7a shows the detector still exhibits good repeatability under continuous 10-cycle switching. The fluctuations observed in the light and dark states may be caused by surface defects generated from ligand exchange processes. Figure 7b demonstrates that τ_r was 2.11 s and τ_f was 1.04 s within one switching cycle.

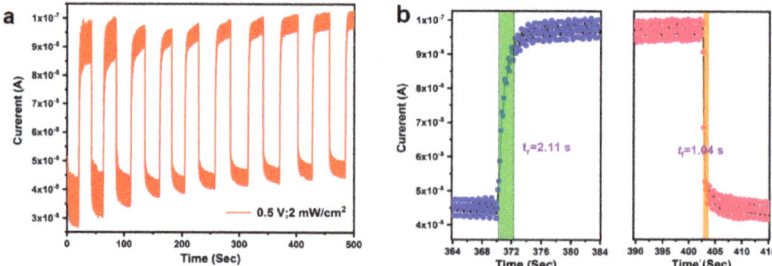

Figure 7. (**a**) Time response of $Ag_2Te_{0.53}S_{0.47}$ QD−based PD at 0.5 V under 1350 nm NIR illumination with power densities of 2 mW/cm^2. (**b**) I−t curve for determining the rise and fall time of $Ag_2Te_{0.53}S_{0.47}$ QD−based PD.

In order to elucidate the electron–hole transport mechanism in the photodetector, the schematic of energy band diagrams of Au/$Ag_2Te_xS_{1−x}$ QDs/Au PD on the glass substrate is shown in Figure 8. Figure 8a shows the energy band diagram of the $Ag_2Te_xS_{1−x}$ QD-based PD under dark and without external bias. The contact between the metal electrode (Au) and $Ag_2Te_xS_{1−x}$ QDs exhibits a slight barrier height (Φ_{SBH}), which proves that the contact between Au and the $Ag_2Te_xS_{1−x}$ QDs is not a simple Ohmic contact but a Schottky contact. This is consistent with the results of the nonlinear I–V curve of the $Ag_2Te_xS_{1−x}$ QD-based PD. Figure 8b shows the energy band diagram of the $Ag_2Te_xS_{1−x}$ QD-based PD under NIR light with a wavelength of 1350 nm illumination and without external bias. The $Ag_2Te_xS_{1−x}$ QDs are excited to produce charge carriers (electron–hole pairs), which are transported between the $Ag_2Te_xS_{1−x}$ QDs due to the quantum tunneling effect. However, only a very small number of carriers can reach the electrode due to the influence of the Schottky barrier. Therefore, no photocurrent can be observed. In addition, under illumination and with external bias, the trap states of $Ag_2Te_xS_{1−x}$ QDs (including surface defect levels and internal defect levels) serve as trapping centers for photogenerated holes [46]. This results in a reduction in the depletion region width and Φ_{SBH}, which offers favorable conditions for carriers tunneling from the $Ag_2Te_xS_{1−x}$ QDs to the metal electrode under external bias, causing increased photocurrent (Figure 8c).

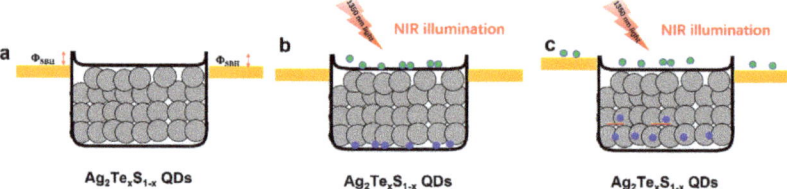

Figure 8. The schematic of energy band diagrams of Au/$Ag_2Te_xS_{1−x}$ QDs/Au PD on the glass substrate: (**a**) under dark and without external bias; (**b**) under NIR light with a wavelength of 1350 nm illumination and without external bias; (**c**) under illumination and with external bias.

4. Conclusions

In summary, we have successfully prepared $Ag_2Te_xS_{1−x}$ ternary QDs via a facile mixture precursor hot-injection method. TEM images and XRD confirmed that the morphology and lattice parameter gradually shifted from Ag_2S to $Ag_2Te_xS_{1−x}$ and finally to Ag_2Te QDs. The ICP-OES results reveal the relationship between the compositions of the precursor and samples, which is consistent with Vegard's equation. In addition, optical characterization confirms the feasibility of $Ag_2Te_xS_{1−x}$ QDs with tunable PL emission and band gap by alloying engineering. The E_{VB} and E_{CB} values of the $Ag_2Te_{0.53}S_{0.47}$ QDs were also calculated using UPS and absorption spectra, and an accurate diagram of the energy band alignment was plotted. The $Ag_2Te_xS_{1−x}$ QD-based PD was fabricated to investigate their photoelectric properties. The PD shows good photoresponse at 1350 nm illumination, D^* is 0.91×10^{10} Jones, and R is 0.48 mA/W under an applied voltage of 0.5 V and power densities of 2 mW/cm^2. The PD maintains a stable response under multiple optical switching cycles, with a rise time of 2.11 s and a fall time of 1.04 s, indicating excellent optoelectronic performance. Therefore, the novel Ag$_2$Te$_x$S$_{1-x}$ ternary alloy QDs extend the photoresponse range of Ag$_2$S QDs, demonstrating promising potential in near-infrared PL emission and photodetection.

Supplementary Materials: The following supporting information can be downloaded at https://www.mdpi.com/article/10.3390/inorganics12010001/s1: Figure S1: The size distribution of Ag$_2$Te$_x$S$_{1-x}$ QDs; Figure S2: ICP-OES data shows the relative amount of Te in the product versus the relative amount of Te in the precursor solution. Figure S3: The schematic band alignment of Ag$_2$Te$_{0.53}$S$_{0.47}$ alloyed QDs. Figure S4: The I-V characteristics of the fabricated Ag$_2$Te$_{0.53}$S$_{0.47}$ QD-based PD in dark and under 1350 nm NIR light with power densities of 2 mW/cm^2.

Author Contributions: Conceptualization, Z.W., H.H. and W.W.; methodology, Z.W.; formal analysis, Z.W.; investigation, Z.W., D.A. and W.W.; re-sources, H.H. and W.W.; data curation, Z.W.; writing—original draft preparation, Z.W.; writing—review and editing, Y.G., D.A., F.L. and W.W.; visualization, Z.W. and D.A.; supervision, H.H. and W.W.; project administration, Y.G., F.L., H.H. and W.W.; funding acquisition, Y.G., F.L., H.H. and W.W. All authors have read and agreed to the published version of the manuscript.

Funding: This work is supported by the National Natural Science Foundation of China (No. 52273242, No. 61975219, No. 22202231), the National Key R&D Program of China (No. 2021YFB2800703, No. 2021YFB2800701) and the Chinese Academy of Sciences.

Data Availability Statement: The data presented in this study are available in article.

Conflicts of Interest: The authors declare no conflict of interest.

References

1. Aiello, R.; Di Sarno, V.; Santi, M.G.D.; De Rosa, M.; Ricciardi, I.; Giusfredi, G.; De Natale, P.; Santamaria, L.; Maddaloni, P. Lamb-dip saturated-absorption cavity ring-down rovibrational molecular spectroscopy in the near-infrared. *Photonics Res.* **2022**, *10*, 1803–1809. [CrossRef]
2. Shanmugam, V.; Selvakumar, S.; Yeh, C.S. Near-infrared light-responsive nanomaterials in cancer therapeutics. *Chem. Soc. Rev.* **2014**, *43*, 6254–6287. [CrossRef]
3. Zhang, Q.Q.; Li, Y.; Jiang, C.H.; Sun, W.B.; Tao, J.W.; Lu, L.H. Near-infrared light-enhanced generation of hydroxyl radical for cancer immunotherapy. *Adv. Healthc. Mater.* **2023**, *12*, 2301502. [CrossRef]
4. Sun, P.; Zhang, M.D.; Dong, F.L.; Feng, L.F.; Chu, W.G. Broadband achromatic polarization insensitive metalens over 950 nm bandwidth in the visible and near-infrared. *Chin. Opt. Lett.* **2022**, *20*, 013601. [CrossRef]
5. Tan, L.L.; Fu, Y.Q.; Kang, S.L.; Wondraczek, L.; Lin, C.G.; Yue, Y.Z. Broadband NIR-emitting Te cluster-doped glass for smart light source towards night-vision and NIR spectroscopy applications. *Photonics Res.* **2022**, *10*, 1187–1193. [CrossRef]
6. Wang, Y.H.; Bai, S.C.; Sun, J.; Liang, H.; Li, C.; Tan, T.Y.; Yang, G.; Wang, J.W. Highly efficient visible and near-infrared luminescence of Sb^{3+}, Tm^{3+} co-doped Cs$_2$NaYCl$_6$ lead-free double perovskite and light emitting diodes. *J. Alloys Compd.* **2023**, *947*, 169602. [CrossRef]
7. Kufer, D.; Nikitskiy, I.; Lasanta, T.; Navickaite, G.; Koppens, F.H.L.; Konstantatos, G. Hybrid 2D-0D MoS$_2$-PbS quantum dot photodetectors. *Adv. Mater.* **2015**, *27*, 176–180. [CrossRef]
8. Zhang, H.; Kumar, S.; Sua, Y.M.; Zhu, S.Y.; Huang, Y.P. Near-infrared 3D imaging with upconversion detection. *Photonics Res.* **2022**, *10*, 2760–2767. [CrossRef]
9. Zeng, X.K.; Wang, C.Y.; Cai, Y.; Lin, Q.G.; Lu, X.W.; Lin, J.H.; Yuan, X.M.; Cao, W.H.; Ai, Y.X.; Xu, S.X. High spatial-resolution biological tissue imaging in the second near-infrared region via optical parametric amplification pumped by an ultrafast vortex pulse. *Chin. Opt. Lett.* **2022**, *20*, 100003. [CrossRef]
10. Konstantatos, G.; Badioli, M.; Gaudreau, L.; Osmond, J.; Bernechea, M.; Garcia de Arquer, F.P.; Gatti, F.; Koppens, F.H.L. Hybrid graphene-quantum dot phototransistors with ultrahigh gain. *Nat. Nanotechnol.* **2012**, *7*, 363–368. [CrossRef]
11. Koppens, F.H.L.; Mueller, T.; Avouris, P.; Ferrari, A.C.; Vitiello, M.S.; Polini, M. Photodetectors based on graphene, other two-dimensional materials and hybrid systems. *Nat. Nanotechnol.* **2014**, *9*, 780–793. [CrossRef]
12. Wang, Z.; Li, J.B.; Huang, F.F.; Hua, Y.J.; Tian, Y.; Zhang, X.H.; Xu, S.Q. Multifunctional optical materials based on transparent inorganic glasses embedded with PbS QDs. *J. Alloys Compd.* **2023**, *942*, 169040. [CrossRef]
13. Semonin, O.E.; Luther, J.M.; Choi, S.; Chen, H.Y.; Gao, J.B.; Nozik, A.J.; Beard, M.C. Peak external photocurrent quantum efficiency exceeding 100% via MEG in a quantum dot solar cell. *Science* **2011**, *334*, 1530–1533. [CrossRef] [PubMed]

14. Chen, M.Y.; Lu, H.P.; Abdelazim, N.M.; Zhu, Y.; Wang, Z.; Ren, W.; Kershaw, S.V.; Rogach, A.L.; Zhao, N. Mercury telluride quantum dot based phototransistor enabling high-sensitivity room-temperature photodetection at 2000 nm. *ACS Nano* **2017**, *11*, 5614–5622. [CrossRef]
15. Wang, R.; Li, X.M.; Zhou, L.; Zhang, F. Epitaxial seeded growth of rare-earth nanocrystals with efficient 800 nm near-infrared to 1525 nm short-wavelength infrared downconversion photoluminescence for in vivo bioimaging. *Angew. Chem. Int. Ed.* **2014**, *53*, 12086–12090. [CrossRef]
16. Kong, J.; Franklin, N.R.; Zhou, C.W.; Chapline, M.G.; Peng, S.; Cho, K.J.; Dai, H.J. Nanotube molecular wires as chemical sensors. *Science* **2000**, *287*, 622–625. [CrossRef]
17. Dong, B.H.; Li, C.Y.; Chen, G.C.; Zhang, Y.J.; Zhang, Y.; Deng, M.J.; Wang, Q.B. Facile synthesis of highly photoluminescent Ag_2Se quantum dots as a new fluorescent probe in the second near-infrared window for in vivo imaging. *Chem. Mater.* **2013**, *25*, 2503–2509. [CrossRef]
18. Gu, Y.P.; Cui, R.; Zhang, Z.L.; Xie, Z.X.; Pang, D.W. Ultrasmall near-infrared Ag_2Se quantum dots with tunable fluorescence for in vivo imaging. *J. Am. Chem. Soc.* **2012**, *134*, 79–82. [CrossRef]
19. Li, C.Y.; Zhang, Y.J.; Wang, M.; Zhang, Y.; Chen, G.C.; Li, L.; Wu, D.M.; Wang, Q.B. In vivo real-time visualization of tissue blood flow and angiogenesis using Ag_2S quantum dots in the NIR-II window. *Biomaterials* **2014**, *35*, 393–400. [CrossRef]
20. Zhao, C.; Chen, C.L.; Wei, R.; Zou, Y.T.; Kong, W.C.; Huang, T.; Yu, Z.; Yang, J.J.; Li, F.; Han, Y.; et al. Laser-assisted synthesis of Ag_2S-quantum-dot-in-perovskite matrix and its application in broadband photodetectors. *Adv. Opt. Mater.* **2022**, *10*, 2101535. [CrossRef]
21. Martin-Garcia, B.; Spirito, D.; Krahne, R.; Moreels, I. Solution-processed silver sulphide nanocrystal film for resistive switching memories. *J. Mater. Chem. C* **2018**, *6*, 13128–13135. [CrossRef]
22. Wang, B.; Zhao, C.; Lu, H.Y.; Zou, T.T.; Singh, S.C.; Yu, Z.; Yao, C.M.; Zheng, X.; Xing, J.; Zou, Y.T.; et al. SERS study on the synergistic effects of electric field enhancement and charge transfer in an Ag_2S quantum dots/plasmonic bowtie nanoantenna composite system. *Photonics Res.* **2020**, *8*, 548–563. [CrossRef]
23. Zhang, Y.; Hong, G.S.; Zhang, Y.J.; Chen, G.C.; Li, F.; Dai, H.J.; Wang, Q.B. Ag_2S quantum dot: A bright and biocompatible fluorescent nanoprobe in the second near-infrared window. *ACS Nano* **2012**, *6*, 3695–3702. [CrossRef]
24. Ji, C.Y.; Zhang, Y.; Zhang, X.Y.; Wang, P.; Shen, H.Z.; Gao, W.Z.; Wang, Y.D.; Yu, W.W. Synthesis and characterization of $Ag_2S_xSe_{1-x}$ nanocrystals and their photoelectrochemical property. *Nanotechnology* **2017**, *28*, 065602. [CrossRef] [PubMed]
25. Bailey, R.E.; Nie, S.M. Alloyed semiconductor quantum dots: Tuning the optical properties without changing the particle size. *J. Am. Chem. Soc.* **2003**, *125*, 7100–7106. [CrossRef] [PubMed]
26. Smith, D.K.; Luther, J.M.; Semonin, O.E.; Nozik, A.J.; Beard, M.C. Tuning the synthesis of ternary lead chalcogenide quantum dots by balancing precursor reactivity. *ACS Nano* **2011**, *5*, 183–190. [CrossRef]
27. Kim, T.; Kim, K.H.; Kim, S.; Choi, S.M.; Jang, H.; Seo, H.K.; Lee, H.; Chung, D.Y.; Jang, E. Efficient and stable blue quantum dot light-emitting diode. *Nature* **2020**, *586*, 385–389. [CrossRef]
28. Qian, H.; Dong, C.; Peng, J.; Qiu, X.; Xu, Y.; Ren, J. High-quality and water-soluble near-infrared photoluminescent CdHgTe/CdS quantum dots prepared by adjusting size and composition. *J. Phys. Chem. C* **2007**, *111*, 16852–16857. [CrossRef]
29. Nakazawa, T.; Kim, D.; Oshima, Y.; Sato, H.; Park, J.; Kim, H. Synthesis and application of $AgBiS_2$ and Ag_2S nanoinks for the production of IR photodetectors. *ACS Omega* **2021**, *6*, 20710–20718. [CrossRef]
30. Grotevent, M.J.; Hail, C.U.; Yakunin, S.; Bachmann, D.; Kara, G.; Dirin, D.N.; Calame, M.; Poulikakos, D.; Kovalenko, M.V.; Shorubalko, I. Temperature-dependent charge carrier transfer in colloidal quantum dot/graphene infrared photodetectors. *ACS Appl. Mater. Interfaces* **2021**, *13*, 848–856. [CrossRef]
31. Bhatt, V.; Kumar, M.; Kim, E.C.; Chung, H.J.; Yun, J.H. Wafer-scale, thickness-controlled p-$CuInSe_2$/n-Si heterojunction for self-biased, highly sensitive, and broadband photodetectors. *ACS Appl. Electron.* **2022**, *4*, 6284–6299. [CrossRef]
32. Hu, H.; Wu, C.; He, C.; Shen, J.; Cheng, Y.; Wu, F.; Wang, S.; Guo, D. Improved photoelectric performance with self-powered characteristics through TiO_2 surface passivation in an α-Ga_2O_3 nanorod array deep ultraviolet photodetector. *ACS Appl. Electron. Mater.* **2022**, *4*, 3801–3806. [CrossRef]
33. Zhang, Y.J.; Yang, H.C.; An, X.Y.; Wang, Z.; Yang, X.H.; Yu, M.X.; Zhang, R.; Sun, Z.Q.; Wang, Q.B. Controlled synthesis of $Ag_2Te@Ag_2S$ core-shell quantum dots with enhanced and tunable fluorescence in the second near-infrared window. *Small* **2020**, *16*, 2001003. [CrossRef]
34. Wang, Z.; Liu, F.H.; Gu, Y.J.; Hu, Y.G.; Wu, W.P. Solution-processed self-powered near-infrared photodetectors of toxic heavy metal-free AgAuSe colloidal quantum dots. *J. Mater. Chem. C* **2022**, *10*, 1097–1104. [CrossRef]
35. Kim, G.; Choi, D.; Eom, S.Y.; Song, H.; Jeong, K.S. Extended short-wavelength infrared photoluminescence and photocurrent of nonstoichiometric silver telluride colloidal nanocrystals. *Nano Lett.* **2021**, *21*, 8073–8079. [CrossRef]
36. Yan, S.G.; Deng, D.Y.; Li, L.; Chen, Y.C.; Song, H.J.; Lv, Y. Glutathione modified Ag_2Te nanoparticles as a resonance Rayleigh scattering sensor for highly sensitive and selective determination of cytochrome C. *Sens. Actuators B Chem.* **2016**, *228*, 458–464. [CrossRef]
37. Chang, P.J.; Cheng, H.Y.; Zhao, F.Y. Photocatalytic reduction of aromatic nitro compounds with Ag/Ag_xS composites under visible light irradiation. *J. Phys. Chem. C* **2021**, *125*, 26021–26030. [CrossRef]
38. Dinda, D.; Ahmed, M.E.; Mandal, S.; Mondal, B.; Saha, S.K. Amorphous molybdenum sulfide quantum dots: An efficient hydrogen evolution electrocatalyst in neutral medium. *J. Mater. Chem. A* **2016**, *4*, 15486–15493. [CrossRef]

39. Denton, A.R.; Ashcroft, N.W. VEGARD LAW. *Phys. Rev. A* **1991**, *43*, 3161–3164. [CrossRef] [PubMed]
40. Kershaw, S.V.; Susha, A.S.; Rogach, A.L. Narrow bandgap colloidal metal chalcogenide quantum dots: Synthetic methods, heterostructures, assemblies, electronic and infrared optical properties. *Chem. Soc. Rev.* **2013**, *42*, 3033–3087. [CrossRef]
41. Du, Y.P.; Xu, B.; Fu, T.; Cai, M.; Li, F.; Zhang, Y.; Wang, Q.B. Near-infrared photoluminescent Ag_2S quantum dots from a single source precursor. *J. Am. Chem. Soc.* **2010**, *132*, 1470–1471. [CrossRef] [PubMed]
42. Zhang, Y.; Zhang, Y.J.; Hong, G.S.; He, W.; Zhou, K.; Yang, K.; Li, F.; Chen, G.C.; Liu, Z.; Dai, H.J.; et al. Biodistribution, pharmacokinetics and toxicology of Ag_2S near-infrared quantum dots in mice. *Biomaterials* **2013**, *34*, 3639–3646. [CrossRef] [PubMed]
43. Chen, C.; He, X.W.; Gao, L.; Ma, N. Cation exchange-based facile aqueous synthesis of small, stable, and nontoxic near-infrared Ag_2Te/ZnS core/shell quantum dots emitting in the second biological window. *ACS Appl. Mater. Interfaces* **2013**, *5*, 1149–1155. [CrossRef]
44. Jeon, J.W.; Jeon, D.W.; Sahoo, T.; Kim, M.; Baek, J.H.; Hoffman, J.L.; Kim, N.S.; Lee, I.H. Effect of annealing temperature on optical band-gap of amorphous indium zinc oxide film. *J. Alloys Compd.* **2011**, *509*, 10062–10065. [CrossRef]
45. Wang, Z.; Gu, Y.J.; Li, X.M.; Liu, Y.; Liu, F.H.; Wu, W.P. Recent progress of quantum dot infrared photodetectors. *Adv. Opt. Mater.* **2023**, *11*, 2300970. [CrossRef]
46. Guo, D.Y.; Wu, Z.P.; An, Y.H.; Guo, X.C.; Chu, X.L.; Sun, C.L.; Li, L.H.; Li, P.G.; Tang, W.H. Oxygen vacancy tuned Ohmic-Schottky conversion for enhanced performance in β-Ga_2O_3 solar-blind ultraviolet photodetectors. *Appl. Phys. Lett.* **2014**, *105*, 023507. [CrossRef]

Disclaimer/Publisher's Note: The statements, opinions and data contained in all publications are solely those of the individual author(s) and contributor(s) and not of MDPI and/or the editor(s). MDPI and/or the editor(s) disclaim responsibility for any injury to people or property resulting from any ideas, methods, instructions or products referred to in the content.

Article
Bipolar Plasticity in Synaptic Transistors: Utilizing HfSe$_2$ Channel with Direct-Contact HfO$_2$ Gate Dielectrics

Jie Lu, Zeyang Xiang, Kexiang Wang, Mengrui Shi, Liuxuan Wu, Fuyu Yan, Ranping Li, Zixuan Wang, Huilin Jin and Ran Jiang *

Faculty of Electrical Engineering and Computer Science, Ningbo University, Ningbo 315211, China
* Correspondence: jiangran@nbu.edu.cn

Abstract: The investigation of dual-mode synaptic plasticity was conducted in thin-film transistors (TFTs) featuring an HfSe$_2$ channel, coupled with an oxygen-deficient (OD)-HfO$_2$ layer structure. In these transistors, the application of negative gate pulses resulted in a notable increase in the post-synaptic current, while positive pulses led to a decrease. This distinctive response can be attributed to the dynamic interplay of charge interactions, significantly influenced by the ferroelectric characteristics of the OD-HfO$_2$ layer. The findings from this study highlight the capability of this particular TFT configuration in closely mirroring the intricate functionalities of biological neurons, paving the way for advancements in bio-inspired computing technologies.

Keywords: plasticity; hafnium dioxide; channel; interlay

1. Introduction

Recently, the exploration into emulating the brain's complex functionality with electronic devices has sparked considerable interest, notably in the realm of neuromorphic engineering [1–5]. At the heart of neuronal functionality is the concept of synaptic plasticity, encompassing both potentiation and depression [6]. Despite the proliferation of diverse electronic devices designed to emulate synaptic plasticity, achieving a faithful replication of both plasticity aspects within a single device remains a formidable challenge, primarily due to the fundamental differences between biological synapses and their electronic counterparts [7]. Consider, for example, the brain's 'profit-and-avoid' characteristic, these properties represent subtle neural responses to external stimuli, intricately dependent on the selective filtering mechanisms within neurons, enabling them to generate two distinct types of responses [8]. In essence, this ability enables neurons to generate two distinct responses. However, electronic devices often simulate a fixed synaptic function with limited single adaptability [9], thereby constraining their effectiveness in accurately mimicking complex biological synaptic processes, such as the brain's 'profit-and-avoid' characteristic. This gap highlights the need for innovative approaches to more accurately model the brain's complex synaptic processes.

In response to this formidable challenge, researchers have turned their attention to HfO$_2$ material, which is known for its high dielectric constant [10]. HfO$_2$ has become a leading material in the semiconductor industry, owing to its seamless integration with silicon-based industrial processes [11–13]. Recent findings have revealed that extremely thin layers of HfO$_2$ demonstrate ferroelectric properties, which has generated considerable interest due to their potential in influencing synaptic behavior [14–17]. This interest stems from the distinct response of ferroelectric materials to electrical stimuli, which opens up new avenues for application. While there are various ferroelectric materials, including perovskites [18], their incompatibility with the semiconductor industry's standard processing methods precludes their use, especially in the fabrication of ultrathin layers, which is crucial for maintaining reliability and precision. In contrast, silicon remains a fundamental element in semiconductor technology, largely attributed to its 1.1 eV band gap enabling operations at low voltage [19]. The function of SiO_2 as a high-quality and inherent insulator has played a crucial role in maintaining silicon's dominance in the industry for over fifty years. Nevertheless, in the past ten years, there has been a transition from using SiO_2 to HfO_2-based high-k insulators in silicon electronics [20]. Despite the benefits of HfO_2, it is not a native oxide of the silicon substrate, leading to various interface challenges and a reduction in ferroelectric effects due to inadequate polarization screening. This situation poses a compelling question: if silicon derives significant benefits from its native SiO_2 insulator, could HfO_2, a well-known high-k dielectric, inherently complement other semiconductors? Layered two-dimensional semiconductors, like $HfSe_2$, are significant in this scenario due to their band gaps, which vary from 0.9 to 1.3 eV. These band gaps cover a range from thicker bulk layers to thinner monolayers. $HfSe_2$ is considered a promising candidate for technological applications due to its compatible bandgap and its ability to work well with HfO_2's dielectric properties. To mimic the 'SiO_2/Si' structure and explore new frontiers in semiconductor technology, a ferroelectric HfO_2/$HfSe_2$ stack has been adopted. This innovative approach potentially marks the beginning of a new era in semiconductor technology, with $HfSe_2$ playing a crucial role in the architecture of transistors, promising advancements in both functionality and efficiency.

In this context, devices containing $HfSe_2$/HfO_2 layers exhibited a dual characteristic in synaptic plasticity under the influence of various pulses [21]. Specifically, it exhibited an enhanced response to negative gate pulses and a subdued response to positive ones, mirroring the selective responsiveness of neurons to favorable or adverse stimuli. In these structures, the magnitude and polarity of voltage pulses emerge as pivotal factors. When negative-polarity pulses are applied, electrons are liberated within the $HfSe_2$ channel, resulting in an augmentation of channel conductivity. Simultaneously, polarization effects within the HfO_2 dielectric layer start to become apparent. The free electrons move rapidly due to the applied voltage, increasing the conductivity of the device. Simultaneously, when positive-polarity pulses are introduced, more charges become confined at the interface or permeate the HfO_2 layer. Consequently, the introduction of positive-polarity pulses during the pulse training process leads to a gradual reduction in response current. Throughout this process, the charge dynamics, influenced by the polarization behavior, mimic the release and reuptake of neurotransmitters. This behavior demonstrates the capacity of these electronic devices to accurately replicate biological synapses, effectively adjusting synaptic plasticity by facilitating both potentiation and depression responses. Furthermore, even when pulses are administered subsequent to the preceding pulse, polarization effects persist. The convergence of these two identical dynamic processes results in an augmentation of current, yielding the effect of increased current, akin to the behavior observed in biological synapses known as paired-pulse facilitation (PPF). Negative-polarity pulses induce electron release and amplify channel conductivity, while positive-polarity pulses prompt charge accumulation and a decline in conductivity. This bidirectional response underscores the potential of electronic devices to faithfully replicate the behavior of biological synapses. The dual nature of synaptic plasticity exhibited by these devices holds immense promise for emulating complex synaptic processes within biological systems.

2. Results and Discussion

Figure 1a presents a representation of a synaptic junction, a part of neural communication. It captures the process of neurotransmitter release from the presynaptic neuron and their reception by the postsynaptic neuron. The synaptic efficacy, the strength of the synaptic signal, is not static; it can be regulated by various factors such as the presence of an activating signal and the dynamic flux of neurotransmitters themselves. This regulation ensures that neural communication is not only robust but also plastic, capable of adapting to different physiological conditions and demands. the gate electrode's voltage pulse is conceptualized as the activating signal, with the consequent current between the source and drain electrodes serving as a proxy for synaptic efficacy. That is, in a fashion analogous to this biological process, the voltage pulse applied to the gate electrode of a transistor can be envisioned as an activating signal. When a voltage pulse is applied, it modulates the conductivity between the source and drain electrodes of the transistor, akin to the way an action potential facilitates the release of neurotransmitters at the synaptic junction. The flow of current that results from this modulation serves as an electrical counterpart to the synaptic efficacy observed in biological systems.

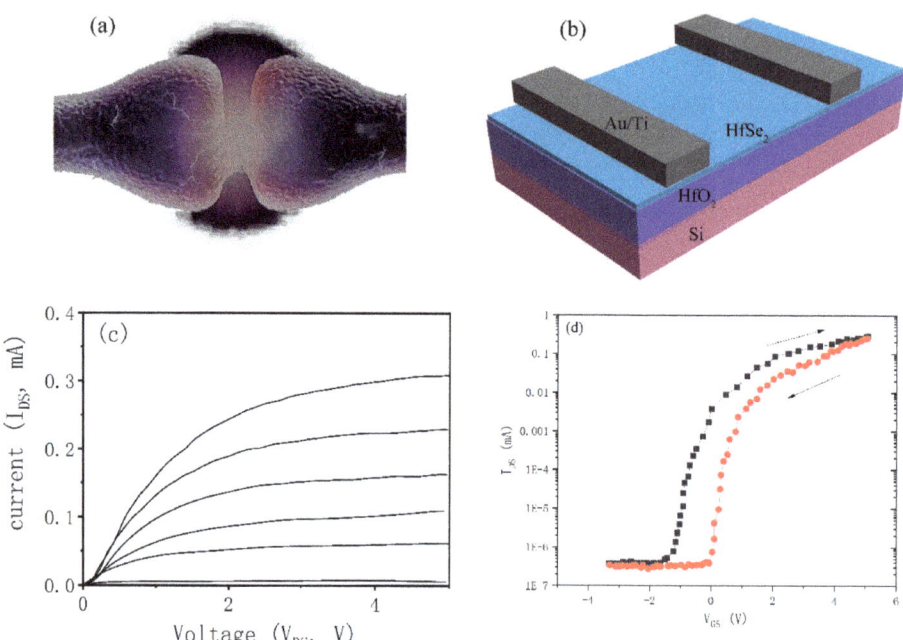

Figure 1. (**a**) Natural synapse architecture. (**b**) Synaptic TFT with HfSe$_2$ channel and HfO$_2$ dielectrics. (**c**) Current–voltage relationship showing drain-source current (I_{DS}) vs. drain-source voltage (V_{DS}). (**d**) Current–voltage behavior of I_{DS} with varying gate-source voltage (V_{GS}) (−3 to 5 V, with a fixed V_{DS} of 5 V).

Figure 1b shows the specific construction of a field-effect transistor (FET) in order to mimic the functions of a synaptic junction. At its core lies the channel made of HfSe$_2$ layers, and close to this channel, there is a layer of HfO$_2$, which is oxygen-deficient, acting as the dielectric material. The presence of oxygen vacancies in HfO$_2$ is critical for the ferroelectrical characteristic which further affects the overall behavior of the FET.

The transfer characteristics of the transistor, which relate the drain-source current (I_{DS}) to the drain-source voltage (V_{DS}), are graphically represented in Figure 1c. This graph shows the I_{DS} response of the transistor to varying gate-source voltages (V_{GS}) in increments of 1 volt, ranging from 1 to 6 volts. Each curve corresponds to a specific V_{DS}, and as V_{DS} increases, there is a noticeable increase in I_{DS} for a given V_{DS}, demonstrating the transistor's ability to modulate current flow in response to changes in gate voltage, much like a neuron's response to different levels of stimuli.

The transistor demonstrates notable operational characteristics, including a threshold voltage of approximately −1.3 V, boasting an ON/OFF ratio exceeding 2.3×10^6, a commendable field-effect mobility of approximately 1.1×10 cm^2/V·s, and an impressive subthreshold swing measuring merely 0.12 V/decade. Figure 1d provides a view of the threshold behavior of the transistor, detailing the subthreshold and above-threshold conduction regions. The graph plots I_{DS} on a logarithmic scale against V_{GS}, illustrating the sharp increase in current as the gate voltage crosses a critical threshold—a behavior that is reminiscent of the all-or-nothing response of a neuron when it fires an action potential. The red and black data points likely represent different measurements sequence that show the consistency of the transistor's behavior in response to varying gate voltages. Within Figure 1d, we present the transfer characteristic curve, which results from a controlled 5 V bias applied across the source and drain electrodes. Interestingly, we observe a subtle hysteresis phenomenon during the gate voltage cycle, spanning from −3 V to 5 V and returning. This behavior hints at the potential occurrence of charge trapping events, either at interface junctions or within the gate dielectric. The presence of hysteresis is further corroborated by the directional arrows noted during the voltage sweep, providing empirical support for its existence and lending weight to the charge trapping [22–24].

Figure 2 provides a detailed visualization of the behavioral patterns of a transistor when subjected to a carefully designed sequence of bipolar voltage pulses. Figure 2a demonstrates the intentional use of a sequential pulse train applied to the transistor's gate. Each pulse is precisely timed with a periodicity of 20 ms, has a fixed amplitude of 1 V, and is delivered with a pulse width of 20 ms. The accurate and regular stimulus is crucial for manipulating the gate of the transistor and imitating synaptic behavior.

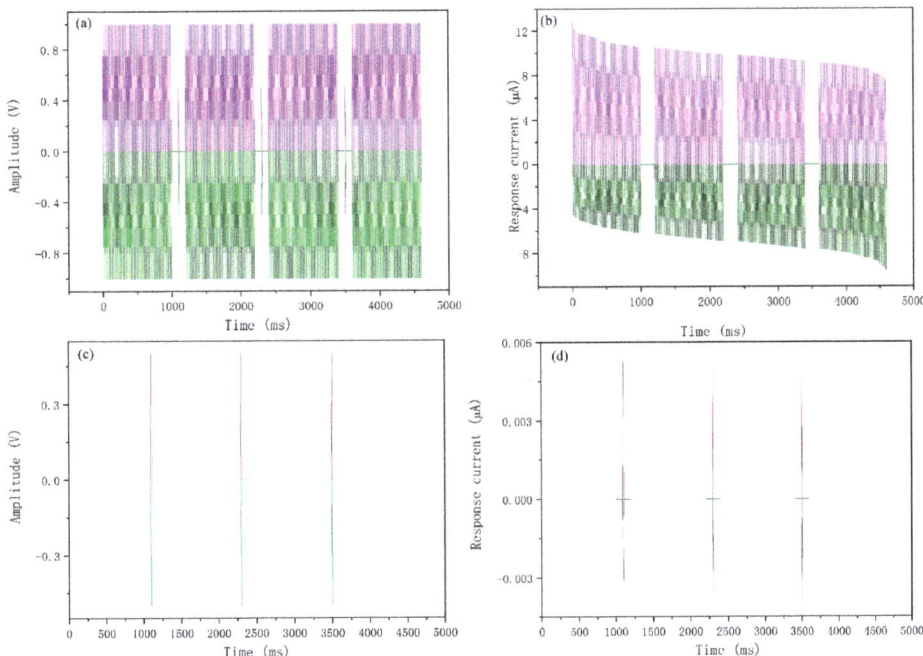

Figure 2. (**a**) Sequence of pulse for different polar voltages. (**b**) The resultant current of the transistor under each voltage condition. (**c**) The read pulses introduced amid pulse trains for different polar voltages. (**d**) The respective synaptic current corresponding to read voltage under alternating voltage conditions.

Figure 2b captures the resultant current traversing the channel at a drain-source voltage (V_{DS}) of 4 V. What is particularly striking in this depiction is the transistor's dynamic current response, which shows a significant potentiation effect in reaction to the negative pulse trains, whereas a clear depression effect is evident with the imposition of positive pulse trains. The contrasting directions of current modulation—increasing with negative pulses and decreasing with positive ones—emphasize the transistor's capability for bidirectional programming. This critical functionality is instrumental in adjusting synaptic weight and closely replicates the bipolar pulse stimuli effect observed in biological synaptic interactions.

The observed phenomena can be attributed to charge trapping and detrapping processes occurring within the structure of the transistor, which further confirms the previously demonstrated trap activity. The initial application of a negative pulse triggers the release of trapped charges. With each subsequent pulse, this release effect is amplified, resulting in a substantial increase in the channel's conductance when a negative pulse train is applied. Conversely, the application of positive pulses achieves the opposite effect, diminishing the channel's conductance. This charge modulation, which dynamically influences synaptic weight, parallels the neurotransmitter dynamics seen in biological synapses, thus heralding a new avenue for semiconductor applications that could mirror the adaptive modulation of synaptic efficacy observed in natural biological processes. What is particularly striking in this depiction is the transistor's dynamic current response, which shows a significant potentiation effect in reaction to the negative pulse trains, whereas a clear depression effect is evident with the imposition of positive pulse trains. The contrasting directions of current modulation—increasing with negative pulses and decreasing with positive ones—emphasize the transistor's capability for bidirectional programming. This critical functionality is instrumental in adjusting synaptic weight and closely replicates the bipolar pulse stimuli effect observed in biological synaptic interactions.

The observed phenomena are attributable to charge trapping and detrapping, corroborated by trap evidence presented in Figure 1d. The initial application of a negative pulse precipitates charge release, with subsequent pulses exacerbating this effect, thereby amplifying the channel's conductance under source/drain (S/D) bias via negative pulse train stimulation, and inversely with positive pulses. Analogous to synaptic neurotransmitter dynamics depicted in Figure 1a, this charge modulation process dynamically alters synaptic weight. This mechanism indicates possible applications that resemble biological synapses, where the effectiveness of synapses is dynamically adjusted by bipolar spike protocols. Furthermore, the enduring characteristic of the response current, which is hypothesized by the use of low-intensity reading pulses (0.2 V, 20 ms) after the pulse train has stopped, is recorded in Figure 2c,d. The response current exhibits a progressive increase or decrease, contingent on the polarity of the applied read pulse. This sustained response, persisting beyond the stimulus signal's duration, suggests an underlying influence beyond mere charge dynamics, potentially linked to the documented ferroelectric properties of HfO_2. Furthermore, Figure 2c,d capture the enduring behavior of the response current, which is meticulously tracked through the use of small amplitude reading pulses of 0.2 V for a duration of 20 ms, administered following the termination of the pulse train. The response current's behavior, as documented, exhibits a consistent and progressive change in magnitude, corresponding directly to the polarity of the read pulse applied. This persistent change, which outlasts the stimulus signal itself, hints at an extended influence that transcends simple charge movement and may be intimately linked to the ferroelectric characteristics of materials like HfO_2, which have been previously noted in other research.

To discern the influence of pulse parameters on the response current in neuromorphic transistors, an exploration was conducted by varying one of the three defining parameters—amplitude, interval, and width—while holding the others constant [25]. Alterations to the other parameters of the pulse sequence, interval or width, did not yield significant shifts in response current. This absence of variation suggests that carrier release and capture are contingent upon reaching a critical energy threshold, principally determined by the voltage amplitude. This finding further corroborates that the pulse conditions applied in this study were sufficient to activate the traps effectively while leaving deeper-level traps unaltered. This observation suggests that the act of carrier release and subsequent capture within the transistor's framework relies on surpassing a pivotal energy boundary, which is predominantly dictated by the voltage amplitude. This unresponsiveness to changes in pulse interval or width thus reinforces the assertion that the pulse conditions selected for this study were optimally chosen to activate the charge traps with high efficacy, while having a negligible effect on deeper-level traps that may not be as readily influenced by the chosen stimulus parameters.

Figure 3a illustrates the synaptic-like behavior of a neuromorphic device in response to a temporal sequence of pulse trains at varying magnitudes. The primary graph reveals the device's response current over an extended time period, with the application of pulse trains ranging from −2 V to +2 V. Each color represents the response to a specific voltage magnitude, indicating how the current either increases or decreases in response to the applied voltage over time. Several key observations can be made: (i) Magnitude-Dependent Responses: There is a clear magnitude-dependent response in the device, with higher absolute voltages eliciting larger currents. This suggests that the device's conductance is sensitive to the voltage magnitude, an important characteristic for emulating the plasticity of biological synapses. (ii) Temporal Dynamics: The response currents display a temporal dependence, highlighting the device's ability to retain a memory of the voltage stimulus. This temporal aspect of the response is critical for mimicking the time-dependent processes found in biological synaptic behavior. (iii) Polarity Sensitivity: The device demonstrates a symmetric response to the polarity of the applied voltages, with positive voltages inducing positive currents and negative voltages inducing negative currents. This symmetry may reflect an intrinsic property of the device's operational mechanism, which is essential for replicating the bidirectional modulation of synaptic strength. (iv) Detailed Observation from Inset: The inset provides a magnified view of the response current at a finer scale (in microamperes) and shows the response after read pulses at various time intervals. This could be instrumental in detecting subtle changes after prolonged sequences, akin to the regulatory mechanisms in biological synapses. (v) Time-Dependent Changes: The response current changes over time suggest a certain degree of potentiation or depression, akin to the learning and memory functions of biological synapses, which may be influenced by the device's charge trapping dynamics.

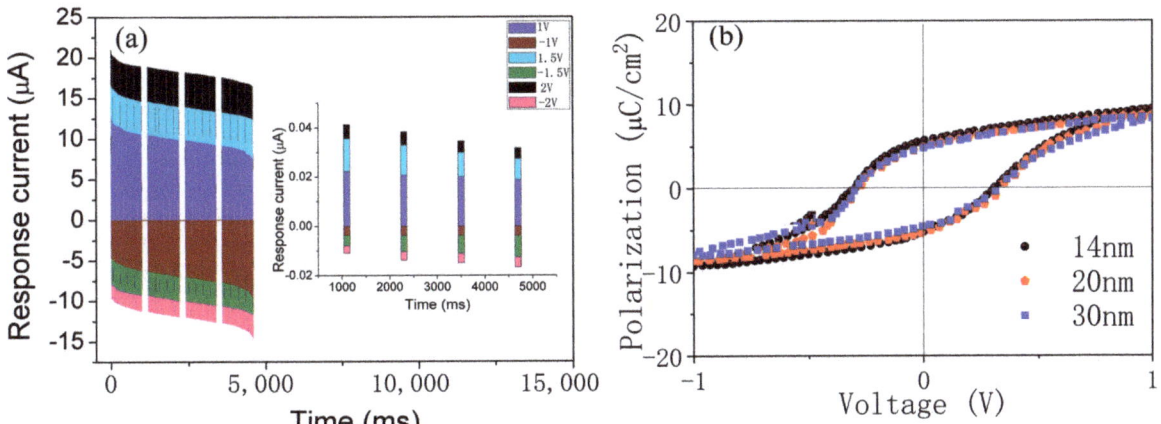

Figure 3. (**a**) Synaptic behavior in relation to the temporal sequence of pulse trains of varying magnitudes. The inset portrays the reading current across pulse trains in correlation with the read intervals. (**b**) Ferroelectric behavior of HfO$_2$ layer with various thickness.

The graph depicts the potential of a neuromorphic device to exhibit synaptic plasticity, with a current response that is highly sensitive to the amplitude and temporal characteristics of voltage pulses. Pulse parameters—amplitude, duration, and periodicity—are pivotal in modulating the neuromorphic transistor's response current. Our study focused on the amplitude's influence while maintaining consistent pulse duration and periodicity. Incrementing the amplitude of negative/positive pulses resulted in an augmented/diminished synaptic response current (refer to Figure 3a), suggesting enhanced efficacy in the charge's release/trapping processes. The augmentation in current under a negative pulse and its reduction under a positive pulse reflect the amplitude's role in controlling the charge dynamics. The supplementary inset of Figure 3a presents the response current continuity post pulse train cessation, utilizing a 0.2 V reading voltage. Notably, this current persists in its incremental/decremental trend under the accumulation of negative (positive) bias voltage, reinforcing the assertion that longer intervals of pulse do not abate the synaptic response's progressive trajectory. This observation alludes to an intrinsic polar electric field's sustained influence, validated through P-V measurements that substantiate the ferroelectric nature of the examined HfO_2 films (as shown in Figure 4b) [26]. Figure 4b illustrates the ferroelectric characteristics of HfO_2 thin films at various thicknesses (14 nm, 20 nm, and 30 nm), as shown by their polarization–voltage (P-V) loops. The P-V loops indicate that the ferroelectric properties of the films are not significantly dependent on the thickness, as all three thicknesses show similar hysteresis loops, which is characteristic of ferroelectric behavior. Notably, even at a reduced thickness of 14 nm, the film exhibits distinct ferroelectricity, which is often not as pronounced in other ferroelectric materials at such thin dimensions.

$HfSe_2$ is widely recognized as an n-type semiconductor. Under the influence of a ferroelectric field and external bias, oxygen vacancies in the $HfSe_2$ channel readily undergo ionization, and the energy barrier for neutralizing these ionized vacancies is remarkably low. Electrons within the $HfSe_2$ channel can traverse the $HfSe_2/HfO_2$ interface and subsequently become trapped within high value of pulse sequence. The observed response exhibits two primary behaviors: (i) The conductivity of $HfSe_2$ strongly depends on its charge carriers. Under negative polar pulse, it induces the liberation of electrons, thereby conductivity promoting within channel. Simultaneously, polarization initiates within the HfO_2 dielectric. The move of these liberated electrons, driven by the voltage (assisted by the polarization within the ferroelectric layer), results in enhanced device conductivity. The above is shown in Figure 4a,b. (ii) Following the cessation of a pulse, the restoration of negative charges is expected. Nonetheless, the inherent polarization of the OD-HfO_2 layer obstructs this recuperative mechanism. As a consequence of this dynamic interaction, it can be observed that, during the intervals between pulse trains, the response current to the modest read pulse (0.2 V) exhibits ongoing fluctuations, which become more pronounced with the introduction of more sparsely distributed read pulses. This phenomenon is graphically illustrated in Figure 4c. (iii) After applying a pulse following another one, polarization persists, and the combination of these identical dynamic processes enhances conduction, leading to an increase in the current. The visual representation of this phenomenon can be observed in Figure 4d. It is worth noting that the postsynaptic current elicited by the second pulse surpasses that of the initial pulse, resembling the behavior observed in paired-pulse facilitation (PPF) within biological synapses. [27].

Paired-pulse facilitation (PPF) is a form of synaptic plasticity characterized by an augmentation in postsynaptic responses when the second spike closely follows the preceding one, reflecting a potentiation phenomenon [28]. Recent reports have extensively covered the ferroelectric characteristics of HfO_2. The emergence of ferroelectricity in HfO_2 films is attributed to the existence of a metastable and non-centrosymmetric orthorhombic phase, characterized by the space group Pca21 [29]. The enhancement and control of ferroelectricity can be achieved by introducing various dopants and applying contact stress. However, when a positive gate voltage is applied, a greater number of charges are either trapped at the interface or pass through the HfO_2 layer. Consequently, the response gradually declines as the pulse trains added. The observation of charge trapping and the retention of polarization post pulse application in both sequences suggest a memory effect within the device, critical for neuromorphic applications. The ability to modulate this effect with the polarity of the applied pulses could be harnessed for simulating synaptic plasticity, thereby emulating the fundamental properties of biological synapses in artificial devices.

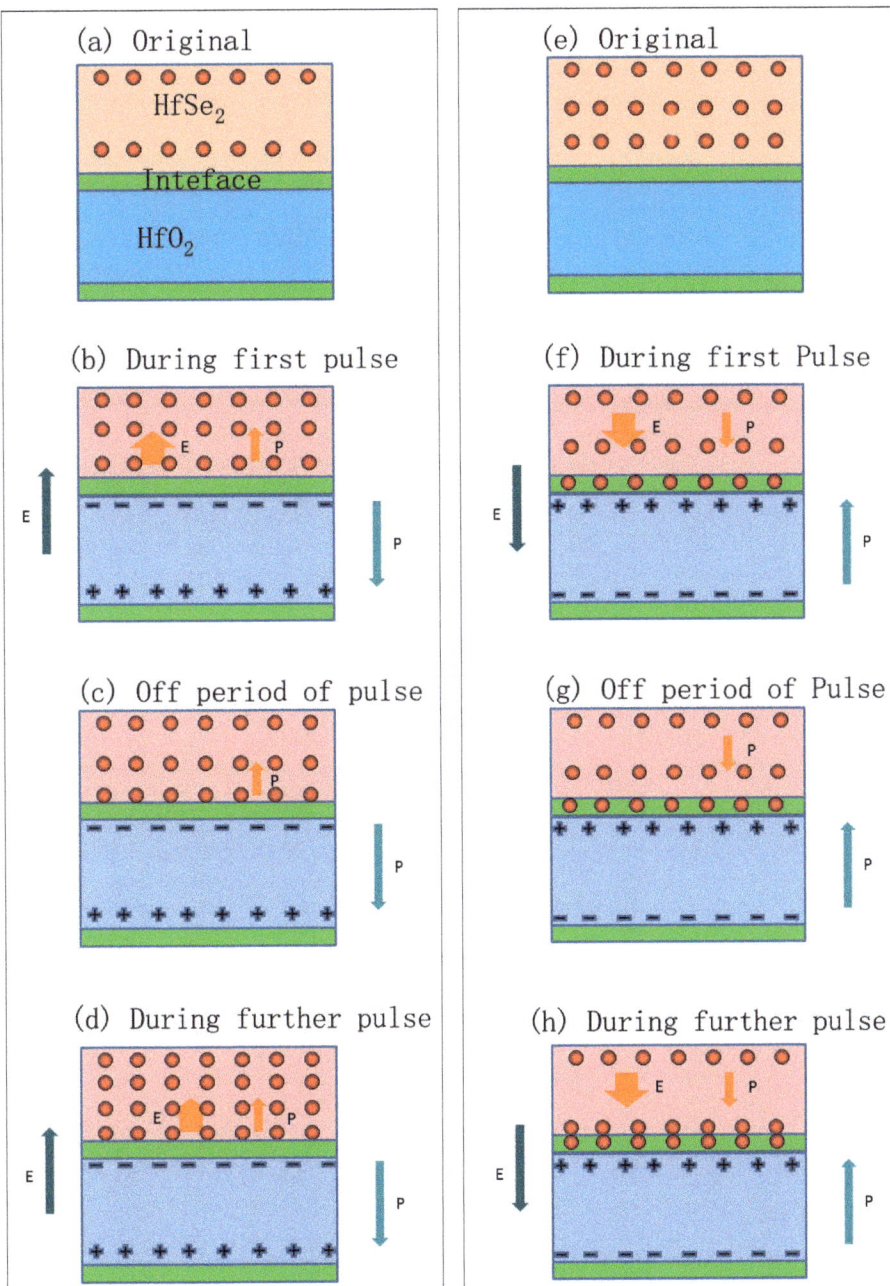

Figure 4. Dynamic sequence within a prototypical pulse train at negative and positive charges.

3. Materials and Methods

In the experimental setup, substrates and bottom electrodes made of n+ silicon were utilized. The initial cleaning process involved submerging the silicon wafers in a dilute hydrofluoric acid solution to effectively strip away any surface oxides and impurities, maintaining a ratio of 1% HF to water. Following this, a thin film of HfO_2, measuring

5 nm in thickness, was deposited onto the silicon substrates. The deposition process under investigation herein was conducted through the utilization of radio frequency (rf) magnetron sputtering as the primary deposition technique. This process entailed the employment of a high-purity Hafnium target, with a purity exceeding 99.99%, which played a pivotal role in ensuring the quality and purity of the deposited material. The deposition took place within a rigorously controlled environment characterized by a blend of Argon and Oxygen gases, affording precise control over the deposition conditions. In order to maintain an atmosphere conducive to the formation of oxygen-deficient oxide layers, the oxygen bias was meticulously regulated to attain a level of 8×10^{-5} Torr, while the Argon bias was set at 2.1×10^{-3} Torr. This fine-tuning of the gas pressures was pivotal in achieving the desired low oxygen environment, which is known to be optimal for the deposition of such layers. The material of interest, $HfSe_2$, was sourced from HQ Graphene in the form of sheet-like crystals, boasting facets measuring several millimeters in size. The synthesis of these $HfSe_2$ layers was carried out through the application of chemical vapor transport (CVT) methodology. Elemental Hafnium and Selenium precursor powders were judiciously employed in the synthesis process, with utmost precision and control, ultimately resulting in the successful production of the desired material. This synthesis approach and the choice of precursor materials were instrumental in ensuring the high quality and purity of the synthesized $HfSe_2$ layers, as demanded by the scientific investigation at hand. These $HfSe_2$ layers were then mechanically exfoliated and carefully transferred onto the prepared HfO_2 layer on the silicon substrate. This transfer was executed with precision, ensuring optimal interface quality between the $HfSe_2$ and HfO_2 layers. After the successful integration of the $HfSe_2$ layer onto the HfO_2, the next step involved the fabrication of electrodes. The experimental procedure encompassed the deposition of two distinct layers of metallic films, specifically 200 nm of gold (Au) and 50 nm of titanium (Ti), functioning as the source and drain electrodes. This deposition process was meticulously executed through electron beam evaporation, ensuring precision and control over the film thickness and material properties. Subsequently, these deposited electrodes underwent a patterning process to achieve specific dimensions. The resulting electrodes were designed to possess lengths of 60 µm and widths measuring 1500 µm, in accordance with the experimental requirements and design specifications. A forming gas anneal (FGA) was applied post-electrode deposition to improve the metal-semiconductor contact. The final stage in the device fabrication was the thermal annealing process. A post-deposition annealing at 300 °C in a nitrogen atmosphere for 120 s was performed to optimize the device properties. The fully assembled device, which included the $HfSe^2$ and HfO^2 layers in conjunction with the meticulously patterned source and drain electrodes, underwent a series of comprehensive electrical characterizations. These characterizations were aimed at assessing its electronic properties and performance. To carry out these assessments, precise current–voltage (I–V) characteristic measurements were conducted. The measurements were conducted using state-of-the-art instrumentation, including a Keithley 4200 SCS system. Additionally, an Agilent B2900 Precision Source/Measure Unit (SMU) was employed to provide precise control over the electrical parameters during the measurements.

4. Conclusions

In this study, we developed thin-film transistors (TFTs) utilizing a $HfSe_2$ channel. These advanced TFTs demonstrated an augmentation in the post-synaptic current upon the application of negative pulse to bottom gate. Conversely, positive bias resulted in a diminution of the post-synaptic current. This phenomenon is attributable to the modulation of carrier concentrations within the $HfSe_2$ layer, with the ferroelectric tendencies of the hafnium oxide layer serving as a facilitator. Notably, the biphasic response behavior of these transistors underlines their immense potential in emulating biological synapses, paving the way for bioinspired neuromorphic applications.

Author Contributions: Conceptualization, R.J.; methodology, J.L.; software, K.W.; formal analysis, Z.X.; investigation, M.S., L.W. and F.Y.; writing—original draft preparation, R.L., H.J. and Z.W. All authors have read and agreed to the published version of the manuscript.

Funding: It is sponsored by the Ningbo Natural Science Foundation (2023J007), the Natural Sciences Fund of Zhejiang Province (LDT23F05015F05) and the National Natural Science Foundation of China under Grant 61774098, 62171242, U1809203, and 61631012. This work was also supported by the Opening Project of Key Laboratory of Microelectronic Devices & Integrated Technology Institute of Microelectronics, Chinese Academy of Sciences.

Data Availability Statement: All the relevant data for this paper can be found in the article.

Conflicts of Interest: The authors declare no conflict of interest.

References

1. Li, Y.; Su, K.; Chen, H.; Zou, X.; Wang, C.; Man, H.; Liu, K.; Xi, X.; Li, T. Research progress of neural synapses based on memristors. *Electronics* **2023**, *12*, 3298. [CrossRef]
2. Rehman, M.M.; Rehman, H.M.M.U.; Gul, J.Z.; Kim, W.Y.; Karimov, K.S.; Ahmed, N. Decade of 2D-materials-based RRAM devices: A review. *Sci. Technol. Adv. Mater.* **2020**, *21*, 147–186. [CrossRef]
3. Duan, X.; Cao, Z.; Gao, K.; Yan, W.; Sun, S.; Zhou, G.; Wu, Z.; Ren, F.; Sun, B. Memristor-Based Neuromorphic Chips. *Adv. Mater.* **2024**, e2310704. [CrossRef] [PubMed]
4. Rehman, M.M.; Rehman, H.M.M.U.; Kim, W.Y.; Sherazi, S.S.H.; Rao, M.W.; Khan, M.; Muhammad, Z. Biomaterial-based nonvolatile resistive memory devices toward ecofriendliness and biocompatibility. *ACS Appl. Electron. Mater.* **2021**, *3*, 2832–2861. [CrossRef]
5. Xu, Z.; Li, Y.; Xia, Y.; Shi, C.; Chen, S.; Ma, C.; Zhang, C.; Li, Y. Organic Frameworks Memristor: An Emerging Candidate for Data Storage, Artificial Synapse, and Neuromorphic Device. *Adv. Funct. Mater.* **2024**, 2312658. [CrossRef]
6. Sumi, T.; Harada, K. Mechanism underlying hippocampal long-term potentiation and depression based on competition between endocytosis and exocytosis of AMPA receptors. *Sci. Rep.* **2020**, *10*, 14711. [CrossRef]
7. Zhao, J.; Zhou, Z.; Zhang, Y.; Wang, J.; Zhang, L.; Li, X.; Zhao, M.; Wang, H.; Pei, Y.; Zhao, Q.; et al. An electronic synapse memristor device with conductance linearity using quantized conduction for neuroinspired computing. *J. Mater. Chem. C* **2019**, *7*, 1298–1306. [CrossRef]
8. Panichello, M.F.; Buschman, T.J. Shared mechanisms underlie the control of working memory and attention. *Nature* **2021**, *592*, 601–605. [CrossRef]
9. Dai, S.; Zhao, Y.; Wang, Y.; Zhang, J.; Fang, L.; Jin, S.; Shao, Y.; Huang, J. Recent advances in transistor-based artificial synapses. *Adv. Funct. Mater.* **2019**, *29*, 1903700. [CrossRef]
10. Zhao, C.; Zhao, C.; Werner, M.; Taylor, S.; Chalker, P. Advanced CMOS gate stack: Present research progress. *Int. Sch. Res. Not.* **2012**, *2012*, 689023. [CrossRef]
11. Schroeder, U.; Park, M.H.; Mikolajick, T.; Hwang, C.S. The fundamentals and applications of ferroelectric HfO_2. *Nat. Rev. Mater.* **2022**, *7*, 653–669. [CrossRef]
12. Mulaosmanovic, H.; Breyer, E.T.; Dünkel, S.; Beyer, S.; Mikolajick, T.; Slesazeck, S. Ferroelectric field-effect transistors based on HfO_2: A review. *Nanotechnology* **2021**, *32*, 502002. [CrossRef]
13. Pathak, S.; Mandal, G.; Das, P.; Dey, A.B. Structural characteristics of HfO_2 under extreme conditions. *Mater. Chem. Phys.* **2020**, *255*, 123633. [CrossRef]
14. Jiang, R.; Ma, P.; Han, Z.; Du, X. Habituation/Fatigue behavior of a synapse memristor based on IGZO–HfO_2 thin film. *Sci. Rep.* **2017**, *7*, 9354. [CrossRef] [PubMed]
15. Jiang, R.; Xie, E.; Chen, Z.; Zhang, Z. Electrical property of HfO_xN_y–HfO_2–HfO_xN_y sandwich-stack films. *Appl. Surf. Sci.* **2006**, *253*, 2421–2424. [CrossRef]
16. Yuan, L.; Zou, X.; Fang, G.; Wan, J.; Zhou, H.; Zhao, X. High-performance amorphous indium gallium zinc oxide thin-film transistors with HfO_xN_y/HfO_2/HfO_xN_y tristack gate dielectrics. *IEEE Electron. Device Lett.* **2011**, *32*, 42. [CrossRef]
17. Jeon, Y.-R.; Kim, D.; Ku, B.; Chung, C.; Choi, C. Synaptic Characteristics of Atomic Layer-Deposited Ferroelectric Lanthanum-Doped HfO_2 (La: HfO_2) and TaN-Based Artificial Synapses. *ACS Appl. Mater. Interfaces* **2023**, *15*, 57359–57368.
18. Zheng, W.; Wang, X.; Zhang, H.; Chen, B.; Suo, H.; Xing, Z.; Wang, Y.; Wei, H.; Chen, J.; Guo, Y.; et al. Emerging Halide Perovskite Ferroelectrics. *Adv. Mater.* **2023**, *35*, 2205410. [CrossRef] [PubMed]
19. Rafin, S.S.H.; Ahmed, R.; Mohammed, O.A. Wide Band Gap Semiconductor Devices for Power Electronic Converters. In Proceedings of the 2023 Fourth International Symposium on 3D Power Electronics Integration and Manufacturing (3D-PEIM), Miami, FL, USA, 1–3 February 2023; IEEE: Piscataway, NJ, USA, 2023; pp. 1–8.
20. Padovani, A.; La Torraca, P. A simple figure of merit to identify the first layer to degrade and fail in dual layer SiO_x/HfO_2 gate dielectric stacks. *Microelectron. Eng.* **2023**, *281*, 112080. [CrossRef]
21. AlMutairi, A.; Yoon, Y. Device Performance Assessment of Monolayer $HfSe_2$: A New Layered Material Compatible with High-κ HfO_2. *IEEE Electron. Device Lett.* **2018**, *39*, 1772–1775. [CrossRef]

22. Ran, J.; Erqing, X.; Zhenfang, W. Interfacial chemical structure of HfO$_2$/Si film fabricated by sputtering. *Appl. Phys. Lett.* **2006**, *89*, 142907.
23. Triska, J.; Conley, J., Jr.; Presley, R.; Wager, J. Bias stress stability of zinc-tin-oxide thin-film transistors with Al$_2$O$_3$ gate dielectrics. *J. Vac. Sci. Technol. B Nanotechnol. Microelectron. Mater. Process. Meas. Phenom.* **2010**, *28*, C5I1–C5I6. [CrossRef]
24. Wu, H.; Wang, J.; He, W.; Shan, C.; Fu, S.; Li, G.; Zhao, Q.; Liu, W.; Hu, C. Ultrahigh output charge density achieved by charge trapping failure of dielectric polymers. *Energy Environ. Sci.* **2023**, *16*, 2274–2283. [CrossRef]
25. Li, X.; Zhong, Y.; Chen, H.; Tang, J.; Zheng, X.; Sun, W.; Li, Y.; Wu, D.; Gao, B.; Hu, X. A memristors-based dendritic neuron for high-efficiency spatial-temporal information processing. *Adv. Mater.* **2023**, *35*, 2203684. [CrossRef] [PubMed]
26. Jaszewski, S.T.; Calderon, S.; Shrestha, B.; Fields, S.S.; Samanta, A.; Vega, F.J.; Minyard, J.D.; Casamento, J.A.; Maria, J.-P.; Podraza, N.J.; et al. Infrared Signatures for Phase Identification in Hafnium Oxide Thin Films. *ACS Nano* **2023**, *17*, 23944–23954. [CrossRef]
27. Lee, K.-C.; Li, M.; Chang, Y.H.; Yang, S.H.; Lin, C.Y.; Chang, Y.M.; Yang, F.-S.; Watanabe, K.; Taniguchi, T.; Ho, C.-H.; et al. Inverse paired-pulse facilitation in neuroplasticity based on interface-boosted charge trapping layered electronics. *Nano Energy* **2020**, *77*, 105258. [CrossRef]
28. Kudryashova, I. Inhibitory control of short-term plasticity during paired pulse stimulation depends on actin polymerization. *Neurochem. J.* **2022**, *16*, 136–146. [CrossRef]
29. Chouprik, A.; Negrov, D.; Tsymbal, E.Y.; Zenkevich, A. Defects in ferroelectric HfO$_2$. *Nanoscale* **2021**, *13*, 11635–11678. [CrossRef]

Disclaimer/Publisher's Note: The statements, opinions and data contained in all publications are solely those of the individual author(s) and contributor(s) and not of MDPI and/or the editor(s). MDPI and/or the editor(s) disclaim responsibility for any injury to people or property resulting from any ideas, methods, instructions or products referred to in the content.

Review

Research and Application Progress of Inverse Opal Photonic Crystals in Photocatalysis

Hongming Xiang, Shu Yang *, Emon Talukder, Chenyan Huang and Kaikai Chen

School of Textiles and Fashion, Shanghai University of Engineering Science, Shanghai 201620, China; 19851543881@163.com (H.X.); emontalukder90@gmail.com (E.T.); 39200002@sues.edu.cn (K.C.)
* Correspondence: shuyang@sues.edu.cn

Abstract: In order to solve the problem of low photocatalytic efficiency in photocatalytic products, researchers proposed a method to use inverse opal photonic crystal structure in photocatalytic materials. This is due to a large specific surface area and a variety of optical properties of the inverse opal photonic crystal, which are great advantages in photocatalytic performance. In this paper, the photocatalytic principle and preparation methods of three-dimensional inverse opal photonic crystals are introduced, including the preparation of basic inverse opal photonic crystals and the photocatalytic modification of inverse opal photonic crystals, and then the application progresses of inverse opal photonic crystal photocatalyst in sewage purification, production of clean energy and waste gas treatment are introduced.

Keywords: inverse opal; photonic crystals; fabrication methods; photocatalytic application

Citation: Xiang, H.; Yang, S.; Talukder, E.; Huang, C.; Chen, K. Research and Application Progress of Inverse Opal Photonic Crystals in Photocatalysis. *Inorganics* **2023**, *11*, 337. https://doi.org/10.3390/inorganics11080337

Academic Editors: Sake Wang, Minglei Sun and Nguyen Tuan Hung

Received: 2 July 2023
Revised: 9 August 2023
Accepted: 14 August 2023
Published: 15 August 2023

Copyright: © 2023 by the authors. Licensee MDPI, Basel, Switzerland. This article is an open access article distributed under the terms and conditions of the Creative Commons Attribution (CC BY) license (https://creativecommons.org/licenses/by/4.0/).

1. Introduction

The world is seriously polluted and short of energy. Clean and renewable resources have attracted many researchers' attention and research. Solar energy is one of the most promising clean energy in the future, and photocatalyst is a material that can convert light energy into chemical energy, which plays an important role in the utilization of solar energy. Researchers have been committed to developing efficient and simple photocatalysts [1,2].

However, photocatalysts usually have a problem of low conversion. The introduction of inverse opal photonic crystal (IOPC) into the photocatalytic system has been proven to be a promising method to solve this problem [3–5]. Inverse opal photonic crystal is a certain kind of photonic crystal. Photonic crystal (PC) is a dielectric structure material with a photonic band gap, which is formed by materials with different dielectric constants and arranged periodically in space. It was independently proposed by John [6] and Yablonovitch [7] in 1987. Among the known photonic crystal structures, the three-dimensional photonic crystal formed by colloidal self-assembly has the same cubic close-packed structure as the natural opal structure, which is called opal photonic crystal; Another kind of three-dimensional ordered porous structure obtained by reverse replication of opal is called reverse opal photonic crystal. Inverse opal photonic crystals have many new physical properties and phenomena, such as slow photon effect [8], photonic band gap [9], photon localization [10–12], etc. Because of its periodic structure and excellent optical properties, inverse opal photonic crystals have been widely used in the field of photocatalysis [13–27].

In detail, the advantages of IOPC in photocatalytic systems are mainly based on the photonic band gap, slow light effect, and high specific surface area.

Photonic band gap: The inverse opal structure has a periodic refractive index and photonic band gap, which can significantly inhibit the propagation of light [28–31]. The periodic ordered porous structure of inverse opal structure makes it selective to incident light. A specific aperture can only allow the light of a specific wavelength to enter, and the incoming photons will be continuously reflected in the "restricted" structure, and the light matching the electronic band gap of the photocatalyst will be absorbed by the photocatalyst material. Photons of other wavelengths cannot be absorbed by the catalyst because they do not match the electron band gap energy of the catalyst and will not affect the photon efficiency of the catalyst.

Slow light effect: The inverse opal structure can also produce structural dispersion and a slow light effect. At the wavelength corresponding to the stop band edge, photons propagate at a strongly reduced group velocity; Therefore, they are called "slow photons." If the energy of the slow photon overlaps with the absorptivity of the material, the photon absorptivity will increase with the increase of the effective optical path length. Curti et al. [32] studied the slow light effect of TiO_2 inverse opal photonic crystal photocatalysis. Their experimental results show that slow photons can enhance the absorption of materials; At the two edges of the stopband, the absorption rate of inverse opal increases sharply. At the red edge of the stopband, the photon absorption capacity of inverse opal is 2.7 times that of opal, while at the blue edge, the value is 1.6. Thomas et al. [33] also studied the slow light effect of inverse opal TiO_2 photonic crystal. They found that due to the slow light effect, the photocatalytic efficiency of the inverse opal photonic structure was seven times that of the opal photonic crystal structure.

High specific surface area: The ordered porous structure of inverse opal photonic crystal improves its specific surface area. The high specific surface area structure of inverse opal photonic crystal makes the active sites on the catalyst surface easier to be exposed, and photo-generated electrons can reach the active sites through the shortest migration path, so as to enhance the transfer of photo-generated charge and reduce the recombination of photo-generated electrons. Reducing the recombination efficiency of photo-generated charge holes is one of the methods to improve photocatalytic efficiency. Huang et al. [34] prepared phosphorus-doped inverse opal structure C_3N_4 and proved that the inverse opal structure could effectively separate the photo-generated charges and reduce the recombination rate of photo-generated charges.

This paper introduces the preparation methods of inverse opal photonic crystals, including the preparation of basic inverse opal photonic crystals and photocatalytic modification of inverse opal photonic crystals. And then, the application progresses of inverse opal photonic crystal photocatalyst in sewage purification, production of clean energy, and waste gas treatment are introduced.

2. Preparation of Inverse Opal Photonic Crystal

At present, a variety of methods have been developed for the preparation of inverse opal photonic crystals, including the template method, micromachining technology, multi-beam laser holographic pattern method, etc. [35,36]. This paper mainly introduces the most common preparation method of inverse opal photonic crystal, namely the template method. The template method can be divided into two-step and three-step methods [37].

The basic process of the two-step method is as follows: the first step is to disperse the colloidal microspheres in the precursor solution and self-assemble into a composite opal structure. The second step is to remove the microsphere template and obtain the inverse opal structure [38,39].

The process of the three-step method is shown in Figure 1. The specific steps are as follows: the first step is to build an opal photonic crystal template using colloid microsphere self-assembly; the second step is to fill the precursor in the resultant material and cure it; and the third step is to remove the template for the inverse opal structure.

Compared with these two methods, it can be seen that the two-step method has obvious limitations on the material, and only the material that can be directly obtained by heating the precursor can be used by the two-step method. So next, the three-step method will be introduced in detail.

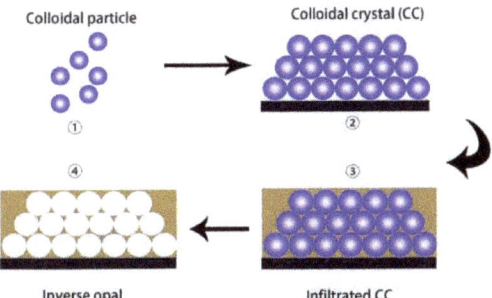

Figure 1. Preparation of inverse opal photonic crystals by three-step method.

2.1. Construction of the Opal Photonic Crystal Template

The construction of the opal template is the basis for the production of inverse opal photonic crystals, and the construction of the opal template includes nanolithography and the self-assembly of colloidal microspheres. Nanolithography, known as the "top-down" method, is expensive and slow. This paper mainly discusses the second method, which is the self-assembly of colloidal microspheres.

Self-assembly methods include the gravity sedimentation method [40], the vertical deposition method [41], etc. The gravity sedimentation method is to disperse the colloidal particle emulsion with uniform particle size and good monodispersity into the solvent according to a certain concentration. With the evaporation of the solvent, the colloidal particles are self-assembled on the material under the action of the gravity field to form a three-dimensional photonic crystal [42], as shown in Figure 2. The gravity sedimentation method has a simple preparation process and low requirements for equipment, but it has strict requirements for the size and density of colloidal particles, many sample defects, and a long preparation cycle. Gao [43] and others used silica nanoparticles (SNP) as materials to obtain silica photonic colloidal crystals via the self-assembly method through gravity sedimentation (Figure 3). In order to solve the problem of the long preparation period of the gravity sedimentation method, centrifugal force or thermal assistance can be introduced on the basis of the gravity sedimentation method to accelerate the deposition of colloidal particles [44]. The gravity sedimentation method also has many disadvantages, such as the long production time and low production efficiency. Because it only relies on the gravity of the ball itself, it is not suitable for large-scale industrial production.

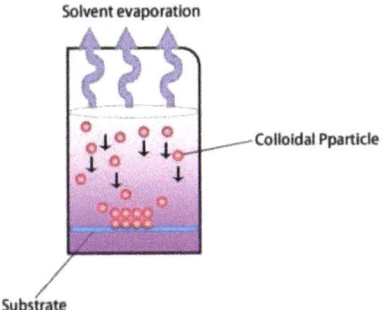

Figure 2. Preparation of inverse opal photonic crystals by gravity sedimentation method.

Vertical deposition is also a widely used self-assembly method. Figure 4 is a schematic diagram of the vertical deposition method; that is, the material is vertically placed in the assembly solution of monodisperse colloidal microspheres. With the evaporation of the solvent, colloidal microspheres gather on both sides of the material under the combined action of capillary force and surface tension, forming a periodic photonic crystal structure [45]. Sinitskii A et al. [46] prepared a polystyrene colloidal opal template using the vertical deposition method, and prepared inverse opals based on different oxide materials (TiO$_2$, SiO$_2$, and Fe$_2$O$_3$). Compared with the gravity deposition method, the vertical deposition method has similar advantages and disadvantages, but it can make photonic crystals adhere to both sides of the material [47,48].

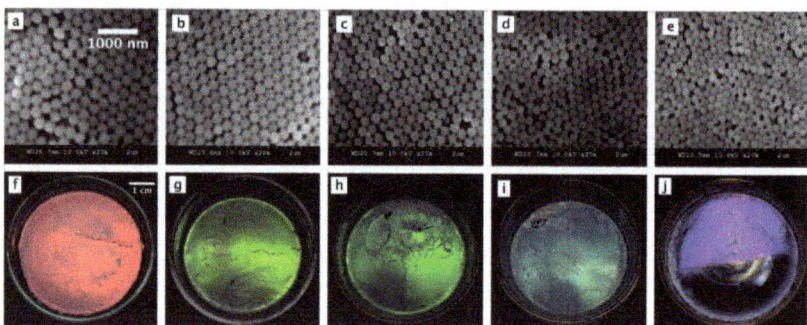

Figure 3. Top view SEM images (**a–e**) and images (**f–j**) of colored CC films with SNPs diameters of 350 nm, 282 nm, 270 nm, 249 nm, and 207 nm, respectively; scale bars are displayed in the first image of each set [43].

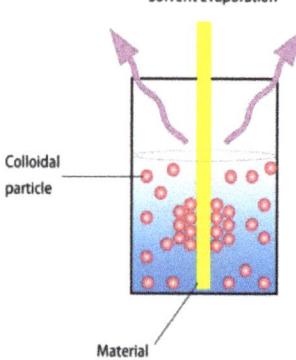

Figure 4. Schematic diagram of vertical deposition self-assembly method.

2.2. Filling of the Precursor

The additional material filling the opal template is called a precursor [49]; the precursor is crucial for the construction of inverse opal photonic crystals. To achieve inverse opal photonic crystals with an excellent morphological structure, the precursor must be uniformly filled in the opal template.

Precursor filling refers to the method of filling liquid precursor into the void of the opal template and curing via chemical reaction under specific conditions (Figure 5), including the sol-gel method [50,51], chemical vapor deposition method [52], atomic layer-by-layer deposition method [53], and electrochemical deposition method [54], etc.

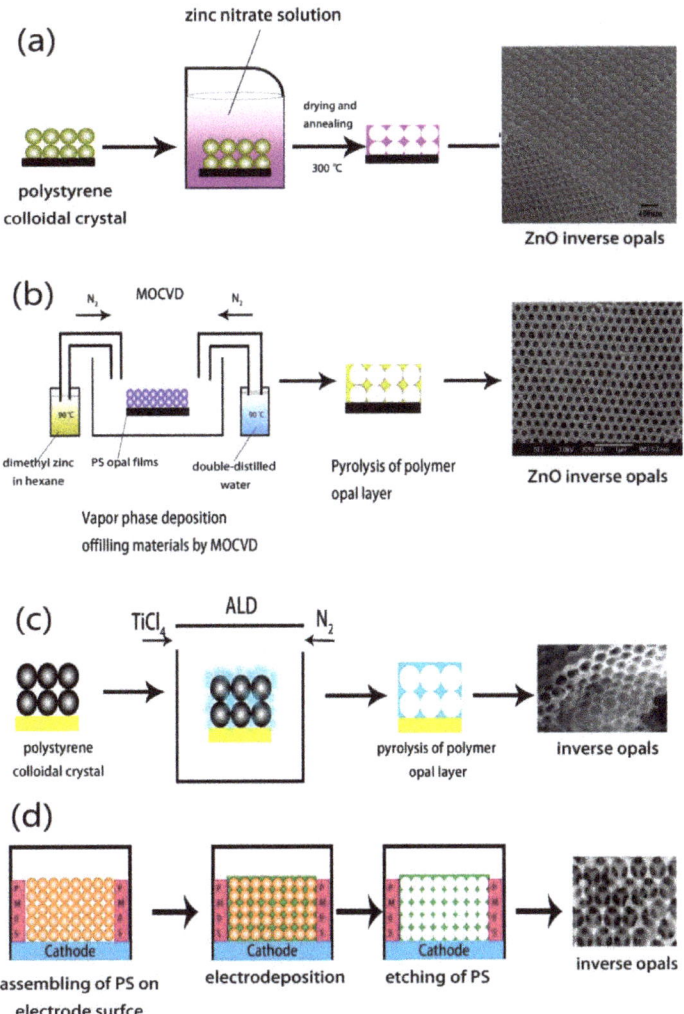

Figure 5. (**a**) Preparation of inverse opal photonic crystals by sol-gel procedure; (**b**) preparation of inverse opal photonic crystals by CVD method; (**c**) preparation of inverse opal photonic crystals by ALD method; and (**d**) preparation of inverse opal photonic crystals by electrodeposition method.

(1) Sol-gel method: The sol-gel method uses hydrolyzing metal alkoxides and other precursors to fill the opal template under appropriate conditions, form a gel, and then calcine to obtain solid oxides. This method is suitable for the filling of most semiconductor oxide materials, but the filling rate is not high, and the volume shrinkage after drying is large. Some researchers have used the sol-gel method to prepare inverse opal photonic crystal photocatalysts [55–57]. Jie Yu [55] developed an inverse opal titanium dioxide photonic crystal photocatalyst to effectively degrade toluene. The catalyst was prepared via the sol-gel method using colloidal photonic crystal as a template. The catalyst was doped with carbon nitride quantum dots (CNQDs) in situ. The catalyst has good photocatalytic performance for toluene degradation. Under simulated sunlight irradiation, the samples were used to degrade the liquid pollutants represented by Rhodamine B (RhB) and phenol. As shown in Figure 6, the degradation rates of dyes and phenol reached more than 97% after 75 min and

100 min of illumination, respectively. Weijie Liu [57] improved the sol-gel permeation method. The prepared high-quality titanium dioxide inverse opal has a dense porous structure. In addition, they studied the optical properties and photocatalytic activity. The photocatalytic degradation of methyl orange was selected to evaluate the photocatalytic activity of the obtained titanium dioxide inverse opal. The conclusion is that compared with the samples prepared before improvement, the obtained titanium dioxide inverse opal has stronger photonic behavior and better photocatalytic activity.

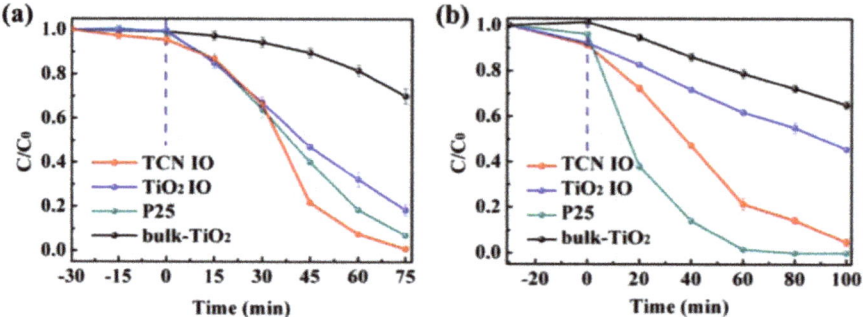

Figure 6. Photocatalytic degradation results in (**a**) 20 mg/L of RhB and (**b**) 10 mg/L phenol over the Nitrogen-doped titanium dioxide inverse opal photonic crystal (TCN IO) under simulated sunlight irradiation [55].

(2) The chemical vapor deposition method: can be used to diffuse the material to be filled in the form of gas precursor, adsorb it in the gap of the orderly microsphere, and then change the temperature or pressure to cause the precursor gas reaction, precipitate the solid material, and deposit it into the pores [58,59].

(3) The atomic layer-by-layer deposition method(ALD): this method is actually a form of chemical vapor deposition; it involves two or more kinds of vapor precursors on the solid surface and deposition to obtain the multilayer film method. The specific process is as follows. The surface or template is deposited in the gas phase of a certain amount of precursor, so that the surface reaches a single-layer saturated adsorption, and then the excess unabsorbed gas extraction is injected into another gas phase precursor. On the surface, two precursors are used to obtain a single-layer thickness film. Repeat the process to obtain a multilayer film with a specific thickness. Some researchers have used this method to prepare inverse opal photonic crystals with excellent photocatalytic performance. László Péter Bakos [60] used the atomic deposition method to fill the carbon nanosphere template to prepare the inverse opal photonic crystal. Their team previously showed that the carbon nanospheres (CS) were the appropriate template for the atomic layer deposition of TiO_2, because they were thermally stable at 300 °C in an inert atmosphere and had oxygen-containing functional groups on their surfaces. The hollow titanium dioxide shell can be prepared by subsequent annealing of the $CS-TiO_2$ composite. They used carbon nanospheres to prepare ordered face-centered colloidal crystals, and used ALD to deposit TiO_2 on them to produce inverse opal structure. They used scanning electron microscopy (SEM), Raman spectroscopy (Raman), X-ray diffraction (XRD), and UV-vis diffuse reflectance spectroscopy (UV-vis diffuse reflectance spectroscopy) to characterize the inverse opal samples, and investigated their photocatalytic degradation activity of methyl orange solution and methylene blue dye dried on the sample surface under UV and visible light irradiation.

Some researchers have also adopted an improved atomic deposition method to prepare inverse opal photonic crystals. Long [61] used O_3 as an oxidant to prepare inverse opal zinc oxide using the ALD method. Because O_3 has higher activity, this method has better performance than the zinc oxide prepared via the H$_2$O atomic deposition method and the electrodeposition method. This is because O$_3$ has higher activity and can produce oxide films with lower concentrations of impurities than films grown with H$_2$O. P. Birnal [62] synthesized TiO$_2$-Au composite inverse opal using the atomic deposition method. While depositing TiO$_2$, they also injected preformed Au nanoparticles. They compared the degradation of methylene blue (MB) by P-TiO$_2$-Au and IO-TiO$_2$-Au composite photocatalysts. The evolution of degradation percentage with time is shown in Figure 7. Two kinds of composite photocatalysts significantly improved the degradation rate of MB. The degradation rate of the P-TiO$_2$-Au flat film was 40% within 2 h of exposure and more than 90% within 14 h of exposure, while the degradation rate of the IO-TiO$_2$ photocatalyst was 95% only within 7 h of exposure to visible light. Their experiments show the potential of the atomic deposition method for preformed nanoparticles to produce complex composite structures.

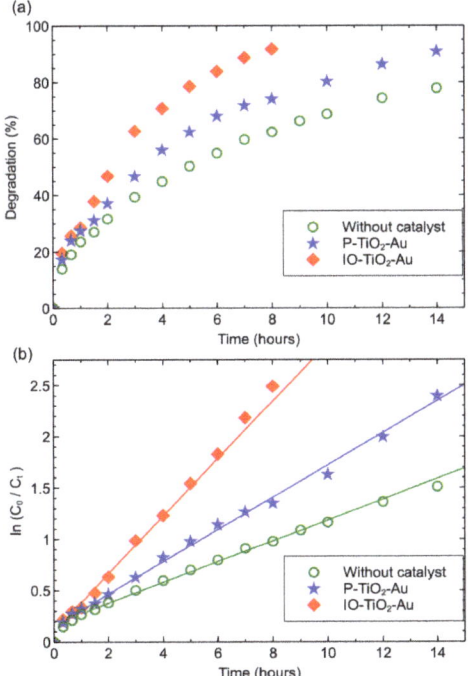

Figure 7. Degradation of methylene blue (1 µmol/L) over time under visible light. Illumination using P-TiO$_2$-Au and IO-TiO$_2$-Au films as photocatalysts, compared to the natural degradation of methylene blue. (**a**) Degradation percentage, (**b**) degradation kinetics in logarithmic scale [62].

(4) Electrochemical deposition method: the electrochemical reaction is used to fill the opal template directly by placing it directly in the cathode of the electrochemical battery. The electrochemical deposition method is characterized by the material filling continuously from the bottom to the top of the opal template, until the filling is relatively complete [63,64].

2.3. Removal of the Opal Template

The common template removal methods include dissolution and pyrolysis.

The dissolution method uses the chemical properties of the opal template material itself, and uses the chemical reagent dissolution to remove the opal template [47]. For example, the silica opal template can be removed using dilute hydrofluoric acid, and the polystyrene opal template can be removed with toluene.

The pyrolysis method is the high-temperature thermal decomposition using the physical properties of opal template material [57]. Min Wu [65] discussed the photocatalytic activity of photodegrading Rhodamine B (RhB) in calcined titania dioxide inverse opal films at different temperatures. The characterization of XRD, SEM, TEM, and HRTEM showed that the structures of inverse opal titanium dioxide photonic crystals were distorted with increasing temperatures. Excessive temperature will reduce the photocatalytic performance of inverse opal photonic crystals.

3. Modification of Photocatalysts for Inverse Opal Photonic Crystals

In order to further improve the photocatalytic efficiency of inverse opal photonic crystals, the researchers proposed the following modification methods, such as the metal modification method, the nonmetal modification method, the self-doping method, and others.

3.1. Metal Modification Method

The metal modification method involves loading the surface of nanomaterials via Au [66], Ag, Pt [67], Pd, and other precious metals, so as to improve the separation efficiency of electron-hole pairs. Some researchers also use metal compounds to improve the performance of inverse opal photonic crystal photocatalysts. Huang et al. [68] developed a new type of CuS-loaded inverse opal g-C_3N_4 photocatalyst (CN) to improve its photocatalytic activity for CO_2 reduction via the surface modification of CuS nanoparticles. In the experiment, inverse opal g-C_3N_4 with a good optical response and pore structure was prepared and characterized. Then, CuS nanoparticles were prepared by hydrothermal method and dispersed in toluene. Then, the CuS nanoparticles solution was added to the surface of inverse opal g-C_3N_4 and calcined at 110 °C for 1 h to obtain the CuS-modified inverse opal g-C_3N_4 photocatalyst. Finally, the photocatalytic reduction of CO_2 was carried out using the catalyst, and its photocatalytic performance and mechanism were studied. The CO generation rate is shown in Figure 8. The results showed that the optimal loading of CuS was 2 wt%, and the corresponding CO-evolution was 13.24 $\mu mol \cdot g^{-1} \cdot h^{-1}$, which is 3.2 times that of IO CN and five times that of bulk CN.

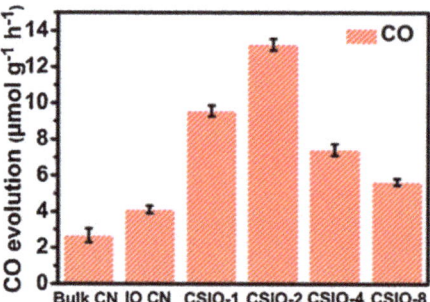

Figure 8. The CO production rate of the photocatalyst samples [68].

In conclusion, metal nanoparticles have a high light absorption rate and catalytic activity, which can effectively improve the photocatalytic efficiency of inverse opal photocrystals. In addition, by selecting different metal nanoparticles or changing the modification mode, the photocatalytic performance of inverse opal photocrystals can be optimized. Its modification method is simple, and its operation is relatively easy. However, it is worth noting that the stability of the metal nanoparticles is poor, and phenomena such as aggregation or dissolution may occur, which may affect the catalytic efficiency and lifetime. The matching between metal nanoparticles and inverse opal photocrystals should be considered in the preparation process to select the appropriate modification mode and conditions. Some metal nanoparticles may pose contamination and toxicity risks to the environment.

3.2. Nonmetal Modification Method

In order to develop an efficient inverse opal photonic crystal photocatalyst, the photocatalyst was modified by doping non-metallic elements such as C, P [69], N [70], and S on the structural basis of inverse opal photonic crystals. Some researchers have used a variety of non-metallic materials mixed with doping to improve the photocatalytic performance of inverse opal photonic crystals. Wenjun Zhang [71] prepared N-CD (nitrogen-doped carbon dot) via the one-step hydrothermal method and then used it to sensitize highly ordered porous TiO_2 IOS (TiO_2 inverse opals) films. Under simulated sunlight irradiation, the photocatalytic energy of N-CD/TiO_2 IOS and CD/TiO_2 IOS were compared. As shown in Figure 9, the photocatalytic degradation rate of MB (methyl blue) of the N-CD/TiO_2 IOS film also reached about 90% (curve 1), while the degradation rate of the CD/TiO_2 IOS film after five cycles was about 80% (curve 2). This fact illustrates that, in the photocatalytic degradation of MB molecules, the N-CD/TiO_2 IOS film has a higher catalytic performance and stability than the CD/TiO_2 IOS, because the n-doped atoms themselves have redox catalytic ability and the N-CD has a wider absorption range (UV and visible light regions).

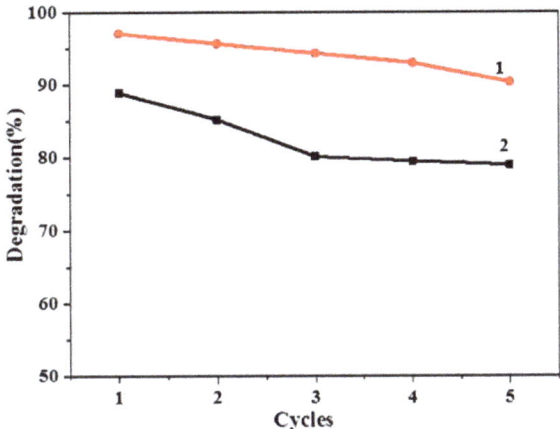

Figure 9. The degradation rate of MB by N-CD/TiO_2 IOS and CD/TiO_2 IOS.

3.3. Self-Doping Methods

Some researchers have utilized self-doping to improve the photocatalytic performance of inverse opal photonic crystals. Qi [72] studied a method to improve the photocatalytic performance of titanium dioxide (TiO_2). They self-doped Ti^{3+} in a titanium dioxide inverse opal photonic crystal and used a slow-light effect to enhance the photocatalytic performance. Their research found that oxygen vacancies and Ti^{3+} are able to narrow the band gap of TiO_2 and induce visible light capture. In order to study whether the synergistic effect of improving visible light absorption can be used to improve the photocatalytic activity driven by visible light, the photodegradation experiment of AO7 was carried out. The photocatalytic activity of pure TiO_2 was the lowest. The photocatalytic activity of doped Ti^{3+} was improved after vacuum activation, which indicated that Ti^{3+} and oxygen vacancy could indeed play a role in visible light photocatalytic activity. On the other hand, the T-170, T-265, and T-355 (170, 265, and 355, respectively, representing the concentrations of Ti^{3+} in titanium dioxide inverse opal photonic crystals) samples showed similar photocatalytic activity. After vacuum activation, the band gap of titanium dioxide was narrowed due to the existence of Ti^{3+} and oxygen vacancies. The inverse opal structure was able to improve the light trapping ability, generate photoelectrons and photoholes, and further improve the photocatalytic activity. V-T-355 (V represents vacuum activation) has the widest absorption and the highest photocatalytic activity in the visible light region. Therefore, the experiment confirmed that the synergistic effect of inverse opals structure and vacuum activation on visible light capture could actually be used to improve photocatalytic activity.

3.4. Other Methods

In some studies, the inverse opal photocatalyst was combined with biomimetic chloroplast to improve the photocatalytic efficiency of inverse opal photonic crystal(Figure 10). Zhou et al. [73] took advantage of this characteristic and combined an inverse opal photonic crystal with bionic chloroplasts to improve its photocatalytic efficiency. Specifically, the researchers first prepared TiO_2 in an inverse opal template. The inverse opal structure, then the surface of the pore in the inverse opal structure, was loaded with chlorophyll (Chl) molecules and ionic liquid (IL) via the cation exchange method. Through the characterization and performance test of the prepared materials, the authors found that the material demonstrated good absorption properties in the UV-visible light region and displayed high photocatalytic activity. The CO_2 conversion rate reached 28.6%. In order to further improve the efficiency of photocatalysis, the author designed a "light-charge separation-transfer" mechanism based on the photosynthesis process in chloroplasts in nature. Under this mechanism, light energy was composed via TiO_2 in the inverse opal structure. The particles absorb and release the electrons, which were taken up by the chlorophyll loaded on the surface, where it was then efficiently excited into the biomass. The slow light effect in the inverse opal structure improved the light absorption and utilization efficiency of the material, enabling more photons to be absorbed and excite the charge, thus improving the photocatalytic efficiency. At the same time, the chlorophyll molecules in the inverse opal structure are able to form the biomimetic chloroplast structure, and the "light-charge separation-transfer" mechanism was adopted so that the light energy could be efficiently converted into biomass (CH_4), which further improved the photocatalytic efficiency. The slow light effect and biomimetic chloroplast structure in inverse opal structure are interrelated and act on the performance of the whole material. The slow light effect improves the light absorption and utilization efficiency, while the bionic chloroplast structure realizes the light-charge separation and efficient transformation.

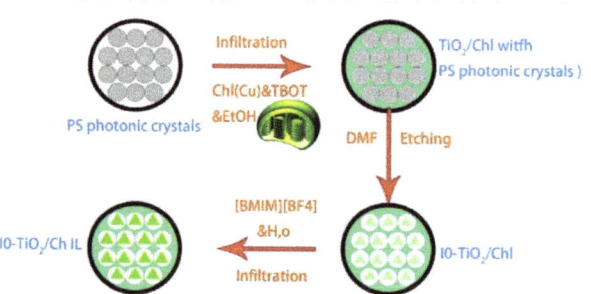

Figure 10. Preparation method of inverse opal photocatalyst combined with biomimetic chloroplast.

In addition to using biomimetic technology, some researchers utilized the heterojunction structure to improve the photocatalytic activity of inverse opal photonic crystals. The so-called heterojunction structure refers to the "S-type" barrier structure based on the built-in electric field. When the materials on both sides of the heterojunction have different conduction bands and valence band energy levels, the built-in electric field will be formed, so that the electrons and holes stimulated under light can move to both sides of the heterojunction, so as to realize effective charge separation. Moreover, the reverse photonic crystal structure can increase the light absorption efficiency and focus the photons near the interface through the photonic localized effect, further promoting the generation and transmission of photo-generated carriers. Therefore, the construction of heterojunctions enables efficient charge separation and rapid transport in photocatalytic reactions, thus increasing the reaction efficiency. Liu et al. [74] studied a novel inverse opal photonic crystal Bi_2WO_6/Bi_2O_3 heterojunction photocatalyst with efficient charge separation and rapid migration to achieve highly active photocatalytic reactions. The results of this heterojunction via the sol-gel method showed that the photocatalysts demonstrated excellent photocatalytic performance under visible light and an obvious effect on grading the organic dye RhB.

4. The Application of Inverse Opal Photonic Crystals in the Field of Photocatalysis

As a photocatalyst, inverse opal photonic crystal is widely used in sewage treatment, clean energy production, and waste gas treatment, as shown in Figure 11.

Figure 11. Application of inverse opal photocatalysis.

4.1. Sewage Treatment

In industrial production, industrial sewage treatment has always been a concern of researchers, who are making efforts to remove organic pollutants and antibiotic pollutants from the water. There are many existing treatment methods, such as filtration, adsorption, precipitation, photodegradation, and biodegradation [75]. Photodegradation, as a new clean and pollution-free organic pollutant treatment technology, has attracted the attention of researchers [76].

When studying water pollution, Rhodamine B (RhB) is generally used as a simulated pollutant. Because RhB is an artificially synthesized dye with a bright peach-red color that is easily soluble in water, it is widely used as a colorant in the textile and food industries and has a carcinogenic effect. Wastewater containing RhB would pollute the environment and harm human health and the growth of animals and plants.

The methods of the photocatalytic degradation of organic pollutants in water are as follows: (1) adsorption: pollutants are adsorbed on the surface of the photocatalyst; (2) electron excitation: due to the stimulation of external light source, when the photon energy obtained by the catalyst is greater than its own band gap, the electron will be excited from the valence band to the conduction band, so that a relatively stable hole will be left in the valence band, thus forming an electron-hole; and (3) redox reaction occurs. Photogenerated holes oxidize the OH^- and H_2O adsorbed on the surface of the catalyst into ·OH radicals with high activity and oxidation. Finally, ·OH oxidizes the organic pollutants adsorbed on the catalyst surface to CO_2 and H_2O [77–79].

Inverse opal photonic crystals are widely used in the field of sewage treatment [80]. Wan et al. [81] successfully prepared an inverse opal TiO_2 photocatalyst which was able to react under visible light via the sol-gel method combined with an opal template. They used the photocatalytic degradation of Rhodamine B to study the photocatalytic performance of the catalyst. They confirmed that the slow photon effect is able to enhance photocatalytic efficiency, and a large specific surface area can reduce the binding rate of photo-generated electron-hole pairs. Figure 12 shows the absorption spectra of RhB photocatalytic degradation and each self-photodegradation. In the process of photocatalytic degradation, the maximum absorption peak of RhB was weakened and blue-shifted. Because the photocatalyst almost completely degraded RhB in 1.5 h, the degradation rate was 92.5% after 1 h, indicating that the photocatalyst has a good photocatalytic performance.

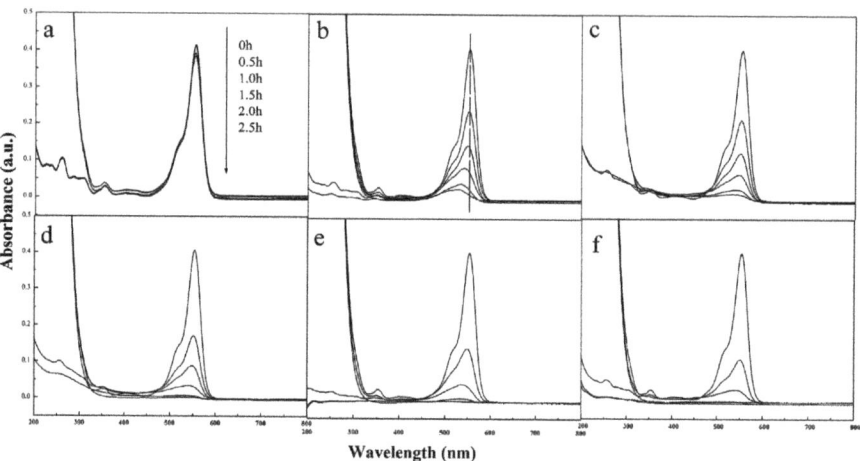

Figure 12. UV-vis absorption spectrum of RhB degraded by as-prepared samples under visible light irradiation; (**a**) self-degradation; (**b**) T-Sol; (**c**) IOT-230; (**d**) IOT-330; (**e**) IOT-440; (**f**) IOT-610 [81].

4.2. Clean Energy Production

As a kind of high-energy clean energy source, hydrogen is widely used in various industrial production activities. However, the synthesis process is complex and has high energy consumption, so how to use solar energy to produce hydrogen has become a concern of researchers. There are two main methods to produce hydrogen from solar energy.

Photoelectric water splitting method [82]: utilizing solar energy to generate electricity, which is used to electrolyze water molecules and decompose water into hydrogen and oxygen. Photocatalytic decomposition method [83]: using a photocatalyst to absorb the solar energy and generate a charge on its surface to start a reaction that decomposes the water into hydrogen and oxygen.

The photocatalytic decomposition method is relatively simple, and the use of photocatalytic decomposition to produce hydrogen does not require traditional energy consumption, and there is no emission of harmful gases such as carbon dioxide, making it an environmentally friendly method. Compared to the traditional electrolytic water method for hydrogen production, the use of photocatalysts can increase the reaction rate and have higher hydrogen production efficiency. And when using the photocatalytic decomposition method to produce hydrogen, the reaction conditions are mild, which can reduce the energy consumption and waste heat generated by the equipment, and improve the overall economy of the process. The most important thing is that the use of photocatalysts can achieve precise control of reaction rate and hydrogen production, which is of great significance for industrial production and scientific research. The principle of photocatalytic hydrogen production is that in the process of the photocatalytic decomposition of water, electrons in the conductive band reduce hydrogen ions to hydrogen, and holes in the valence band oxidize oxygen ions to oxygen.

Many researchers have conducted a lot of research on the preparation of hydrogen using inverse opal photonic crystal photocatalyst. Fiorenza et al. [84] prepared TiO_2-$BiVO_4$

and TiO$_2$-CuO samples with inverse opal structure, characterized them, and carried out photocatalytic water decomposition experiments under ultraviolet light and solar light irradiation. Figure 13 shows the H$_2$ production of the inverse opal titanium dioxide (I.O.TiO$_2$) system under UV and sunlight. Under UV irradiation, it can be seen that the hydrogen production rate of pure I.O.TiO$_2$ is higher than that of the TiO$_2$ photocatalyst purchased from the market (Figure 13A,B). In addition, it can be seen that the presence of BiVO$_4$ (Figure 13A) leads to a moderate increase in H$_2$ production. Under visible light irradiation, I.O.TiO$_2$ showed a five-time higher activity than the commercially available TiO$_2$ (Figure 13C,D). In this case, the I.O.TiO$_2$ 25% BiVO4 sample showed the best performance (purple line in Figure 13C). Lv et al. [85] prepared an efficient photocatalytic material called the inverse opal structure (IO)-TiO$_2$-MoO$_3$-x, which is able to catalyze H$_2$ generation and Rhodamine B dye degradation simultaneously. The material is made up of titanium dioxide (TiO$_2$), and the molybdate (MoO$_3$-x) composition has an inverse opal structure and plasma enhancement effect. The experimental results show that the photocatalytic activity of the material is not only seven times that of pure titanium dioxide photocatalyst but also has an effective catalytic effect in the visible light range. Liu et al. [86] identified an efficient photocatalyst, namely Ag (silver) modified onto g-C$_3$N$_4$ (melamine), an inverse opal photonic crystal structure, known as an Ag/g-C$_3$N$_4$ 3D inverse opal photonic crystal. The experimental results show that when using Ag-CN IO (Ag/g-C$_3$N$_4$ 3D inverse opal photonic crystal) as the photocatalyst, the H$_2$ release rate under ultraviolet light irradiation was 4.93 µmol·g^{-1}·h^{-1}, and because Ag nanoparticles enhance the optical absorption and the photonic crystal effect of the g-C$_3$N$_4$ inverse opal structure, the photocatalyst has excellent hydrogen production and stability.

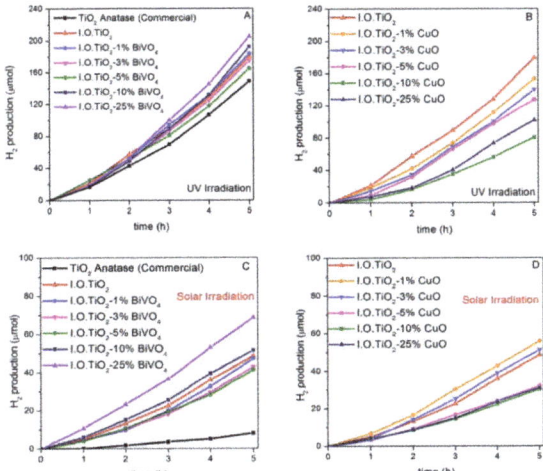

Figure 13. H$_2$ production of the inverse opal TiO$_2$ (I.O.TiO$_2$) system under UV and sunlight irradiation. (**A**) Inverse Opal TiO2-BiVO4 and (**B**) Inverse Opal TiO2-CuO composites under UV irradiation; (**C**) Inverse Opal TiO2-BiVO4 and (**D**) Inverse Opal TiO2-CuO under solar light irradiation [84].

4.3. Waste Gas Treatment

With the development of global industrialization, the emission of industrial waste gas has become one of the main factors endangering the earth's environment. As a waste gas that is widely emitted in industrial production, the decomposition of CO$_2$ has become a popular research direction among researchers. Converting carbon dioxide into organic compounds, such as methane or methanol, can both reduce the concentration of carbon dioxide in the atmosphere and also provide a new and convenient energy storage method [87]. In the conversion treatment of CO$_2$, due to its stable chemical properties, there are multiple problems, such as the low conversion efficiency of traditional catalysts and many reaction by-products. Research has shown that photonic crystals exhibit high catalytic performance in the photo-reduction of CO_2.

Many researchers have conducted a great amount of research to degrade carbon dioxide by using inverse opal photonic crystal photocatalysts. Wherein Wei et al. [88] added Au nanoparticles to titanium dioxide inverse opal photonic crystal to prepare inverse opal photonic crystal with core-shell structure, and used it as a photocatalyst for CO_2 reduction reaction. The catalyst has the highest photocatalytic activity and CO_2 reduction selectivity, and the experimental results indicate that CH_4 generation rate of 41.6 $\mu mol \cdot g^{-1} \cdot h^{-1}$ and 98.6% selectivity for CO_2 reduction to generate CH_4. This composite has highly efficient photocatalytic CO_2 conversion properties under visible light. Xu et al. [89] prepared photocatalytic experiments to build rhenium doped into inverse opal SnO_2/TiO_2-x and reducing CO_2, and the CO_2 yield using the finally obtained catalyst was 16.59 $\mu mol \cdot g^{-1} \cdot h^{-1}$, which was about 1.21, 2.14 and 7.44 times obtained using inverse opal SnO_2/TiO_2-x, inverse opal TiO_2-x, and SnO_2, respectively.

Some examples of photocatalyst applications of the inverse opal structure are summarized in Table 1.

Table 1. Some examples of the photocatalytic applications of IOPCs.

IOPC Materials	Photocatalysis Application	Result	Ref
Mg-TiO_2	Sewage treatment	The Mg-TiO_2 system exhibits much higher activity than its counterpart due to the reduced band gap, which is due to the doping of Mg^{2+} in the system. By adding Mg^{2+}, the sterilization rate under visible light can reach 100%.	[80]
g-C_3N_4-BiOBr	Sewage treatment	It provides a new idea for the preparation of a new visible light-driven Z-type photocatalyst and a new idea for the study of wastewater treatment methods.	[90]
TiO_2-$BiVO_4$ TiO_2-CuO	H_2 production	The synthesis of inverse opal material combined with TiO_2 structure modification and chemical modification through the addition of $BiVO_4$ or CuO can improve H_2 production via the photocatalytic decomposition of water.	[84]
TiO_2-MoO_3-x	H_2 production	The results show that compared with a single control factor, the composite of IO structure and plasma material (MoO_3) has higher light capture ability and carrier separation and transfer efficiency, which can significantly improve the photocatalytic activity of RhB degradation and hydrogen evolution.	[85]
Ag-C_3N_4	H_2 production	The results showed that the hydrogen evolution performance of Ag-CN IO was the best among all the tested samples.	[86]
TiO_2-ZrO_2	Carbon dioxide conversion	This study improved the oxidation-reduction ability of the material because the construction of inverse opal core-shell structure promoted the nano-crystallization of the material.	[55]
Au@CdS/IO-TiO_2	Carbon dioxide conversion	Under simulated sunlight irradiation, the Au@CdS/IO-TiO_2 displayed excellent photocatalytic performance for the CO_2 reduction of CH_4.	[88]
TiO_2-x/SnO_2	Carbon dioxide conversion	A strategy for constructing a new S-type heterojunction structure in visible light photocatalysts is proposed, which provides an ideal method for improving photocatalytic activity to treat organic pollutants and renewable energy production.	[89]

5. Summary

The inverse opal photonic crystal, because of its own excellent characteristics—large surface area, interrelated channel structure, more active catalytic sites, quicker electron transmission speeds, a slow photonic effect able to enhance the absorption of light, good mass transfer properties, and a three-dimensional order within a multiple scattering structure, has great research value for improving photocatalytic efficiency.

However, there are still many challenges in the application of inverse opal photonic crystals in photocatalysis. For example, inverse opal photonic crystals cannot be produced on a large scale, their service life is not long enough, and the photocatalytic efficiency of inverse opal photonic crystals still needs to be improved. In the near future, researchers may find solutions to these problems and make significant progress in the application of inverse opal photonic crystals in the field of photocatalysis.

Author Contributions: Conceptualization, S.Y.; writing—original draft preparation, H.X.; writing—review and editing, E.T.; visualization, C.H.; funding acquisition, K.C. All authors have read and agreed to the published version of the manuscript.

Funding: This research was funded by the Natural Science Foundation of Zhejiang Province [LY20E030009], the National Natural Science Foundation of China [51403078 and 52103035], China Scholarship Council [201508330017].

Institutional Review Board Statement: Not applicable.

Informed Consent Statement: Not applicable.

Data Availability Statement: Not applicable.

Conflicts of Interest: The authors declare no conflict of interest.

References

1. Kumar, C.M.S.; Singh, S.; Gupta, M.K.; Nimdeo, Y.M.; Raushan, R.; Deorankar, A.V.; Kumar, T.M.A.; Rout, P.K.; Chanotiya, C.S.; Pakhale, V.D.; et al. Solar energy: A promising renewable source for meeting energy demand in Indian agriculture applications. *Sustain. Energy Technol. Assess.* **2023**, *55*, 102905.
2. Kasaeian, A.; Javidmehr, M.; Mirzaie, M.R.; Fereidooni, L. Integration of solid oxide fuel cells with solar energy systems: A review. *Appl. Therm. Eng.* **2023**, *224*, 120117. [CrossRef]
3. Zong, X.; Wang, L. Ion-exchangeable semiconductor materials for visible light-induced photocatalysis. *J. Photochem. Photobiol. C Photochem. Rev.* **2014**, *18*, 32–49. [CrossRef]
4. Wu, X.; Lan, D.; Zhang, R.; Pang, F.; Ge, J. Fabrication of Opaline ZnO Photonic Crystal Film and Its Slow-photonic Effect on Photoreduction of Carbon Dioxide. *Langmuir ACS J. Surf. Colloids* **2019**, *35*, 194–202. [CrossRef] [PubMed]
5. Likodimos, V. Photonic crystal-assisted visible light activated TiO_2 photocatalysis. *Appl. Catal. B Environ.* **2018**, *230*, 269–303. [CrossRef]
6. John, S. Strong localization of photons in certain disordered die lectric superlattices. *Phys. Rev. Lett.* **1987**, *58*, 2486–2489. [CrossRef] [PubMed]
7. Yablonovitch, E. Inhibited spontaneous emission in solid state physics and electronics. *Phys. Rev. Lett.* **1987**, *58*, 2059–2062. [CrossRef]
8. Zhou, L.; Lei, J.; Wang, L.; Liu, Y.; Zhang, J. Highly efficient photo-Fenton degradation of methyl orange facilitated by slow light effect and hierarchical porous structure of Fe_2O_3-SiO_2 photonic crystals. *Appl. Catal. B Environ.* **2018**, *237*, 1160–1167. [CrossRef]
9. Yablonovitch EJ, J.B. Photonic band gap structures. *J. Opt. Soc. Am. B* **1993**, *10*, 283–295. [CrossRef]
10. Peigen, N. Progress in the fabrication and application of photonic crystals. *Acta Phys. Sin. Chin. Ed.* **2010**, *59*, 340–350.
11. Gumus, M.; Giden, I.H.; Akcaalan, O.; Turduev, M.; Kurt, H. Enhanced superprism effect in symmetry reduced photonic crystals. *Appl. Phys. Lett.* **2018**, *113*, 131103. [CrossRef]
12. Parimi, P.V.; Lu, W.T.; Vodo, P.; Sokoloff, J.; Derov, J.S.; Sridhar, S. Negative refraction and left-handed electromagnetism in microwave photonic crystals. *Phys. Rev. Lett.* **2004**, *92*, 127401. [CrossRef] [PubMed]
13. Zhu, K.; Fang, C.; Pu, M.; Song, J.; Wang, D.; Zhou, X. Recent advances in photonic crystal with unique structural colors: A review. *J. Mater. Sci. Technol.* **2023**, *141*, 78–99. [CrossRef]
14. Li, L.; Li, J.; Xu, J.; Liu, Z. Recent advances of polymeric photonic crystals in molecular recognition. *Dye. Pigment.* **2022**, *205*, 110544. [CrossRef]
15. Li, T.; Liu, G.; Kong, H.; Yang, G.; Wei, G.; Zhou, X. Recent advances in photonic crystal-based sensors. *Coord. Chem. Rev.* **2023**, *475*, 214909. [CrossRef]
16. Butt, M.A.; Khonina, S.N.; Kazanskiy, N.L. Recent advances in photonic crystal optical devices: A review. *Opt. Laser Technol.* **2021**, *142*, 107265. [CrossRef]
17. Liu, W.; Ma, H.; Walsh, A. Advance in photonic crystal solar cells. *Renew. Sustain. Energy Rev.* **2019**, *116*, 109436. [CrossRef]
18. Fathi, F.; Monirinasab, H.; Ranjbary, F.; Nejati-Koshki, K. Inverse opal photonic crystals: Recent advances in fabrication methods and biological applications. *J. Drug Deliv. Sci. Technol.* **2022**, *72*, 103377. [CrossRef]
19. Schroden, R.C.; Al-Daous, M.; Blanford, C.F.; Stein, A. Optical properties of inverse opal photonic crystals. *Chem. Mater.* **2002**, *14*, 3305–3315. [CrossRef]

20. Fathi, F.; Chaghamirzaei, P.; Allahveisi, S.; Ahmadi-Kandjani, S.; Rashidi, M.-R. Investigation of optical and physical property in opal films prepared by colloidal and freeze-dried microspheres. *Colloids Surf. A Physicochem. Eng. Asp.* **2021**, *611*, 125842. [CrossRef]
21. Asoh, H.; Ono, S. Fabrication of inverse opal structure of silica by si anodization. In *Proceedings of the 206th Meeting of the Electrochemical Society: Third International Symposium on Pits and Pores: Formation, Properties, and Significance for Advanced Materials, Honolulu, HW, USA, 3–8 October 2004*; The Electrochemical Society: Honolulu, HI, USA, 2006.
22. Zhou, Z.; Zhao, X.S. Opal and inverse opal fabricated with a flow-controlled vertical deposition method. *Langmuir* **2005**, *21*, 4717–4723. [CrossRef]
23. Borel, P.I.; Harpøth, A.; Frandsen, L.H.; Kristensen, M.; Shi, P.; Jensen, J.S.; Sigmund, O. Topology optimization and fabrication of photonic crystal structures. *Opt. Express* **2004**, *12*, 1996–2001. [CrossRef] [PubMed]
24. Waterhouse, G.I.N.; Waterland, M.R. Opal and inverse opal photonic crystals: Fabrication and characterization. *Polyhedron* **2007**, *26*, 356–368. [CrossRef]
25. Qi, M.; Lidorikis, E.; Rakich, P.T.; Johnson, S.G.; Joannopoulos, J.D.; Ippen, E.P.; Smith, H.I. A three-dimensional optical photonic crystal with designed point defects. *Nature* **2004**, *429*, 538–542. [CrossRef] [PubMed]
26. von Freymann, G.; Kitaev, V.; Lotsch, B.V.; Ozin, G.A. Bottom-up assembly of photonic crystals. *Chem. Soc. Rev.* **2013**, *42*, 2528–2554. [CrossRef]
27. Klimonsky, S.O.; Abramova, V.V.; Sinitskii, A.S.; Tretyakov, Y.D. Photonic crystals based on opals and inverse opals: Synthesis and structural features. *Russ. Chem. Rev.* **2011**, *80*, 1191–1207. [CrossRef]
28. Blanco, A.; Chomski, E.; Grabtchak, S.; Ibisate, M.; John, S.; Leonard, S.W.; Lopez, C.; Meseguer, F.; Miguez, H.; Mondia, J.P.; et al. Large-scale synthesis of a silicon photonic crystal with a complete three-dimensionalband gap near 1.5 micrometres. *Nature* **2000**, *405*, 437–440. [CrossRef]
29. Noda, S. Full three-dimensional photonicband gap crystals at near-infrared wavelengths. *Science* **2000**, *289*, 604–606. [CrossRef]
30. Baba, T. Control of light emission and propagation in photonic crystals. In Proceedings of the 2008 International Nano-Optoelectronics Workshop, Tokyo, Japan, 2–15 August 2008.
31. Lodahl, P.; Floris van Driel, A.; Nikolaev, I.S.; Irman, A.; Overgaag, K.; Vanmaekelbergh, D.; Vos, W.L. Controlling the dynamics of spontaneous emission from quantum dots by photonic crystals. *Nature* **2004**, *430*, 654–657. [CrossRef]
32. Curti, M.; Zvitco, G.; Grela, M.A.; Mendive, C.B. Angle dependence in slow photonic photocatalysis using TiO_2 inverse opals. *Chem. Phys.* **2018**, *502*, 33–38. [CrossRef]
33. Madanu, T.L.; Mouchet, S.R.; Deparis, O.; Liu, J.; Li, Y.; Su, B.-L. Tuning and transferring slow photons from TiO_2 photonic crystals to $BiVO_4$ nanoparticles for unprecedented visible light photocatalysis. *J. Colloid Interface Sci.* **2023**, *634*, 290–299. [CrossRef] [PubMed]
34. Huang, X.; Gu, W.; Hu, S.; Hu, Y.; Zhou, L.; Lei, J.; Wang, L.; Liu, Y.; Zhang, J. Phosphorus-doped inverse opal g-C_3N_4 for efficient and selective CO generation from photocatalytic reduction of CO_2. *Catal. Sci. Technol.* **2020**, *10*, 3694–3700. [CrossRef]
35. Waterhouse, G.I.; Chen, W.T.; Chan, A.; Sun-Waterhouse, D. Achieving Color and Function with Structure: Optical and Catalytic Support Properties of ZrO_2 Inverse Opal Thin Films. *ACS Omega* **2018**, *3*, 9658–9674. [CrossRef]
36. Fathi, F.; Rashidi, M.R.; Pakchin, P.S.; Ahmadi-Kandjani, S.; Nikniazi, A. Photonic Crystal Based Biosensors: Emerging Inverse Opals for Biomarker Detection. *Talanta* **2020**, *221*, 121615. [CrossRef]
37. Chen, X.; Wang, L.; Wen, Y.; Zhang, Y.; Wang, J.; Song, Y.; Jiang, L.; Zhu, D. Fabrication of closed-cell polyimide inverse opal photonic crystals with excellent mechanical properties and thermal stability. *J. Mater. Chem.* **2008**, *18*, 2262–2267. [CrossRef]
38. Quan, L.N.; Jang, Y.H.; Stoerzinger, K.A.; May, K.J.; Jang, Y.J.; Kochuveedu, S.T.; Shao-Horn, Y.; Kim, D.H. Soft-template-carbonization route to highly textured mesoporous carbon-TiO_2 inverse opals for efficient photocatalytic and photoelectrochemical applications. *Phys. Chem. Chem. Phys.* **2014**, *16*, 9023–9030. [CrossRef]
39. Luo, D.; Chen, Q.; Qiu, Y.; Liu, B.; Zhang, M. Carbon Dots-Decorated Bi_2WO_6 in an Inverse Opal Film as a Photoanode for Photoelectrochemical Solar Energy Conversion under Visible-Light Irradiation. *Materials* **2019**, *12*, 1713. [CrossRef]
40. Liu, L.; Karuturi, S.K.; Su, L.T.; Tok, A.I.Y. TiO_2 inverse-opal electrode fabricated by atomic layer deposition for dye-sensitized solar cell applications. *Energy Environ. Sci.* **2010**, *4*, 209–215. [CrossRef]
41. Xia, Y.N.; Gates, B.; Yin, Y.D.; Lu, Y.J.A.M. Monodispersed Colloidal Spheres: Old Materials with New Applications. *Adv. Mater.* **2000**, *12*, 693–713. [CrossRef]
42. Zhou, L.; Wu, Y.; Liu, G.; Li, Y.; Fan, Q.; Shao, J. Fabrication of high-quality silica photonic crystals on polyester fabrics by gravitational sedimentation self-assembly. *Color. Technol.* **2016**, *131*, 413–423. [CrossRef]
43. Gao, W.; Rigout, M.; Owens, H. Self-assembly of silica colloidal crystal thin films with tuneable structural colours over a wide visible spectrum. *Appl. Surf. Sci.* **2016**, *380*, 12–15. [CrossRef]
44. Yu, J.L.; Lee, C.H.; Kan, C.W.; Jin, S. Fabrication of structural-coloured carbon fabrics by thermal assisted gravity sedimentation method. *Nanomaterials* **2020**, *10*, 1133. [CrossRef] [PubMed]
45. Liu, G.J.; Zhou, L.; Fan, Q.G.; Chai, L.; Shao, J. The vertical deposition selfassembly process and the formation mechanism of poly photonic crystals on polyester fabrics. *J. Mater. Sci.* **2016**, *51*, 2859–2868. [CrossRef]
46. Sinitskii, A.; Abramova, V.; Grigorieva, N.; Grigoriev, S.; Snigirev, A.; Byelov, D.V.; Petukhov, A.V. Revealing stacking sequences in inverse opals by microradian X-ray diffraction. *EPL Europhys. Lett.* **2010**, *89*, 14002. [CrossRef]

47. Wu, Y.; Ren, S.; Chang, X.; Hu, J.; Wang, X. Effect of the combination of inverse opal photonic superficial areaic crystal and heterojunction on active free radicals and its effect on the mechanism of photocatalytic removal of organic pollutants. *Ceram. Int.* **2023**, *49*, 27107–27116. [CrossRef]
48. Xia, Y.; Yin, Y.; Lu, Y.; McLellan, J. Template-Assisted Self-Assembly of Spherical Colloids into Complex and Controllable Structures. *Adv. Funct. Mater.* **2010**, *13*, 907–918. [CrossRef]
49. Pham, K.; Temerov, F.; Saarinen, J.J. Multicompound inverse opal structures with gold nanoparticles for visible light photocatalytic activity. *Mater. Des.* **2020**, *194*, 108886. [CrossRef]
50. Yang, Z.; Ji, Z.; Huang, X.; Xie, Q.; Fu, M.; Li, B.; Li, L. Preparation and photonicband gap properties of $Na_{1/2}Bi_{1/2}TiO_3$ inverse opal photonic crystals. *J. Alloys Compd.* **2009**, *471*, 241–243. [CrossRef]
51. Engelken, R.D.; Mccloud, H.E.; Lee, C.; Slayton, M.; Ghoreishi, H. ChemInform Abstract: Low Temperature Chemical Precipitation and Vapor Deposition of SnxS Thin Films. *ChemInform* **1988**, *19*, 50269–50278. [CrossRef]
52. Míguez, H.; Chomski, E.; García-Santamaría, F.; Ibisate, M.; John, S.; López, C.; Meseguer, F.; Mondia, J.P.; Ozin, G.A.; Toader, O.; et al. Photonicband gap Engineering in Germanium Inverse Opals by Chemical Vapor Deposition. *Adv. Mater.* **2010**, *13*, 1634–1637. [CrossRef]
53. Liu, Y.R.; Hsueh, Y.C.; Perng, T.P. Fabrication of TiN inverse opal structure and Pt nanoparticles by atomic layer deposition for proton exchange membrane fuel cell. *Int. J. Hydrogen Energy* **2017**, *42*, 10175–10183. [CrossRef]
54. Chung, Y.W.; Leu, I.C.; Lee, J.H.; Hon, M.-H. Filling behavior of ZnO nanoparticles into opal template via electrophoretic deposition and the fabrication of inverse opal. *Electrochim. Acta* **2009**, *54*, 3677–3682. [CrossRef]
55. Yu, J.; Caravaca, A.; Guillard, C.; Vernoux, P.; Zhou, L.; Wang, L.; Lei, J.; Zhang, J.; Liu, Y. Carbon Nitride Quantum Dots Modified TiO_2 Inverse Opal Photonic Crystal for Solving Indoor VOCs Pollution. *Catalysts* **2021**, *11*, 464. [CrossRef]
56. Tan, T.; Xie, W.; Zhu, G.; Shan, J.; Xu, P.; Li, L.; Wang, J. Fabrication and photocatalysis of $BiFeO_3$ with inverse opal structure. *J. Porous Mater.* **2015**, *22*, 659–663. [CrossRef]
57. Liu, W.; Zou, B.; Zhao, J.; Cui, H. Optimizing sol-gel infiltration for the fabrication of high-quality titania inverse opal and its photocatalytic activity. *Thin Solid Film.* **2010**, *518*, 4923–4927. [CrossRef]
58. Juarez, B.H.; Garcia, P.D.; Golmayo, D.; Blanco, A.; López, C. *ZnO Inverse Opals by Chemical Vapor Deposition*; Wiley: New York, NY, USA, 2005; pp. 2761–2765.
59. Caicedo, J.; Taboada, E.; Hrabovský, D.; López-García, M.; Herranz, G.; Roig, A.; Blanco, A.; López, C.; Fontcuberta, J. Facile route to magnetophotonic crystals by infiltration of 3D inverse opals with magnetic nanoparticles. *J. Magn. Magn. Mater.* **2010**, *322*, 1494–1496. [CrossRef]
60. Bakos, L.P.; Karajz, D.; Katona, A.; Hernadi, K.; Parditka, B.; Erdélyi, Z.; Lukács, I.; Hórvölgyi, Z.; Szitási, G.; Szilágyi, I.M. Carbon nanosphere templates for the preparation of inverse opal titania photonic crystals by atomic layer deposition—ScienceDirect. *Appl. Surf. Sci.* **2020**, *504*, 144443. [CrossRef]
61. Long, J.; Fu, M.; Li, C.; Sun, C.; He, D.; Wang, Y. High-quality ZnO inverse opals and related heterostructures as photocatalysts produced by atomic layer deposition. *Appl. Surf. Sci.* **2018**, *454*, 112–120. [CrossRef]
62. Birnal, P.; de Lucas, M.M.; Pochard, I.; Herbst, F.; Heintz, O.; Saviot, L.; Domenichini, B.; Imhoff, L. Visible-light photocatalytic degradation of dyes by TiO_2-Au inverse opal films synthesized by Atomic Layer Deposition. *Appl. Surf. Sci.* **2023**, *609*, 155213. [CrossRef]
63. Teh, L.K.; Wong, C.C.; Yang, H.Y.; Lau, S.P.; Yu, S.F. Lasing in electrodeposited ZnO inverse opal. *Appl. Phys. Lett.* **2007**, *91*, 485. [CrossRef]
64. Yan, H.; Yang, Y.; Fu, Z.; Yang, B.; Xia, L.; Xu, Y.; Fu, S.; Li, F. Cathodic Electrodeposition of Ordered Porous Titania Films by Polystyrene Colloidal Crystal Templating. *Chem. Lett.* **2006**, *35*, 864–865. [CrossRef]
65. Wu, M.; Liu, J.; Jin, J.; Wang, C.; Huang, S.; Deng, Z.; Li, Y.; Su, B.-L. Probing significant light absorption enhancement of titania inverse opal films for highly exalted photocatalytic degradation of dye pollutants. *Appl. Catal. B Environ.* **2014**, *150–151*, 411–420. [CrossRef]
66. Chung, W.A.; Hung, P.S.; Wu, C.J.; Guo, W.Q.; Wu, P.W. Fabrication of composite Cu_2O/Au inverse opals for enhanced detection of hydrogen peroxide: Synergy effect from structure and sensing mechanism. *J. Alloys Compounds* **2021**, *886*, 161243–161255. [CrossRef]
67. Chen, J.I.L.; Loso, E.; Ebrahim, N.; Ozin, G.A. Synergy of Slow photonic and Chemically Amplified Photochemistry in Platinum Nanocluster-Loaded Inverse Titania Opals. *J. Am. Chem. Soc.* **2008**, *130*, 5420–5421. [CrossRef]
68. Huang, X.; Hu, Y.; Zhou, L.; Lei, J.; Wang, L.; Zhang, J. Fabrication of CuS-modified inverse opal g-C_3N_4 photocatalyst with enhanced performance of photocatalytic reduction of CO_2. *J. CO2 Util.* **2021**, *54*, 101779. [CrossRef]
69. Sun, Q.; Zhang, B.; He, Y.; Sun, L.; Hou, P.; Gan, Z.; Yu, L.; Dong, L. Design and synthesis of black phosphorus quantum dot sensitized inverse opal TiO_2 photonic crystal with outstanding photocatalytic activities. *Appl. Surf. Sci.* **2023**, *609*, 155442. [CrossRef]
70. Hu, Z.; Xu, L.; Wang, L.; Huang, Y.; Xu, L.; Chen, J. One-step fabrication of N-doped TiO_2 inverse opal films with visible light photocatalytic activity. *Catal. Commun.* **2013**, *40*, 106–110. [CrossRef]
71. Zhang, W.J.; Zhang, X.; Zhang, Z.; Wang, W.; Xie, A.; Xiao, C.; Zhang, H.; Shen, Y. A Nitrogen-Doped Carbon Dot-Sensitized TiO_2 Inverse Opal Film: Preparation, Enhanced Photoelectrochemical and Photocatalytic Performance. *J. Electrochem. Soc.* **2015**, *162*, H638–H644. [CrossRef]

72. Qi, D.; Lu, L.; Xi, Z.; Wang, L.; Zhang, J. Enhanced photocatalytic performance of TiO$_2$ based on synergistic effect of Ti^{3+} self-doping and slow light effect. *Appl. Catal. B Environ.* **2014**, *160–161*, 621–628. [CrossRef]
73. Zhou, L.; He, H.; Tao, M.; Muhammad, Y.; Gong, W.; Liu, Q.; Zhao, Z.; Zhao, Z. Chloroplast-inspired microenvironment engineering of inverse opal structured IO-TiO$_2$/Chl/IL for highly efficient CO$_2$ photolytic reduction to CH$_4$. *Chem. Eng. J.* **2023**, *464*, 142685. [CrossRef]
74. Liu, W.; Li, X.; Qi, K.; Wang, Y.; Wen, F.; Wang, J. Novel inverse opal Bi$_2$WO$_6$/Bi$_2$O$_3$ S-scheme heterojunction with efficient charge separation and fast migration for high activity photocatalysis. *Appl. Surf. Sci.* **2023**, *607*, 155085. [CrossRef]
75. Fang, J.; Fan, H.; Li, M.; Long, C. Nitrogen self-doped graphitic carbon nitride as efficient visible light photocatalyst for hydrogen evolution. *J. Mater. Chem. A* **2015**, *3*, 13819–13826. [CrossRef]
76. Prabhu, S.; Pudukudy, M.; Sohila, S.; Harish, S.; Navaneethan, M.; Navaneethan, D.; Ramesh, R.; Hayakawa, Y. Synthesis, structural and optical properties of ZnO spindle/reduced graphene oxide composites with enhanced photocatalytic activity under visible light irradiation. *Opt. Mater.* **2018**, *79*, 186–195. [CrossRef]
77. Hu, J.; Zhang, P.; An, W.; Liu, L.; Liang, Y.; Cui, W. In-situ Fe-doped g-C$_3$N$_4$ heterogeneous catalyst via photocatalysis-Fenton reaction with enriched photocatalytic performance for removal of complex wastewater. *Appl. Catal. B Environ.* **2019**, *245*, 130–142. [CrossRef]
78. Li, J.; Peng, T.; Zhang, Y.; Zhou, C.; Zhu, A. Polyaniline modified SnO$_2$ nanoparticles for efficient photocatalytic reduction of aqueous Cr(VI) under visible light. *Sep. Purif. Technol.* **2018**, *210*, 120–129. [CrossRef]
79. Ahmed, A.; Siddique, M.N.; Alam, U.; Ali, T.; Tripathi, P. Improved photocatalytic activity of Sr doped SnO$_2$ nanoparticles: A role of oxygen vacancy. *Appl. Surf. Sci.* **2019**, *463*, 976–985. [CrossRef]
80. Zhang, Y.; Wang, L.; Liu, D.; Gao, Y.; Song, C.; Shi, Y.; Qu, D.; Shi, J. Morphology effect of honeycomb-like inverse opal for efficient photocatalytic water disinfection and photodegradation of organic pollutant. *Mol. Catal.* **2018**, *444*, 42–52. [CrossRef]
81. Wan, Y.; Wang, J.; Wang, X.; Xu, H.; Yuan, S.; Zhang, Q.; Zhang, M. Preparation of inverse opal titanium dioxide for photocatalytic performance research. *Opt. Mater.* **2019**, *96*, 109287. [CrossRef]
82. Ogden, J.M.; Williams, R.H. Electrolytic hydrogen from thin-film solar cells. *Int. J. Hydrogen Energy* **1990**, *15*, 155–169. [CrossRef]
83. Aydin, E.B.; Ates, S.; Sigircik, G. CuO-TiO$_2$ nanostructures prepared by chemical and electrochemical methods as photo electrode for hydrogen production. *Int. J. Hydrogen Energy* **2022**, *47*, 6519–6534. [CrossRef]
84. Fiorenza, R.; Bellardita, M.; Scirè, S.; Palmisano, L. Photocatalytic H$_2$ production over inverse opal TiO$_2$ catalysts. *Catal. Today* **2019**, *321–322*, 113–119. [CrossRef]
85. Lv, C.; Wang, L.; Liu, X.; Zhao, L.; Lan, X.; Shi, J. An efficient inverse opal (IO)-TiO$_2$-MoO$_{3-x}$ for photocatalytic H$_2$ evolution and RhB degradation—The synergy effect of IO structure and plasmonic MoO$_{3-x}$. *Appl. Surf. Sci.* **2020**, *527*, 146726. [CrossRef]
86. Liu, Y.; Xu, G.; Ma, D.; Li, Z.; Yan, Z.; Xu, A.; Zhong, W.; Fang, B. Synergistic effects of g-C$_3$N$_4$ three-dimensional inverse opals and Ag modification toward high-efficiency photocatalytic H$_2$ evolution. *J. Clean. Prod.* **2021**, *328*, 129745. [CrossRef]
87. Habisreutinger, S.N.; Schmidt-Mende, L.; Stolarczyk, J.K. Photocatalytic reduction of CO$_2$ on TiO$_2$ and other semiconductors. *Angew. Chem.* **2013**, *52*, 7372–7408. [CrossRef]
88. Wei, Y.; Jiao, J.; Zhao, Z.; Liu, J.; Li, J.; Jiang, G.; Wang, Y.; Duan, A. Fabrication of inverse opal TiO$_2$-supported Au@CdS core-shell nanoparticles for efficient photocatalytic CO$_2$ conversion. *Appl. Catal. B. Environ. Int. J. Devoted Catal. Sci. Its Appl.* **2015**, *179*, 422–423. [CrossRef]
89. Jiating, X.; Xiaohan, Z. Efficient photocatalytic reduction of CO$_2$ by a rhenium-doped TiO$_2$-x/SnO$_2$ inverse opal S-scheme heterostructure assisted by the slow-phonon effect. *Separat. Purificat. Technol.* **2021**, *277*, 119431. [CrossRef]
90. Chen, B.; Zhou, L.; Tian, Y.; Yu, J.; Lei, J.; Wang, L.; Liu, Y.; Zhang, J. Z-scheme Inverse Opal CN/BiOBr Photocatalyst for Highly Efficient Degradation of Antibiotics. *Phys. Chem. Chem. Phys.* **2019**, *21*, 12818–12825. [CrossRef]

Disclaimer/Publisher's Note: The statements, opinions and data contained in all publications are solely those of the individual author(s) and contributor(s) and not of MDPI and/or the editor(s). MDPI and/or the editor(s) disclaim responsibility for any injury to people or property resulting from any ideas, methods, instructions or products referred to in the content.

Review

Recent Progress in Source/Drain Ohmic Contact with β-Ga₂O₃

Lin-Qing Zhang [1], Wan-Qing Miao [1], Xiao-Li Wu [1], Jing-Yi Ding [1], Shao-Yong Qin [1], Jia-Jia Liu [1], Ya-Ting Tian [1], Zhi-Yan Wu [1,*], Yan Zhang [1], Qian Xing [2,*] and Peng-Fei Wang [3]

[1] College of Electronic and Electrical Engineering, Henan Normal University, No. 46 East of Construction Road, Xinxiang 453007, China; 14110720069@fudan.edu.cn (L.-Q.Z.); lxjhjy0815@163.com (J.-J.L.)
[2] College of Intelligent Engineering, Henan Institute of Technology, No. 699 Pingyuan Road (East Section), Xinxiang 453003, China
[3] State Key Laboratory of ASIC and System, School of Microelectronics, Fudan University, No. 220 Han Dan Road, Shanghai 200433, China
* Correspondence: 2018010@htu.edu.cn (Z.-Y.W.); xingqian8911@163.com (Q.X.)

Abstract: β-Ga₂O₃, with excellent bandgap, breakdown field, and thermal stability properties, is considered to be one of the most promising candidates for power devices including field-effect transistors (FETs) and for other applications such as Schottky barrier diodes (SBDs) and solar-blind ultraviolet photodetectors. Ohmic contact is one of the key steps in the β-Ga₂O₃ device fabrication process for power applications. Ohmic contact techniques have been developed in recent years, and they are summarized in this review. First, the basic theory of metal–semiconductor contact is introduced. After that, the representative literature related to Ohmic contact with β-Ga₂O₃ is summarized and analyzed, including the electrical properties, interface microstructure, Ohmic contact formation mechanism, and contact reliability. In addition, the promising alternative schemes, including novel annealing techniques and Au-free contact materials, which are compatible with the CMOS process, are discussed. This review will help our theoretical understanding of Ohmic contact in β-Ga₂O₃ devices as well as the development trends of Ohmic contact schemes.

Keywords: β-Ga₂O₃; Ohmic contact; ion implantation; interface; annealing temperature

1. Introduction

Si-based devices are the dominant devices used for power applications. However, with the increasing demand for much faster and more convenient network communication, Si-based device techniques cannot meet these requirements due to their physical properties. Thus, new-material devices should be investigated for operating at high temperatures, at high power, and in harsh environments. In recent years, wide-bandgap semiconductors including GaN (3.4 eV) and SiC (3.25 eV) have been developed, and they have replaced Si-based techniques in many fields due to their advantages in terms of their material properties [1–5]. Recently, β-Ga₂O₃, which is mostly thermally and chemically stable in five polymorphs [6–8], has attracted more and more attention for power applications because β-Ga₂O₃ has a wide bandgap of 4.6–4.9 eV and a breakdown field strength as high as 8 MV/cm [9–11]. In addition, for the Baliga figure and Johnson's figure of merit, when evaluating its application potential in power devices, β-Ga₂O₃ exhibits the best performance [9–11]. The basic physical properties and figures of merit (FOM) of commonly used semiconductor materials are shown in Table 1. For this reason, researchers have obtained plenty of results related to β-Ga₂O₃-based FETs [12–14], SBDs [15–18], and solar-blind ultraviolet photodetectors [19–21]. In 2023, Wang et al. [13] demonstrated a metal–heterojunction composite field-effect transistor that exhibited a breakdown voltage (BV) of around 2160 V. In addition, the corresponding $R_{ON,SP}$ was 6.35 mΩ·cm². So far, the power figure of merit (P-FOM) achieved the highest value of 0.73 GW/cm² for e-mode β-Ga₂O₃ devices. For SBDs, Hao et al. [17] used an optimized p-type NiO (with a hole concentration of 10^{17} cm^{-3}) junction-termination extension (JTE) technique that exhibited a BV and $R_{ON,SP}$ of 2.11 kV and 2.9 m$\Omega\cdot$cm^2, respectively. For this reason, the P-FOM was as high as 1.54 GW/cm^2. For the junction barrier Schottky (JBS) diode, Wu et al. [18] fabricated a device with a well-designed field plate to suppress the crowding effect of the electric field. The forward current and BV could reach 5.1 A and 1060 V, respectively. At the circuit level, the hybrid circuit exhibited more efficiency compared with the Si-based one. An R_{ON}-BV benchmark comparison of β-Ga$_2$O$_3$-based devices with other published results was also presented in their work. The R_{ON}-BV characteristics for β-Ga$_2$O$_3$-based devices were comparable with GaN-based ones [22]. To fully exploit β-Ga$_2$O$_3$'s potential in power electronics applications, the material quality, device structure, and process details should be further optimized. By embedding indium tin oxide (ITO) electrodes, Zhang et al. [19] fabricated a fully transparent MSM-structured solar-blind UV photodetector with an excellent dark current, normalized photocurrent-to-dark-current ratio (NPDR), responsivity, rejection ratio, and specific detectivity characteristics. Another advantage of β-Ga$_2$O$_3$ material is that single large β-Ga$_2$O$_3$ crystals can be cost-effectively mass produced using melt–growth methods, such as EFG [23], FZ [24,25], VB [26,27], and CZ [28,29] methods. Additionally, a high-quality β-Ga$_2$O$_3$ epilayer can be realized using MOCVD [30,31], MBE [32,33], HVPE [34,35], and MOVPE [36,37] methods to form a well-controlled n-type doping using Si, Ge, and Sn. However, the p-type doping technique is still challenging because the activation energy of the acceptors and the self-trapping energy of the holes are large [38,39]. For the purpose of achieving p-type Ga$_2$O$_3$, great efforts have been taken by researchers from all over the world [40–51].

Table 1. Physical properties and FOMs of the commonly used semiconductors.

Parameters	Si	GaAs	4H-SiC	GaN	β-Ga$_2$O$_3$
Bandgap, E_G (eV)	1.12	1.43	3.25	3.4	4.6–4.9
Breakdown field, E_{br} (MV/cm)	0.3	0.4	2.5	3.3	8
Electron mobility, μ (cm^2 V^{-1} s^{-1})	1480	8400	1000	2000 (2DEG)	300
Saturation velocity, V_s (10^7 cm/s)	1	1.2	2	2.5	1.8–2
BFOM, $\varepsilon\mu E_{br}^3$	1	14.7	317	846	2000–3000
JFOM, $E_{br}^2 V_s^2/(4\pi^2)$	1	1.8	278	1089	2844

For power applications, Ohmic contact is one of the key steps in β-Ga$_2$O$_3$ device fabrication processes. Ohmic contact resistance (R_C), specific contact resistance (ρ_c), and thermal stability are important indexes of contact quality. A lower R_C can reduce voltage drop across the contact region and power loss. For GaN-based devices, Au-free low-temperature Ohmic contact techniques are proposed to realize CMOS-compatible and gate-first techniques [52,53]. Until now, because of the wide-bandgap property of β-Ga$_2$O$_3$ and Fermi-level pinning [54–56], Ohmic contact methods for Ga$_2$O$_3$-based devices have remained challenging. The metal schemes, annealing conditions (the annealing temperature, durations, and atmosphere), and doping concentration of Ga$_2$O$_3$ have been investigated and optimized to obtain low-R_C contact. In this review, we will first give a brief introduction of metal–semiconductor contact theory. After that, the state-of-the-art advances in Ohmic contact techniques for β-Ga$_2$O$_3$ will be presented and discussed, including metal electrodes, surface treatments, ion implantation, epitaxial regrowth, and adding an interlayer. Finally, we will give some perspectives for further studies on Ohmic contact with β-Ga$_2$O$_3$ in the future.

2. Basic Metal–Ga$_2$O$_3$ Contact Physical Theory

Metal–semiconductor contact is a critical part of β-Ga$_2$O$_3$ power devices. A device's performance is mainly limited by the Ohmic contact property. Two types of contacts (Schottky and Ohmic) can be formed due to the differences in the work functions of contact metals [57–59]. For wide-bandgap β-Ga$_2$O$_3$, the contacts always exhibit Schottky behavior. When metal and Ga$_2$O$_3$ come into contact, the energy band of the Ga$_2$O$_3$ side bends up to make their Fermi levels equal. As shown in Figure 1a, the Schottky barrier height from the metal side (Φ_B) can be described as

$$\Phi_B = \Phi_m - \chi,$$

where χ represents the semiconductor's electron affinity (4 eV for Ga_2O_3 in our case [60]) and Φ_m represents the metal work function. Therefore, it is desirable to select a metal with a Φ_m lower than 4 eV to realize a negative Φ_B, which allows electrons to flow freely across it to form an Ohmic contact. Unfortunately, the lack of suitable metal materials with lower work functions makes Ohmic contact formation challenging. Generally, researchers have proposed Ohmic contact schemes to form a lower Φ_B or an n^+-doped Ga_2O_3 region for electron tunneling. When a semiconductor is heavily doped ($N_D > 10^{18}$ cm^{-3}), field emission (FE) dominates the electron tunneling [61,62]. In order to obtain a low R_C or ρ_c, a higher N_D is expected.

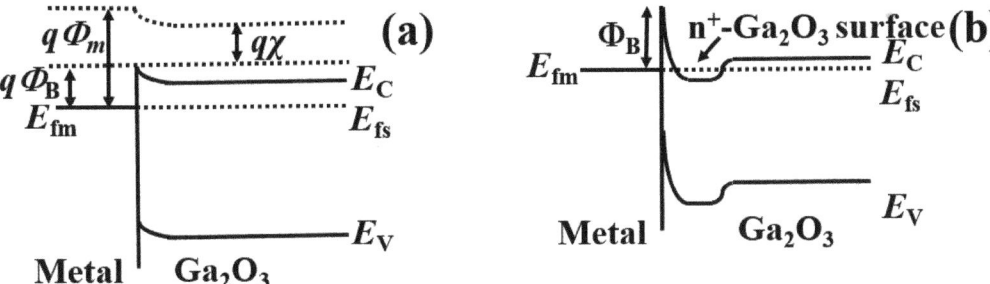

Figure 1. Energy-band diagrams of metal–Ga_2O_3 Schottky contacts with (**a**) a lower Φ_{bn} and (**b**) an n^+-doped Ga_2O_3 region.

The Ohmic contact resistance (R_C, measured in $\Omega \cdot$mm) and specific contact resistance (ρ_c, measured in $\Omega \cdot$cm^{-2}) are always determined using the transmission line model (TLM) method [63,64]. Details concerning the TLM measurement technique can be seen in the references mentioned above.

The metal work function, metal schemes, interfacial reactions between metal and a semiconductor during the annealing process, and the doping concentration of Ga_2O_3 in the source/drain region are significant influencing factors for the Ohmic contact property. Until now, researchers from universities and research institutes have proposed Ohmic contact schemes involving optimizing the metal materials, annealing condition, source/drain doping method and concentration, and source/drain etching as well as adding an interlayer in the source/drain region.

3. Approaches to Metal–Ga_2O_3 Ohmic Contact

3.1. Metal Electrode

From the metal–semiconductor contact theory, the work function of the selected metal material crucially affects the Ohmic contact quality. Thus, in the early period, Yao et al. [65] investigated the Ohmic contact properties and surface morphologies of nine metal materials, including Ti, In, Ag, Sn, W, Mo, Sc, Zn, and Zr with Sn-doped ($\bar{2}01$) β-Ga_2O_3. From their results, the work function is not the main factor influencing the contact quality. Sc, with the lowest work function, cannot form Ohmic contacts under different annealing conditions. Ti/Au metal schemes with a 400 °C annealing process exhibited the lowest R_C values. Cross-sectional transmission electron microscopy (TEM) and energy-dispersive X-ray spectroscopy (EDX) mapping showed that Ga and O diffused into the Ti layer during the annealing process. They concluded that interfacial reactions during the annealing process played a crucial part in Ohmic contact formation. Otherwise, the ultra-wide bandgap property leads to a pinning effect due to defects and surface states that lie in the mid-gap, which are not beneficial for forming an Ohmic contact.

Other groups have reported Mg/Au and Cr/Au metal schemes for Ohmic contact formation [66,67]. In the Mg (3.66 eV)/Au method, the ρ_c of 2.2×10^{-4}~2.1×10^{-5} $\Omega \cdot cm^{-2}$ was achieved with an annealing process at temperatures varying from 300 °C to 500 °C. Until now, the most common metallization schemes of Ohmic contact for β-Ga$_2$O$_3$ have used Ti/Au. Ti is used as an adhesion layer with a low work function. Au serves as a cap layer to prevent the oxidation of metal stacks during the high-temperature process. For the purpose of understanding the mechanism, Lee et al. [68] deposited Ti/Au (20/80 nm) on a Sn-doped β-Ga$_2$O$_3$ (010) substrate and carried out a 470 °C rapid thermal annealing (RTA) process for 1 min to form an Ohmic contact. Scanning transmission electron microscopy (STEM), high-resolution transmission electron microscopy (HRTEM), and EDX measurements were taken to understand the interfacial reactions and components. They found that a defective β-Ga$_2$O$_3$ layer (3–5 nm), a Ti–TiO$_x$ layer (3–5 nm), and an intermixed Au–Ti layer containing Ti-rich nanocrystalline inclusions were formed sequentially, as shown in Figure 2. They deduced that the Ti–TiO$_x$ layer (3–5 nm) with a small bandgap could provide an efficient path for the electron flow. In addition, the lattice matching between the defective β-Ga$_2$O$_3$ layer and the β-Ga$_2$O$_3$ substrate could enhance the carrier mobility by reducing the collision probability, resulting in a lower R_C. Before this work, Higashiwaki et al. [69] showed TEM results for an interface and deduced that interface reactions help improve contact quality.

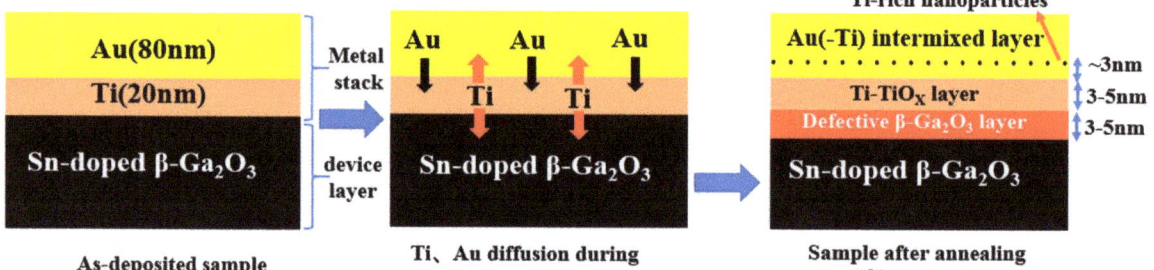

Figure 2. Schematic illustrations of Ti/Au metallization layers on Sn-doped β-Ga$_2$O$_3$ with a 1 min 470 °C N$_2$ annealing process. Reproduced from Ref. [68].

Also, multilayer metal contact schemes were proposed for obtaining lower R_C values, such as Ti/Al/Au [70,71], Ti/Al/Ni/Au [72,73], and Ti/Au/Ni [32,33]. As can be seen in Figure 3, Krishnamoorthy et al. formed a δ-doped β-Ga$_2$O$_3$ structure in the source/drain region to form a heavily doped contact area. After Ti/Au/Ni deposition, a 470 °C RTA process was employed for 1 min for Ohmic contact formation. The extracted R_C and ρ_c were 0.35 $\Omega \cdot mm$ and 4.3×10^{-6} $\Omega \cdot cm^{-2}$, respectively. In addition, the fabricated β-Ga$_2$O$_3$ FET exhibited excellent I_D and g_m results. By using Ti/Al/Au contact metals, Zhou et al. [71] achieved a low R_C of 0.75 $\Omega \cdot mm$ by adopting a highly Sn-doped channel. For AlGaN/GaN HEMT, Ti/Al/Ni/Au is one of the most mature metal schemes for Ohmic contact formation [74,75]. For β-Ga$_2$O$_3$ devices, Chen et al. [73] deposited Ti/Al/Ni/Au multilayer metal stacks and carried out an RTA process with the temperature at 470 °C for 70 s. By analyzing the X-ray photoelectron spectroscopy (XPS) results, as shown in Figure 4, they concluded that the use of Al can lead to the formation of a Ti–Al phase with a low work function, which is beneficial for oxygen vacancy generation at the interface. In n-type β-Ga$_2$O$_3$, the vacancies act as donors, enhancing the electron flow and realized Ohmic contact.

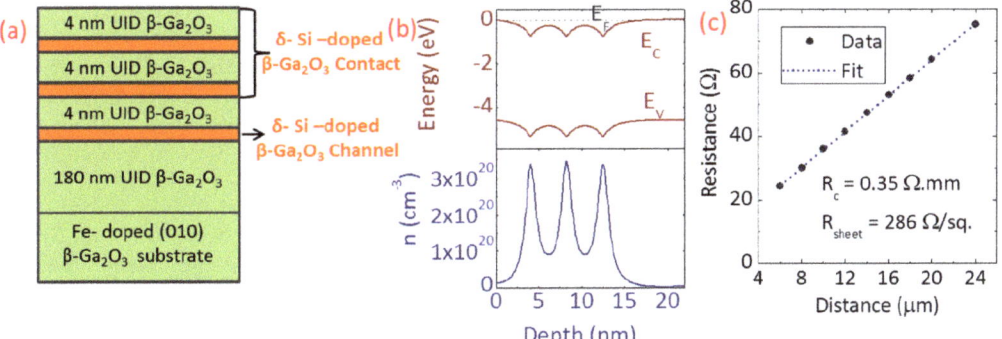

Figure 3. (**a**) Device structure, (**b**) equilibrium band diagram and charge profile, and (**c**) TLM results. Reproduced from Ref. [33].

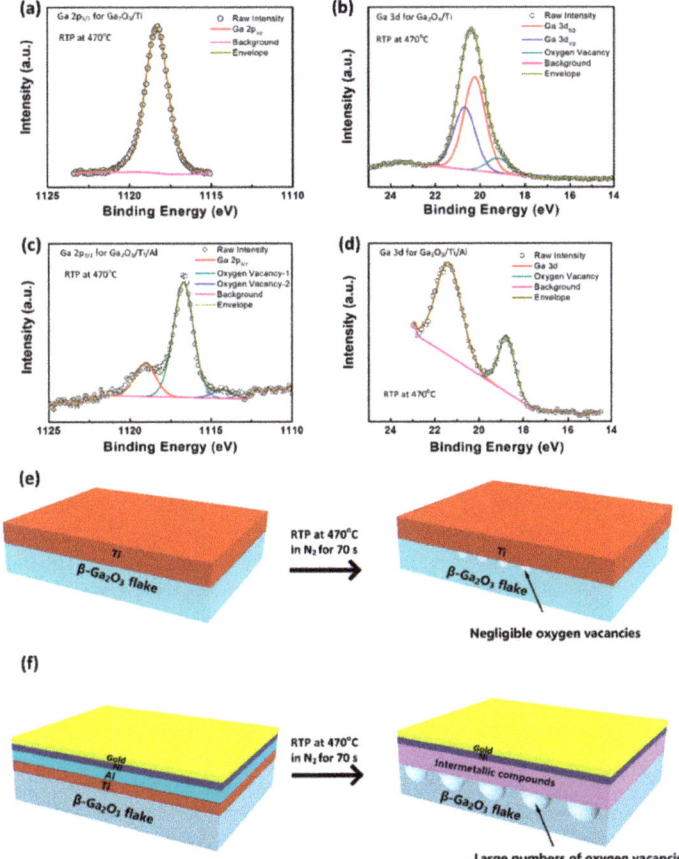

Figure 4. XPS results of (**a**) Ga $2p_{3/2}$ and (**b**) Ga 3d core-level spectra from the Ti-coated (~2.5 nm) β-Ga_2O_3 sample. (**c**) Ga $2p_{3/2}$ and (**d**) Ga 3d core-level spectra for the Ti/Al-coated (2/2 nm) β-Ga_2O_3 sample. (**e**) Schematic diagram of the role of Ti in the generation of oxygen vacancies. (**f**) Schematic of the formation process of oxygen vacancies at the interface of β-Ga_2O_3 and metal. Reproduced from Ref. [73].

In 2022, Tetzner et al. [76] used a TiW alloy instead of the traditional Ti/Au metal schemes, and a low ρ_c of 1.5×10^{-5} $\Omega \cdot cm^{-2}$ was extracted after a 700 °C RTA process. The temperature was 200 °C higher than that of the Ti/Au schemes. To understand the Ohmic contact formation mechanism, STEM HAADF and EDX were employed. The STEM HAADF image showed that a 3–5 nm TiO$_X$ interlayer was formed, which was confirmed with the STEM EDX. They suspected that vacancies, defects, or Ga impurities that exist in the interlayer are beneficial for electrons flowing freely to reduce the R_C.

Thermal stability is another important index of contact quality. For Ti/Au electrodes, the most commonly used annealing temperatures are between 400 °C to 500 °C. Above 500 °C, Yao et al. [65] found that the Ohmic contacts degraded in their results. In 2022, Lee et al. [77] systematically investigated the influence of temperature on Ohmic contact quality. In their results, when the annealing temperature increased from 470 °C to 520 °C, aggressive Au diffused into the interface and reacted with Ga that diffused out, resulting in a much thicker Ti–TiO$_x$ layer due to GaAu$_2$ formation, which accounted for the contact degradation. In Kim's [78] results, the R_C increased when the temperature changed from 400 °C to 500 °C or 600 °C. They deduced that this could have been due to an increased amount of Ti oxide. Related investigations have also been conducted and reported [79–81]. Therefore, more research into interfacial reactions for Ti/Au schemes and alternative metallization schemes, including Au-free electrodes, should be proposed to solve the instability issue of the Ti/β-Ga$_2$O$_3$ interface using Ti/Au metal schemes. It should be noted that excellent Ohmic contact cannot be achieved just by selecting metal materials. Combined with other techniques, including surface treatment, ion implantation, epitaxial regrowth, adding an interlayer, etc., the contact quality can be improved and optimized.

3.2. Surface Treatment

A surface treatment before metal deposition can also help improve the Ohmic contact property (dry etching, plasma bombardment, etc.). In 2012, Higashiwaki et al. [82] compared the I–V results of Ga$_2$O$_3$ devices with and without the RIE treatment. The RIE process was implemented by using a BCl$_3$/Ar mixing gas for 1 min before Ti/Au (20/230 nm) deposition. The samples with the RIE treatment exhibited Ohmic contact characteristics, while without the RIE process the samples showed Schottky contact features. They speculated that the Ohmic contact formation was due to the large number of oxygen-vacancy surface defects formed during the RIE process. The defects acted as donors for Ohmic contact realization. Combined with Si ion implantation [83], they achieved a ρ_c of 4.6×10^{-6} $\Omega \cdot cm^{-2}$ with a doping concentration of 5×10^{19} cm^{-3}. In addition, Zhou et al. [70] performed an Ar plasma bombardment process and optimized the duration of 30 s for generating oxygen vacancies, which are good for n-type surface doping. The mechanism was similar to that of the BCl$_3$/Ar RIE process. The R_C values in their results were as low as 0.95 $\Omega \cdot mm$. Related results have also been reported by other groups [33,69,71,84–86].

Also, the annealing temperature and atmosphere may affect the interfacial reactions that dominate Ohmic contact formation. Bae et al. [87] compared the electrical results of the fabricated β-Ga$_2$O$_3$ nanobelts under different atmospheres with various temperatures. The samples treated under a N$_2$ atmosphere exhibited better characteristics that the ones treated in an air environment. Under an Ar atmosphere, Li et al. [88] reduced the R_C to 0.387 $\Omega \cdot mm$ by optimizing the annealing temperature and the durations. In their results, a large drain current density of ~3.1 mA/μm (V_{ds} = 100 V) was achieved due to the low R_C. To fully understand the influence of an Ar atmosphere on improving the β-Ga$_2$O$_3$ device's I–V characteristics, XPS was used to show the material changes during the RTA process. From the results, as can be seen in Figure 5, they deduced that Ti reduced β-Ga$_2$O$_3$ and generated large numbers of oxygen vacancies at the interface during the annealing process, which served as effective electron donors. For this reason, the depletion layer was narrower, resulting in Ohmic behavior and a low R_C for β-Ga$_2$O$_3$ FETs. The annealing temperature is another element that affects the interface reactions to determine the Ohmic contact property. In 2022, Lee et al. [77] systematically investigated the influence of temperature (from 370 °C to 520 °C) on the Ohmic contact property. The lowest R_C occurred when temperature was 420 °C. When the temperature increased, the R_C increased as well. To investigate the reason for the degradation of the R_C with an increasing temperature, cross-sectional S/TEM was employed for a sample with an annealing temperature of 520 °C. Their results show that the thickness of the Ti–TiO$_x$ layer (25–30 nm) increased due to the formation of GaAu$_2$ inclusions, which was caused by Au aggressively diffusing in and its reaction with Ga that had diffused out. In their early results [68], a thin Ti–TiO$_x$ layer was beneficial for electron transport. The degradation of contact quality was the result of the increasing Ti–TiO$_x$ layer thickness. Also, in earlier results, the degradation of contact characteristics was observed when the annealing process was performed above 500 °C [65]. Yao et al. speculated that Ti reduces Ga$_2$O$_3$, possibly forming an insulating oxide layer at the interface, which would account for the Ohmic contact degradation. In their results, the optimized annealing temperature was 400 °C, which achieved the lowest R_C value.

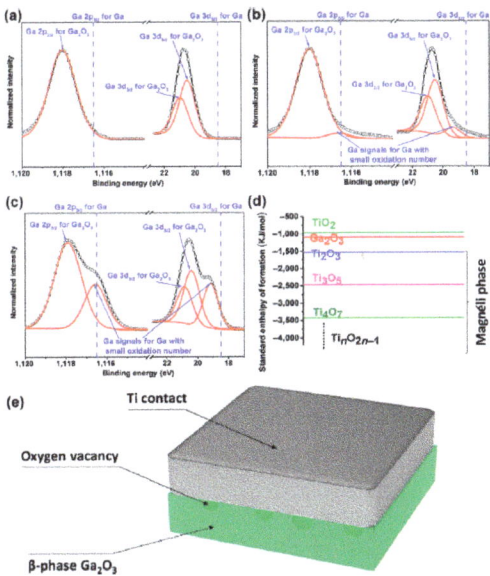

Figure 5. XPS results from β-Ga$_2$O$_3$. (**a**) Normalized Ga 2p$_{3/2}$ XPS spectra and Ga 3d XPS spectra from pure β-Ga$_2$O$_3$, (**b**) β-Ga$_2$O$_3$ after annealing in argon at 300 °C for 180 min, and (**c**) Ti-coated (1 nm) β-Ga$_2$O$_3$ after annealing in argon at 300 °C for 180 min. Black dots show experimental data, and red curves show simulated fitting curves. (**d**) Free energy scheme of different metal oxides. (**e**) Schematic diagram of the proposed oxygen vacancy model at the Ti/β-Ga$_2$O$_3$ interface. Reproduced from Ref. [88].

Surface treatment, including BCl$_3$/Ar RIE and Ar plasma bombardment, before metal deposition can help to reduce the R_C to a degree. During these processes, the accelerated high-energy ions react with Ga$_2$O$_3$ via physical and chemical methods, creating defects to form a highly damaged surface, which enables high recombination rates. However, excellent Ohmic contact cannot be achieved only using such methods. Techniques, including RIE, ion implantation, RTA, etc., are always used together to improve the Ohmic contact quality. In addition, the RIE technique is not always reproducible or practically applicable due to the undesired damage induced during semiconductor processing.

3.3. Ion Implantation

The ion-implantation doping technique (including Si, Sn, etc.) is another effective way for Ga$_2$O$_3$ to realize low-contact-resistance Ohmic electrodes by forming a heavily doped n$^+$ region that facilitates electron flow. In 2013, Sasaki et al. [83] successfully fabricated Ohmic electrodes with a low contact resistance via Si implantation, which requires the MOVPE method in β-Ga$_2$O$_3$. They optimized the Si doping concentration to 5×10^{19} cm^{-3}, which was activated by annealing at a temperature of 950 °C. The R$_C$ and ρ$_c$ in their results were as low as 1.4 mΩ·cm and 4.6×10^{-6} Ω·cm^{-2}, respectively. In other results, Zhou et al. [70] doped β-Ga$_2$O$_3$ with Sn at a concentration of 2.7×10^{18} cm^{-3}. Combined with Ar plasma bombardment, the R$_C$ was dramatically reduced to 0.95 Ω·mm. The fabricated devices also exhibited an excellent on/off ratio and output characteristics and a low SS value. In addition, Ge and Sn were also studied for doping β-Ga$_2$O$_3$ [89]. In that study, the samples were treated under the same annealing condition (925 °C for 30 min). The efficiencies of Sn, Ge, and Si were calculated to be 28.2%, 40.3%, and 64.7%, respectively, using SIMS measurements. The same activation annealing condition for Ge and Sn with Si resulted in low activation efficiencies for Ge and Sn. The heavier Ge and Sn ions also created more implant damage than the Si ions due to the greater momentum transfer required to achieve the same implant depth, likely contributing to decreased implant activation and increasing both the contact and sheet resistances. In 2023, Tetzner et al. [90] analyzed the optimized annealing temperature for the activation of Ge-implanted β-Ga$_2$O$_3$ from 900 to 1200 °C using a pulsed RTA technique. The lowest recorded ρ$_c$ value of 4.8×10^{-7} Ω cm^{-2} was achieved after a pulsed RTA at 1100 °C using 40 pulses. The activation efficiency was 14.2%. The measured R$_C$ and ρ$_c$ values at various annealing temperatures can be seen in Figure 6. Also, other representative studies related to the ion implantation technique have been reported [91–94].

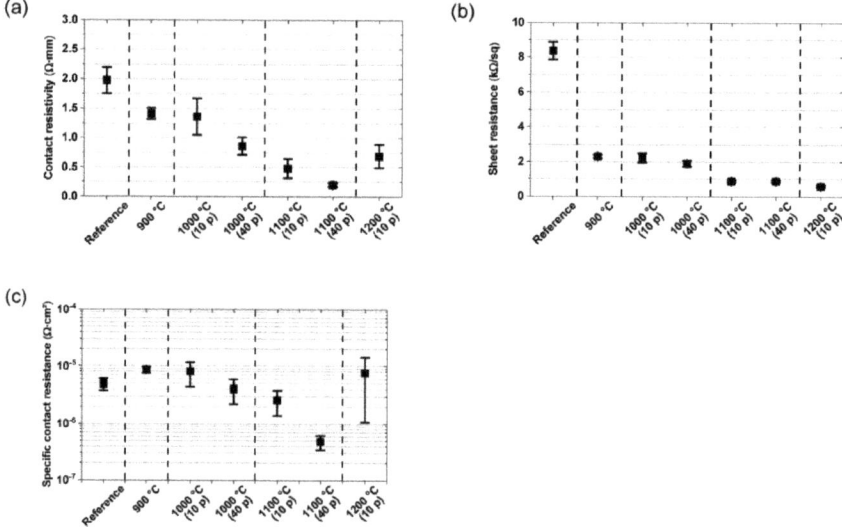

Figure 6. Measured contact resistivity (**a**), R$_C$ (**b**), and specific contact resistance (**c**) as a function of the annealing conditions. Reproduced from Ref. [90].

Considering the high cost of ion implantation, the complicated steps, and the potential damage-induced diffusion of species, Zeng et al. [95] successfully proposed a Sn spin-on-glass (SOG) technique for β-Ga$_2$O$_3$ doping. A Sn-doped epitaxial layer with a doping density of 1×10^{18} cm^{-3} was formed on a Ga$_2$O$_3$ substrate. The obtained ρ$_c$ was determined to be $2.1 \pm 1.4 \times 10^{-5}$ Ω·cm^{-2} in their results. As shown in Figure 7, the fabricated devices also exhibited improved output current, peak transconductance, on/off ratio, and breakdown voltage values. The SOG technique is an effective alternative to the simple, low-cost doping technique to make low-R$_C$ Ohmic contact. Thus, based on the existing results using doping techniques, it is possible to form a heavily doped Ga$_2$O$_3$ layer and obtain ρ$_c$ values from 10^{-5} to 10^{-7} Ω·cm^{-2}.

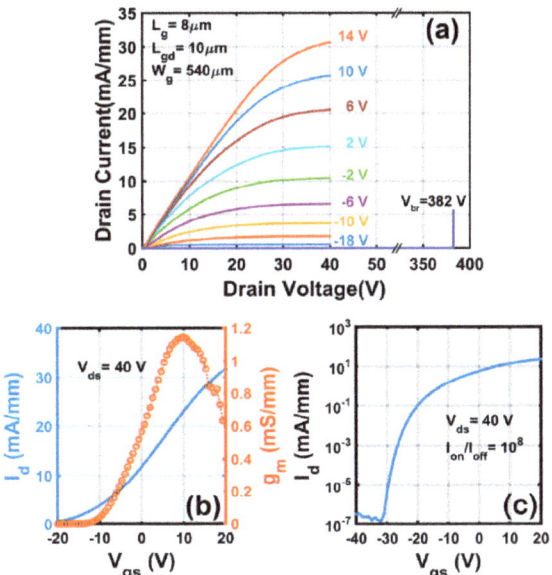

Figure 7. (**a**) Output characteristics of the SOG-doped MOSFET and (**b**,**c**) linear and log-scale transfer characteristics of the same device. Reproduced from Ref. [95].

Ion implantation, surface treatment, and post-RTA annealing are always used together to obtain a low R_C. Ion implantation can form a heavily doped interface to enhance electron tunneling. A surface treatment combined with RTA can generate oxygen vacancies that act as donors in Ga_2O_3, resulting in a low R_C. For ion implantation, a high-temperature post-anneal is required to activate the implanted donor impurity and recover the induced crystalline damage. During the high-temperature process, dopant redistribution, residuals, crystalline defects, and incomplete activation should be noticed and optimized.

3.4. Epitaxial Regrowth

To further reduce the contact resistance, regrown contacts have been reported to fabricate Ohmic contacts. Ion implantation and spin-on-glass techniques need a high annealing temperature around 900–1200 °C and potentially deteriorate the material quality in the active region. However, the regrowth process, which is performed at a much lower temperature of about 600 °C can avoid this potential problem. In 2018, Xia et al. [32] used a molecular-beam epitaxy (MBE) method to form a heavily doped n-type Ga_2O_3 with a doping concentration of 2×10^{20} cm^{-3}. The device's structure can be seen in Figure 8. An extracted R_C of 1.5 Ω·mm was obtained from the TLM structure. The regrowth technique avoids gate recessing and potential damage associated with etching, which may degrade the carrier mobility. The fabricated devices exhibited a peak drain current of 140 mA/mm and an excellent transconductance of 34 mS/mm. Considering the advantage of high room-temperature electron mobility values (close to the theoretical limit) grown using metalorganic vapor phase epitaxy (MOVPE), Bhattacharyya et al. [36] proposed an MOVPE epitaxy approach to realize low-resistance regrown S/D contacts in a Ga_2O_3 lateral MESFET for the first time. As shown in Figure 9, the heavily Si-doped (~1.8×10^{20} cm^{-3}) Ga_2O_3 was grown using MOVPE at a relatively low temperature of 600 °C. After that, an Ohmic metal stack of Ti/Au/Ni (20 nm/100 nm/30 nm) was evaporated, followed by 470 °C annealing in N_2. From their testing results, an ultralow R_C of 80 mΩ·mm and a ρ_c of 8.3×10^{-7} Ω·cm^{-2} were achieved. In order to systematically study the mechanism of heavily doped β-Ga_2O_3 using MOVPE to achieve low contact resistance, in 2022 Alema et al. [96] optimized the doping concentration to 3.23×10^{20} cm^{-3}, and the R_C and ρ_c values were as low as 1.62×10^{-7} Ω·cm^{-2} and 0.023 Ω·mm. The ultralow contact characteristics had a significant impact, improving the RF devices' performance.

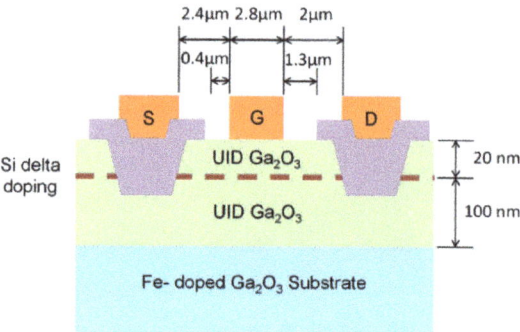

Figure 8. Device schematic of delta-doped β-Ga$_2$O$_3$ MESFET. Reproduced from Ref. [32].

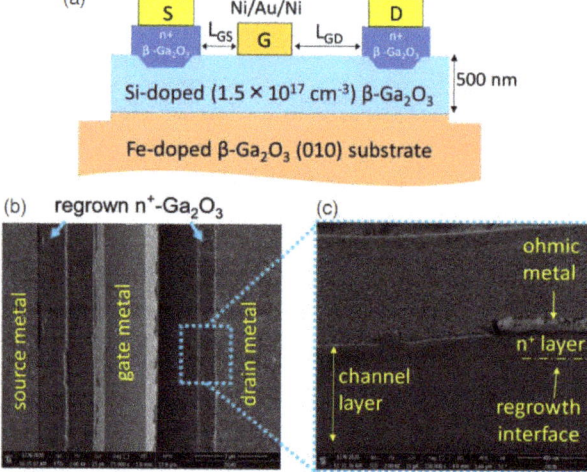

Figure 9. (**a**) Schematic of the fully MOVPE-grown Ga$_2$O$_3$ MESFET with regrown Ohmic contacts. (**b**) Top-view SEM image of the MESFET showing the regrown access regions. (**c**) Cross-sectional SEM image of the contact region showing the estimated regrowth interface. Reproduced from Ref. [36].

The existing results that have been reported in recent years demonstrate that regrown contact is an effective approach to achieve an ultralow R_C. Epitaxial regrowth obtains a high-quality crystalline film and can be versatile. However, there are also several constraints, such as low throughput, high expense, strict material compatibility, and the need for selective growth or subsequent etchings, as mentioned in Refs. [97,98].

3.5. Adding the Interlayer

The Ti/Au schemes always form a 3–5 nm interlayer, which facilitates electron transport for Ohmic contact formation. Another approach is inserting an intermediate semiconductor layer (ISL) with a low work function and a narrower bandgap. In 2016, Oshima et al. [99] proved the insertion of indium tin oxide (ITO) for forming Ohmic contact with β-Ga$_2$O$_3$. In their results, as shown in Figure 10, the ITO method exhibited Ohmic behavior at temperatures from 900 °C to 1150 °C. However, Pt/β-Ga$_2$O$_3$ maintained Schottky contact, even at the RTA temperature of 500 °C. They also confirmed the existence of an intermediate semiconductor layer (ISL) at the interface using TEM and EDS analyses.

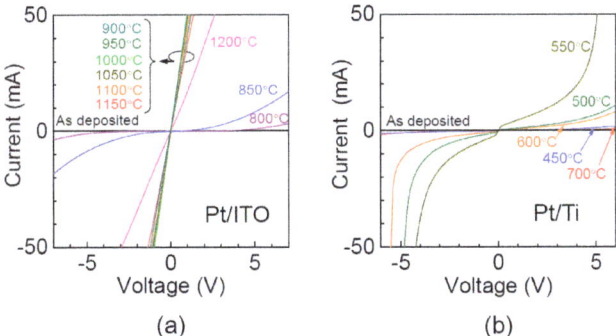

Figure 10. Typical I–V characteristics of (**a**) Pt/ITO and (**b**) Pt/Ti electrodes annealed at various temperatures. Reproduced from Ref. [99].

In 2017, considering the excellent conductivity property of ITO, by depositing an ITO layer before the metal deposition, Carey et al. created a Au/Ti/ITO/Si-doped Ga_2O_3 structure to form low-R_C contact. As shown in Figure 11, by optimizing the annealing temperature at 600 °C, the minimum R_C and ρ_c were determined to be 0.6 Ω·mm and 6.3×10^{-5} Ω·cm^{-2}. A schematic of the band offset for Au/Ti/ITO on Ga_2O_3 and Au/Ti on Ga_2O_3 can be seen in Figure 12 [100]. The insertion of an ITO interlayer allows for reduced conduction band discontinuity between Ti and Ga_2O_3, which is beneficial for reducing R_C values. By inserting an aluminum zinc oxide (AZO) interlayer [101], Au/Ti/AZO/Ga_2O_3 schemes exhibit the minimum R_C and specific contact resistance values of 0.42 Ω·mm and 2.82×10^{-5} cm^{-2}, respectively. The optimized annealing temperature is 400 °C, as shown in Figure 13. In their results, samples without an AZO interlayer did not exhibit Ohmic I–V characteristics when varying the annealing temperature. The use of a thin layer of AZO, with a bandgap of 3.2 eV, can lower the barrier for electron transport and achieve a low R_C. A corresponding schematic of the band offset for AZO on Ga_2O_3 can be seen in Figure 14. Other related results have also been presented by other researchers [102–105]. It should be noted that different metal layers capping ITO layers are needed to prevent the degradation of the surface morphology.

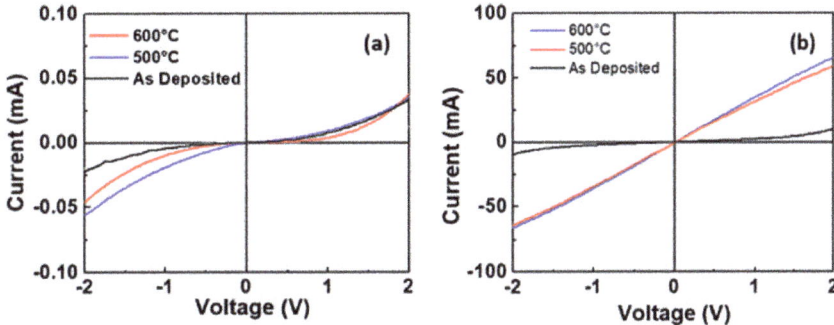

Figure 11. I–V curves of (**a**) Au/Ti/Ga_2O_3 and (**b**) Au/Ti/ITO/Ga_2O_3 contact stacks as a function of annealing temperature. Reproduced from Ref. [100].

Other elements such as substrate orientation have also been reported to influence the Ohmic contact property. To form Ohmic contact, Ti/Au contacts were deposited, followed by an RTA process at 450 °C for 5 min, which was employed on both ($\bar{2}$01) and (010) Sn-doped Ga_2O_3. The former sample exhibited Ohmic characteristics compared to the rectifying behavior of the (010) sample. Related content has also been investigated and reported by other groups [106–109].

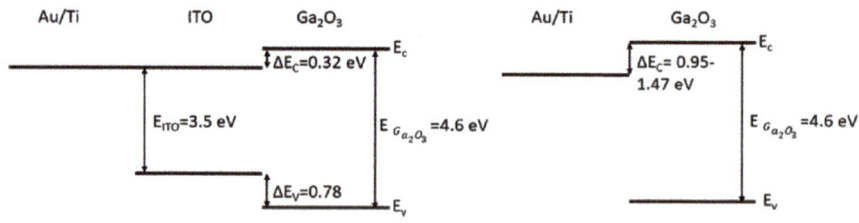

Figure 12. Schematics of band offset for Au/Ti/ITO on Ga$_2$O$_3$ and Au/Ti on Ga$_2$O$_3$. Reproduced from Ref. [100].

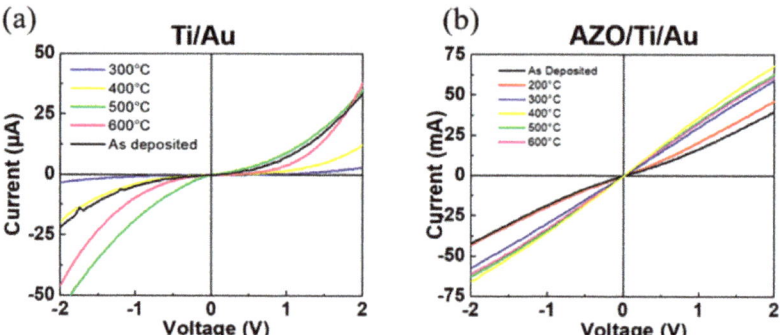

Figure 13. I–V curves of (**a**) Au/Ti/Ga$_2$O$_3$ and (**b**) Au/Ti/AZO/Ga$_2$O$_3$ contact stacks as a function of annealing temperature from as-deposited samples (black lines) to 600 °C (purple lines). The 200 °C data were similar to those of the as-deposited samples, and the contact resistance decreased with temperature in the AZO-based contacts. Reproduced from Ref. [101].

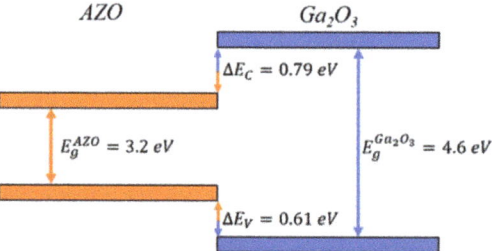

Figure 14. Schematic of band offset for AZO on Ga$_2$O$_3$. Reproduced from Ref. [101].

In recent years, researchers have investigated and optimized the Ohmic contact property of β-Ga$_2$O$_3$ by choosing a metal with a proper work function and investigating metal schemes, interfacial reactions between metal and semiconductors during the annealing process, and the doping concentration of Ga$_2$O$_3$ in the source/drain region, and they have achieved excellent results. Representative results with excellent Ohmic contact quality are summarized in Figure 15 [32,33,36,37,70,83,88,90,96,100]. Despite the significant improvement in the Ohmic contact techniques for β-Ga$_2$O$_3$, there are also some questions that need to be solved before commercializing the devices. (1) For power device applications, contact performance in high-temperature, -current, and -voltage environments is another concern. Failure analyses of the electrical stress/cycling of Ohmic electrodes have been investigated for other WBG semiconductor systems [110–112], while for β-Ga$_2$O$_3$, the research is lacking and efforts should be made to understand the degradation mechanism of Ohmic contacts under electrical stress. (2) To realize the integration of β-Ga$_2$O$_3$ semiconductors into Si CMOS technology, Au-free metal schemes should be investigated and proposed. The commonly used Ti/Au layer for Ga_2O_3 is not CMOS-compatible due to the existence of Au, which is a contaminant for Si fabrication lines [113,114]. Au is used for oxidation protection. Some oxidation-resistant capping materials, such as TiN, which has been proven to realize low-R_C Ohmic contact in AlGaN/GaN HEMT [53], can be substitutes for Ohmic contact realization in β-Ga_2O_3 devices. Related investigations should be carried out to prove the feasibility of Au-free schemes. (3) RIE, ion implantation, and epitaxial regrowth are used to achieve a low R_C. For the RIE process, the influences of plasma gas (including BCl_3/Ar, Ar, and CF_4 [115]), plasma power, bias power, etc., should be fully understood. For ion implantation, the high-temperature annealing used for impurity activation and damage recovery may cause unwanted effects, which should be noticed and further studied. In addition, the effect of substrate orientation should also be investigated. Other annealing techniques can also be used for Ohmic contact formation in β-Ga_2O_3 devices [116,117].

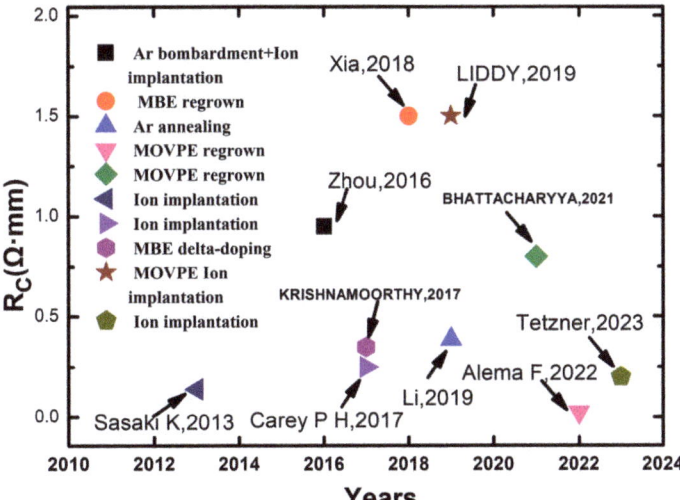

Figure 15. Research progress in source/drain Ohmic contact of β-Ga_2O_3 devices [32,33,36,37,70,83, 88,90,96,100].

4. Conclusions

In this work, the β-Ga_2O_3 Ohmic contact technique has been discussed comprehensively, including the selected metal stack, surface treatment, ion implantation, epitaxial regrowth, adding the interlayer, etc. Although state-of-the-art methods for forming Ohmic contacts with β-Ga_2O_3 have been proposed and summarized in this work, there is still significant room for exploration to improve Ohmic contact, and related prospects have been proposed. In summary, Ohmic contacts with β-Ga_2O_3 will continue to be a research focus for power application in the future. We believe the content presented in this work will be beneficial for understanding and achieving high-performance β-Ga_2O_3 devices with low R_C values.

Author Contributions: L.-Q.Z.: writing—original draft, investigation, data curation, and conceptualization; W.-Q.M., X.-L.W., J.-Y.D., S.-Y.Q., J.-J.L. and Y.-T.T.: writing—original draft, investigation, data curation, and conceptualization; Z.-Y.W., Q.X., Y.Z. and P.-F.W.: writing—review and editing, visualization, and validation. All authors have read and agreed to the published version of the manuscript.

Funding: This work was supported by the Henan Province Key Research and Development and Promotion of Special Scientific and Technological Research Project (Grant No. 222102320283), key scientific research projects of higher education institutions in Henan Province (Grant No. 22B510009), and in part by the Key Laboratory of Optoelectronic Sensing Integrated Application of Hennan Province.

Data Availability Statement: The authors declare that the data supporting the findings of this study are available within the article.

Conflicts of Interest: The authors declare no conflict of interest.

References

1. Sonnenberg, T.; Verploegh, S.; Pinto, M.; Popović, Z. W-Band GaN HEMT Frequency Multipliers. *IEEE Trans. Microw. Theory Tech.* **2023**, *71*, 4327–4336. [CrossRef]
2. Akso, E.; Collins, H.; Clymore, C.; Li, W.; Guidry, M.; Romanczyk, B.; Wurm, C.; Liu, W.; Hatui, N.; Hamwey, R.; et al. First Demonstration of Four-Finger N-polar GaN HEMT Exhibiting Record 712 mW Output Power With 31.7% PAE at 94 GHz. *IEEE Microw. Wirel. Technol. Lett.* **2023**, *33*, 683–686. [CrossRef]
3. Han, L.; Tang, X.; Wang, Z.; Gong, W.; Zhai, R.; Jia, Z.; Zhang, W. Research Progress and Development Prospects of Enhanced GaN HEMTs. *Crystals* **2023**, *13*, 911. [CrossRef]
4. Li, H.; Zhao, S.; Wang, X.; Ding, L.; Mantooth, A.H. Parallel Connection of Silicon Carbide MOSFETs Challenges, Mechanism, and Solutions. *IEEE Trans. Power Electron.* **2023**, *38*, 9731–9749. [CrossRef]
5. Lyu, G.; Sun, J.; Wei, J.; Chen, K.J. Static and Dynamic Characteristics of a 1200 V/22 mΩ Normally-Off SiC/GaN Cascode Device Built with Parallel-Connected SiC JFETs Controlled by a Single GaN HEMT. *IEEE Trans. Power Electron.* **2023**, *38*, 12648–12658. [CrossRef]
6. Roy, R.; Hill, V.G.; Osborn, E.F. Polymorphism of Ga_2O_3 and the system Ga_2O_3-H_2O. *J. Am. Chem. Soc.* **1952**, *74*, 719–722. [CrossRef]
7. Yoshioka, S.; Hayashi, H.; Kuwabara, A.; Oba, F.; Matsunaga, K.; Tanaka, I. Structures and energetics of Ga_2O_3 polymorphs. *J. Phys. Condens. Matter* **2007**, *19*, 346211. [CrossRef]
8. Bechstedt, F.; Furthmüller, J. Influence of screening dynamics on excitons in Ga_2O_3 polymorphs. *Appl. Phys. Lett.* **2019**, *114*, 122101. [CrossRef]
9. Pearton, S.J.; Yang, J.; Cary, P.H.; Ren, F.; Kim, J.; Tadjer, M.J.; Mastro, M.A. A review of Ga_2O_3 materials, processing, and devices. *Appl. Phys. Rev.* **2018**, *5*, 011301. [CrossRef]
10. Wang, C.; Zhang, J.; Xu, S.; Zhang, C.; Feng, Q.; Zhang, Y.; Ning, J.; Zhao, S.; Zhou, H.; Hao, Y. Progress in state-of-the-art technologies of Ga_2O_3 devices. *J. Phys. D Appl. Phys.* **2021**, *54*, 243001. [CrossRef]
11. Sheoran, H.; Kumar, V.; Singh, R. A comprehensive review on recent developments in ohmic and Schottky contacts on Ga_2O_3 for device applications. *ACS Appl. Electron. Mater.* **2022**, *4*, 2589–2628. [CrossRef]
12. Abedi Rik, N.; Orouji, A.A.; Madadi, D. 500 V breakdown voltage in β-Ga_2O_3 laterally diffused metal-oxide-semiconductor field-effect transistor with 108 MW/cm^2 power figure of merit. *IET Circuits Devices Syst.* **2023**, *17*, 199–204. [CrossRef]
13. Wang, X.; Lu, X.; He, Y.; Liu, P.; Shao, Y.; Li, Y.; Yang, Y.; Li, Y.; Hao, Y.; Ma, X. An E-mode β-Ga_2O_3 metal-heterojunction composite field effect transistor with a record high P-FOM of 0.73 GW/cm2. In Proceedings of the 2023 35th International Symposium on Power Semiconductor Devices and ICs (ISPSD), Hong Kong, 28 May–1 June 2023; pp. 390–393. [CrossRef]
14. Liu, H.; Wang, Y.; Lv, Y.; Han, S.; Han, T.; Dun, S.; Guo, H.; Bu, A.; Feng, Z. 10-kV Lateral β-Ga_2O_3 MESFETs with B ion Implanted Planar Isolation. *IEEE Electron. Device Lett.* **2023**, *44*, 1048–1051. [CrossRef]
15. Guo, W.; Han, Z.; Zhao, X.; Xu, G.; Long, S. Large-area β-Ga_2O_3 Schottky barrier diode and its application in DC-DC converters. *J. Semicond.* **2023**, *44*, 072805. [CrossRef]
16. Wei, Y.; Peng, X.; Jiang, Z.; Sun, T.; Wei, J.; Yang, K.; Hao, L.; Luo, X. Low Reverse Conduction Loss β-Ga_2O_3 Vertical FinFET with an Integrated Fin Diode. *IEEE Trans. Electron. Devices* **2023**, *70*, 3454–3461. [CrossRef]
17. Hao, W.; Wu, F.; Li, W.; Xu, G.; Xie, X.; Zhou, K.; Guo, W.; Zhou, X.; He, Q.; Zhao, X.; et al. Improved Vertical β-Ga_2O_3 Schottky Barrier Diodes with Conductivity-Modulated p-NiO Junction Termination Extension. *IEEE Trans. Electron. Devices* **2023**, *70*, 2129–2134. [CrossRef]
18. Wu, F.; Wang, Y.; Jian, G.; Xu, G.; Zhou, X.; Guo, W.; Du, J.; Liu, Q.; Dun, S.; Yu, Z.; et al. Superior Performance β-Ga_2O_3 Junction Barrier Schottky Diodes Implementing p-NiO Heterojunction and Beveled Field Plate for Hybrid Cockcroft–Walton Voltage Multiplier. *IEEE Trans. Electron. Devices* **2023**, *70*, 1199–1205. [CrossRef]
19. Zhang, C.; Liu, K.; Ai, Q.; Sun, X.; Chen, X.; Yang, J.; Zhu, Y.; Cheng, Z.; Li, B.; Liu, L.; et al. High-performance fully transparent Ga_2O_3 solar-blind UV photodetector with the embedded indium-tin-oxide electrodes. *Mater. Today Phys.* **2023**, *33*, 101034. [CrossRef]
20. Li, Y.; Deng, C.; Huang, B.; Yang, S.; Xu, J.; Zhang, G.; Hu, S.; Wang, D.; Liu, B.; Ji, Z.; et al. High-Performance Solar-Blind UV Phototransistors Based on ZnO/Ga_2O_3 Heterojunction Channels. *ACS Appl. Mater. Interfaces* **2023**, *15*, 18372–18378. [CrossRef]
21. Zeng, G.; Zhang, M.R.; Chen, Y.C.; Li, X.; Chen, D.; Shi, C.; Zhao, X.; Chen, N.; Wang, T.; Zhang, W.; et al. A solar-blind photodetector with ultrahigh rectification ratio and photoresponsivity based on the $MoTe_2$/Ta: β-Ga_2O_3 pn junction. *Mater. Today Phys.* **2023**, *33*, 101042. [CrossRef]
22. Matys, M.; Kitagawa, K.; Narita, T.; Uesugi, T.; Suda, J.; Kachi, T. Mg-implanted vertical GaN junction barrier Schottky rectifiers with low on resistance, low turn-on voltage, and nearly ideal nondestructive breakdown voltage. *Appl. Phys. Lett.* **2022**, *121*, 203507. [CrossRef]

23. Aida, H.; Nishiguchi, K.; Takeda, H.; Aota, N.; Sunakawa, K.; Yaguchi, Y. Growth of β-Ga$_2$O$_3$ single crystals by the edge-defined, film fed growth method. *Jpn. J. Appl. Phys.* **2008**, *47*, 8506. [CrossRef]
24. Ohira, S.; Suzuki, N.; Arai, N.; Tanaka, M.; Sugawara, T.; Nakajima, K.; Shishido, T. Characterization of transparent and conducting Sn-doped β-Ga$_2$O$_3$ single crystal after annealing. *Thin Solid Film.* **2008**, *516*, 5763–5767. [CrossRef]
25. Víllora, E.G.; Shimamura, K.; Yoshikawa, Y.; Aoki, K.; Ichinose, N. Large-size β-Ga$_2$O$_3$ single crystals and wafers. *J. Cryst. Growth* **2004**, *270*, 420–426. [CrossRef]
26. Hoshikawa, K.; Ohba, E.; Kobayashi, T.; Yanagisawa, J.; Miyagawa, C.; Nakamura, Y. Growth of β-Ga$_2$O$_3$ single crystals using vertical Bridgman method in ambient air. *J. Cryst. Growth* **2016**, *447*, 36–41. [CrossRef]
27. Nikolaev, V.I.; Maslov, V.; Stepanov, S.I.; Pechnikov, A.I.; Krymov, V.; Nikitina, I.P.; Guzilova, L.I.; Bougrov, V.E.; Romanov, A.E. Growth and characterization of β-Ga$_2$O$_3$ crystals. *J. Cryst. Growth* **2017**, *457*, 132–136. [CrossRef]
28. Tomm, Y.; Reiche, P.; Klimm, D.; Fukuda, T. Czochralski grown Ga$_2$O$_3$ crystals. *J. Cryst. Growth* **2000**, *220*, 510–514. [CrossRef]
29. Galazka, Z.; Uecker, R.; Irmscher, K.; Albrecht, M.; Klimm, D.; Pietsch, M.; Brützam, M.; Bertram, R.; Ganschow, S.; Fornari, R. Czochralski growth and characterization of β-Ga$_2$O$_3$ single crystals. *Cryst. Res. Technol.* **2010**, *45*, 1229–1236. [CrossRef]
30. Alema, F.; Seryogin, G.; Osinsky, A.; Osinsky, A. Ge doping of β-Ga$_2$O$_3$ by MOCVD. *APL Mater.* **2021**, *9*, 091102. [CrossRef]
31. Seryogin, G.; Alema, F.; Valente, N.; Fu, H.; Steinbrunner, E.; Neal, A.; Mou, S.; Fine, A.; Osinsky, A. MOCVD growth of high purity Ga$_2$O$_3$ epitaxial films using trimethylgallium precursor. *Appl. Phys. Lett.* **2020**, *117*, 262101. [CrossRef]
32. Xia, Z.; Joishi, C.; Krishnamoorthy, S.; Bajaj, S.; Zhang, Y.; Brenner, M.; Lodha, S.; Rajan, S. Delta Doped β-Ga$_2$O$_3$ Field Effect Transistors with Regrown Ohmic Contacts. *IEEE Electron. Device Lett.* **2018**, *39*, 568–571. [CrossRef]
33. Krishnamoorthy, S.; Xia, Z.; Bajaj, S.; Brenner, M.; Rajan, S. Delta-doped β-gallium oxide field-effect transistor. *Appl. Phys. Express* **2017**, *10*, 051102. [CrossRef]
34. Oshima, Y.; Víllora, E.G.; Shimamura, K. Halide vapor phase epitaxy of twin-free α-Ga$_2$O$_3$ on sapphire (0001) substrates. *Appl. Phys. Express* **2015**, *8*, 055501. [CrossRef]
35. Murakami, H.; Nomura, K.; Goto, K.; Sasaki, K.; Kawara, K.; Thieu, Q.; Togashi, R.; Kumagai, Y.; Higashiwaki, M.; Kuramata, A.; et al. Homoepitaxial growth of β-Ga$_2$O$_3$ layers by halide vapor phase epitaxy. *Appl. Phys. Express* **2014**, *8*, 015503. [CrossRef]
36. Bhattacharyya, A.; Roy, S.; Ranga, P.; Shoemaker, D.; Song, Y.; Lundh, J.; Choi, S.; Krishnamoorthy, S. 130 mA mm^{-1} β-Ga$_2$O$_3$ metal semiconductor field effect transistor with low-temperature metalorganic vapor phase epitaxy-regrown ohmic contacts. *Appl. Phys. Express* **2021**, *14*, 076502. [CrossRef]
37. Liddy, K.J.; Green, A.; Hendricks, N.; Heller, E.; Moser, N.; Leedy, K.; Popp, A.; Lindquist, M.; Tetlak, S.; Wagner, G.; et al. Thin channel β-Ga$_2$O$_3$ MOSFETs with self-aligned refractory metal gates. *Appl. Phys. Express* **2019**, *12*, 126501. [CrossRef]
38. Varley, J.B.; Janotti, A.; Franchini, C.; Walle, C. Role of self-trap in luminescence and p-type conductivity of wide-band-gap oxides. *Phys. Rev. B* **2012**, *85*, 081109. [CrossRef]
39. Lyons, J.L. A survey of acceptor dopants for β-Ga$_2$O$_3$. *Semicond. Sci. Technol.* **2018**, *33*, 05LT02. [CrossRef]
40. Gallagher, J.C.; Koehler, A.D.; Tadjer, M.J.; Mahadik, A.; Anderson, T.; Budhathoki, S.; Law, K.; Hauser, A.; Hobart, K.; Kub, F. Demonstration of CuI as a P-N heterojunction to β-Ga$_2$O$_3$. *Appl. Phys. Express* **2019**, *12*, 104005. [CrossRef]
41. Budde, M.; Splith, D.; Mazzolini, P.; Tahraoui, A.; Feld, J.; Ramsteiner, M.; Wenckstern, H.; Grundmann, M.; Bierwage, O. SnO/β-Ga$_2$O$_3$ vertical pn heterojunction diodes. *Appl. Phys. Lett.* **2020**, *117*, 252106. [CrossRef]
42. Watahiki, T.; Yuda, Y.; Furukawa, A.; Yamamuka, M.; Takiguchi, Y.; Miyajima, S. Heterojunction p-Cu$_2$O/n-Ga$_2$O$_3$ diode with high breakdown voltage. *Appl. Phys. Lett.* **2017**, *111*, 222104. [CrossRef]
43. Lu, X.; Zhou, X.; Jiang, H.; Ng, K.; Chen, Z.; Pei, Y.; Lau, K.; Wang, G. 1-kV Sputtered p-NiO/n-Ga$_2$O$_3$ Heterojunction Diodes With an Ultra-Low Leakage Current Below 1μA/cm^2. *IEEE Electron. Device Lett.* **2020**, *41*, 449–452. [CrossRef]
44. Wang, Y.G.; Gong, H.H.; Lv, Y.J.; Fu, X.; Dun, S.; Han, T.; Liu, H.; Zhou, X.; Liang, S.; Zhang, R.; et al. 2.41 kV vertical P-NiO/n-Ga$_2$O$_3$ heterojunction diodes with a record Baliga's figure-of-merit of 5.18 GW/cm^2. *IEEE Trans Power Electron.* **2022**, *37*, 3743. [CrossRef]
45. Nakagomi, S.; Hiratsuka, K.; Kakuda, Y.; Yoshihiro, K. Beta-gallium oxide/SiC heterojunction diodes with high rectification ratios. *ECS J. Solid State Sci. Technol.* **2016**, *6*, Q3030. [CrossRef]
46. Zhang, Y.C.; Li, Y.F.; Wang, Z.Z.; Guo, R.; Xu, S.; Liu, C.; Zhao, S.; Zhang, J.; Hao, Y. Investigation of β-Ga$_2$O$_3$ films and β-Ga$_2$O$_3$/GaN heterostructures grown by metal organic chemical vapor deposition. *Sci. China Phys. Mech. Astron.* **2020**, *63*, 117311. [CrossRef]
47. Jaquez, M.; Specht, P.; Yu, K.M.; Walukiewicz, W.; Dubon, D. Amorphous gallium oxide sulfide: A highly mismatched alloy. *J. Appl. Phys.* **2019**, *126*, 105708. [CrossRef]
48. Cai, X.; Sabino, F.P.; Janotti, A.; Wei, S. Approach to achieving a p-type transparent conducting oxide: Doping of bismuth-alloyed Ga$_2$O$_3$ with a strongly correlated band edge state. *Phys. Rev. B* **2021**, *103*, 115205. [CrossRef]
49. Kaneko, K.; Masuda, Y.; Kan, S.; Takahashi, I.; Kato, Y.; Shinohe, T.; Fujita, S. Ultra-wide bandgap corundum-structured p-type α-(Ir, Ga)$_2$O$_3$ alloys for α-Ga$_2$O$_3$ electronics. *Appl. Phys. Lett.* **2021**, *118*, 102104. [CrossRef]
50. Li, Z.H.; Egbo, K.O.; Lv, X.H.; Wang, Y.; Yu, K. Electronic structure and properties of Cu$_{2-x}$S thin films: Dependence of phase structures and free-hole concentrations. *Appl. Surf. Sci.* **2022**, *572*, 151530. [CrossRef]
51. Ezeh, C.V.; Egbo, K.O.; Musah, J.D.; Yu, K.M. Wide gap p-type NiO-Ga$_2$O$_3$ alloy via electronic band engineering. *J. Alloys Compd.* **2023**, *932*, 167275. [CrossRef]

52. Fan, M.Y.; Yang, G.Y.; Zhou, G.N.; Jiang, Y.; Li, W.; Jiang, Y.; Yu, H. Ultra-low Contact Resistivity of< 0.1 Ω mm for Au-free Ti$_x$Al$_y$ Alloy Contact on Non-recessed i-AlGaN/GaN. *IEEE Electron. Device Lett.* **2020**, *41*, 143–146. [CrossRef]
53. Zhang, J.; Kang, X.; Wang, X.; Huang, S.; Chen, C.; Kei, K.; Zheng, Y.; Zhou, Q.; Chen, W.; Zhang, B.; et al. Ultralow-Contact-Resistance Au-Free Ohmic Contacts With Low Annealing Temperature on AlGaN/GaN Heterostructures. *IEEE Electron. Device Lett.* **2018**, *39*, 847–850. [CrossRef]
54. Xu, R.; Lin, N.; Jia, Z.; Liu, Y.; Wang, H.; Yu, Y.; Zhao, X. First principles study of Schottky barriers at Ga$_2$O$_3$(100)/metal interfaces. *RSC Adv.* **2020**, *10*, 14746–14752. [CrossRef] [PubMed]
55. Lovejoy, T.C.; Chen, R.; Zheng, X.; Villora, E.; Shimamura, K.; Yoshikawa, H.; Yamashita, Y.; Ueda, S.; Kobayashi, K.; Dunham, S.; et al. Band bending and surface defects in β-Ga$_2$O$_3$. *Appl. Phys. Lett.* **2012**, *100*, 181602. [CrossRef]
56. Yao, Y.; Gangireddy, R.; Kim, J.; Das, K.; Davis, R.; Porter, L. Electrical behavior of β-Ga$_2$O$_3$ Schottky diodes with different Schottky metals. *J. Vac. Sci. Technol. B* **2017**, *35*, 03D113. [CrossRef]
57. Mott, N.F. Note on the contact between a metal and an insulator or semi-conductor. In *Mathematical Proceedings of the Cambridge Philosophical Society*; Cambridge University Press: Cambridge, UK, 1938; pp. 568–572.
58. Sze, S.M.; Ng, K.K. *Physics of Semiconductor Devices*; John Wiley & Sons: Hoboken, NJ, USA, 2006.
59. Neamen, D. *Semiconductor Physics and Devices*; McGraw-Hill, Inc.: New York, NY, USA, 2002.
60. Mohamed, M.; Irmscher, K.; Janowitz, C.; Galazka, Z.; Manzke, R.; Fornari, R. Schottky barrier height of Au on the transparent semiconducting oxide β-Ga$_2$O$_3$. *Appl. Phys. Lett.* **2012**, *101*, 132106. [CrossRef]
61. Sze, S.M. *Physics of Semiconductor Devices*, 2nd ed.; Wily: New York, NY, USA, 1982.
62. Rideout, V.L. A review of the theory and technology for ohmic contacts to group III-V compound semiconductors. *Solid-State Electron.* **1975**, *18*, 541–550. [CrossRef]
63. Reeves, G.K.; Harrison, H.B. Obtaining the specific contact resistance from transmission line model measurements. *IEEE Electron. Device Lett.* **1982**, *3*, 111–113. [CrossRef]
64. Schroder, D.K. *Contact Resistance and Schottky Barriers*; Wiley: New York, NY, USA, 2006.
65. Yao, Y.; Davis, R.F.; Porter, L.M. Investigation of different metals as ohmic contacts to β-Ga$_2$O$_3$: Comparison and analysis of electrical behavior, morphology, and other physical properties. *J. Electron. Mater.* **2017**, *46*, 2053–2060. [CrossRef]
66. Shi, J.; Xia, X.; Liang, H.; Abbas, Q.; Liu, J.; Zhang, H.; Liu, Y. Low resistivity ohmic contacts on lightly doped n-type β-Ga$_2$O$_3$ using Mg/Au. *J. Mater. Sci. Mater. Electron.* **2019**, *30*, 3860–3864. [CrossRef]
67. Ma, J.; Yoo, G. Low Subthreshold Swing Double-Gate β-Ga$_2$O$_3$ Field-Effect Transistors with Polycrystalline Hafnium Oxide Dielectrics. *IEEE Electron. Device Lett.* **2019**, *40*, 1317–1320. [CrossRef]
68. Lee, M.-H.; Peterson, R.L. Interfacial reactions of titanium/gold ohmic contacts with Sn-doped β-Ga$_2$O$_3$. *APL Mater.* **2019**, *7*, 022524. [CrossRef]
69. Higashiwaki, M.; Sasaki, K.; Kamimura, T.; Wong, M.; Krishnamurthy, D.; Kuramata, A.; Masui, T.; Yamakoshi, S. Depletion-mode Ga$_2$O$_3$ metal-oxide-semiconductor field-effect transistors on β-Ga$_2$O$_3$ (010) substrates and temperature dependence of their device characteristics. *Appl. Phys. Lett.* **2013**, *103*, 123511. [CrossRef]
70. Zhou, H.; Si, M.; Alghamdi, S.; Qiu, G.; Yang, L.; Ye, P. High-Performance Depletion/Enhancement-mode β-Ga$_2$O$_3$ on Insulator (GOOI) Field-Effect Transistors with Record Drain Currents of 600/450 mA/mm. *IEEE Electron. Device Lett.* **2016**, *38*, 103–106. [CrossRef]
71. Zhou, H.; Maize, K.; Qiu, G.; Shakouri, A.; Ye, P. β-Ga$_2$O$_3$ on insulator field-effect transistors with drain currents exceeding 1.5 A/mm and their self-heating effect. *Appl. Phys. Lett.* **2017**, *111*, 092102. [CrossRef]
72. Chabak, K.D.; Moser, N.; Green, A.; Jr, D.; Tetlak, S.; Heller, E.; Crespo, A.; Fitch, R.; McCandless, J.; Leedy, K.; et al. Enhancement-mode Ga$_2$O$_3$ wrap-gate fin field-effect transistors on native (100) β-Ga$_2$O$_3$ substrate with high breakdown voltage. *Appl. Phys. Lett.* **2016**, *109*, 213501. [CrossRef]
73. Chen, J.; Li, X.; Ma, H.; Huang, W.; Ji, Z.; Xia, C.; Lu, H.; Zhang, W. Investigation of the mechanism for Ohmic contact formation in Ti/Al/Ni/Au contacts to β-Ga$_2$O$_3$ nanobelt field-effect transistors. *ACS Appl. Mater. Interfaces* **2019**, *11*, 32127–32134. [CrossRef]
74. Gong, R.; Wang, J.; Liu, S.; Dong, Z.; Yu, M.; Wen, C.; Cai, Y.; Zhang, B. Analysis of surface roughness in Ti/Al/Ni/Au ohmic contact to AlGaN/GaN high electron mobility transistors. *Appl. Phys. Lett.* **2010**, *97*, 062115. [CrossRef]
75. Fontserè, A.; Tomás, A.; Placidi, M.; Llobet, J.; Baron, N.; Chenot, S.; Cordier, Y.; Moreno, J.; Gammon, P.; Jennings, M.; et al. Micro and nano analysis of 0.2 Ω mm Ti/Al/Ni/Au ohmic contact to AlGaN/GaN. *Appl. Phys. Lett.* **2011**, *99*, 213504. [CrossRef]
76. Tetzner, K.; Schewski, R.; Popp, A.; Anooz, S.; Chou, T.; Ostermay, I.; Kirmse, H.; Würfl, J. Refractory metal-based ohmic contacts on β-Ga$_2$O$_3$ using TiW. *APL Mater.* **2022**, *10*, 071108. [CrossRef]
77. Lee, M.H.; Chou, T.S.; Bin Anooz, S.; Galazka, Z.; Popp, A.; Peterson, R. Effect of post-metallization anneal on (100) Ga$_2$O$_3$/Ti-Au ohmic contact performance and interfacial degradation. *APL Mater.* **2022**, *10*, 091105. [CrossRef]
78. Kim, Y.; Kim, M.K.; Baik, K.H.; Jang, S. Low-resistance Ti/Au ohmic contact on (001) plane Ga$_2$O$_3$ Crystal. *ECS J. Solid State Sci. Technol.* **2022**, *11*, 045003. [CrossRef]
79. Porter, L.M.; Hajzus, J.R. Perspectives from research on metal-semiconductor contacts: Examples from Ga$_2$O$_3$, SiC, (nano) diamond, and SnS. *J. Vac. Sci. Technol. A* **2020**, *38*, 031005. [CrossRef]
80. Lyle, L.A.M.; Back, T.C.; Bowers, C.T.; Green, A.; Chabak, K.; Dorsey, D.; Heller, E.; Porter, L.M. Electrical and chemical analysis of Ti/Au contacts to β-Ga$_2$O$_3$. *APL Mater.* **2021**, *9*, 061104. [CrossRef]

81. Guo, D.Y.; Wu, Z.P.; An, Y.H.; Guo, X.C.; Chu, X.L.; Sun, C.L.; Li, L.H.; Li, P.G.; Tang, W.H. Oxygen vacancy tuned Ohmic-Schottky conversion for enhanced performance in β-Ga$_2$O$_3$ solar-blind ultraviolet photodetectors. *Appl. Phys. Lett.* **2014**, *105*, 023507. [CrossRef]
82. Higashiwaki, M.; Sasaki, K.; Kuramata, A.; Masui, T.; Yamakoshi, S. Gallium oxide (Ga$_2$O$_3$) metal-semiconductor field-effect transistors on single-crystal β-Ga$_2$O$_3$ (010) substrates. *Appl. Phys. Lett.* **2012**, *100*, 013504. [CrossRef]
83. Sasaki, K.; Higashiwaki, M.; Kuramata, A.; Masui, T.; Yamakoshi, S. Si-ion implantation doping in β-Ga$_2$O$_3$ and its application to fabrication of low-resistance ohmic contacts. *Appl. Phys. Express* **2013**, *6*, 086502. [CrossRef]
84. Oshima, T.; Okuno, T.; Arai, N.; Suzuki, N.; Ohira, S.; Fujita, S. Vertical solar-blind deep-ultraviolet Schottky photodetectors based on β-Ga$_2$O$_3$ substrates. *Appl. Phys. Express* **2008**, *1*, 011202. [CrossRef]
85. Wong, M.H.; Sasaki, K.; Kuramata, A.; Yamakoshi, S.; Higashiwaki, M. Field-plated Ga$_2$O$_3$ MOSFETs with a breakdown voltage of over 750 V. *IEEE Electron. Device Lett.* **2015**, *37*, 212–215. [CrossRef]
86. Higashiwaki, M.; Kuramata, A.; Murakami, H.; Kumagai, Y. State-of-the-art technologies of gallium oxide power devices. *J. Phys. D Appl. Phys.* **2017**, *50*, 333002. [CrossRef]
87. Bae, J.; Kim, H.Y.; Kim, J. Contacting mechanically exfoliated β-Ga$_2$O$_3$ nanobelts for (opto) electronic device applications. *ECS J. Solid State Sci. Technol.* **2016**, *6*, Q3045. [CrossRef]
88. Li, Z.; Liu, Y.; Zhang, A.; Liu, Q.; Shen, C.; Wu, F.; Xu, C.; Chen, M.; Fu, H.; Zhou, C. Quasi-two-dimensional β-Ga$_2$O$_3$ field effect transistors with large drain current density and low contact resistance via controlled formation of interfacial oxygen vacancies. *Nano Res.* **2019**, *12*, 143–148. [CrossRef]
89. Spencer, J.A.; Tadjer, M.J.; Jacobs, A.G.; Mastro, M.A.; Lyons, J.L.; Freitas, J.A., Jr.; Gallagher, J.C.; Thieu, Q.T.; Sasaki, K.; Kuramata, A.; et al. Activation of implanted Si, Ge, and Sn donors in high-resistivity halide vapor phase epitaxial β-Ga$_2$O$_3$: N with high mobility. *Appl. Phys. Lett* **2022**, *121*, 192102. [CrossRef]
90. Tetzner, K.; Thies, A.; Seyidov, P.; Chou, T.; Rehm, J.; Ostermay, J.; Galazka, Z.; Fiedler, A.; Popp, A.; Würfl, J. Ge-ion implantation and activation in (100) β-Ga$_2$O$_3$ for ohmic contact improvement using pulsed rapid thermal annealing. *J. Vac. Sci. Technol. A* **2023**, *41*, 043102. [CrossRef]
91. Tadjer, M.; Lyons, J.; Nepal, N.; Freitas, J.A., Jr.; Koehler, A.; Foster, G. Review-Theory and characterization of doping and defects in β-Ga$_2$O$_3$. *ECS J. Solid State Sci. Technol.* **2019**, *8*, Q3187–Q3194. [CrossRef]
92. Nikolskaya, A.; Okulich, E.; Korolev, D.; Stepanov, A.; Nikolichev, D.; Mikhaylov, A.; Tetelbaum, D.; Almaev, A.; Bolzan, C.A.; Buaczik, A., Jr.; et al. Ion implantation in β-Ga$_2$O$_3$: Physics and technology. *J. Vac. Sci. Technol. A* **2021**, *39*, 030802. [CrossRef]
93. Huang, H.L.; Chae, C.; Johnson, J.M.; Senckowski, A.; Sharma, S.; Singisetti, U.; Wong, M.; Hwang, J. Atomic scale defect formation and phase transformation in Si implanted β-Ga$_2$O$_3$. *APL Mater.* **2023**, *11*, 061113. [CrossRef]
94. Lee, M.H.; Chou, T.S.; Bin Anooz, S.; Galazka, Z.; Popp, A.; Peterson, R. Exploiting the nanostructural anisotropy of β-Ga$_2$O$_3$ to demonstrate giant improvement in titanium/gold ohmic contacts. *ACS Nano* **2022**, *16*, 11988–11997. [CrossRef]
95. Zeng, K.; Wallace, J.S.; Heimburger, C.; Sasaki, K.; Kuramata, A.; Masui, T.; Gardella, J.; Singisetti, U. Ga$_2$O$_3$ MOSFETs using spin-on-glass source/drain doping technology. *IEEE Electron. Device Lett.* **2017**, *38*, 513–516. [CrossRef]
96. Alema, F.; Peterson, C.; Bhattacharyya, A.; Roy, S.; Krishnamoorthy, S.; Osinsky, A. Low resistance Ohmic contact on epitaxial MOVPE grown β-Ga$_2$O$_3$ and β-(Al$_x$Ga$_{1-x}$)$_2$O$_3$ films. *IEEE Electron. Device Lett.* **2022**, *43*, 1649–1652. [CrossRef]
97. Lee, M.H.; Peterson, R.L. Process characterization of ohmic contacts for beta-phase gallium oxide. *J. Mater. Res.* **2021**, *36*, 4771–4789. [CrossRef]
98. Kim, H. Control and understanding of metal contacts to β-Ga$_2$O$_3$ single crystals: A review. *SN Appl. Sci.* **2022**, *4*, 27. [CrossRef]
99. Oshima, T.; Wakabayashi, R.; Hattori, M.; Hashiguchi, A.; Kawano, N.; Sasaki, K.; Masui, T.; Kuramata, A.; Yamakoshi, S.; Yoshimatsu, K. Formation of indium-tin oxide ohmic contacts for β-Ga$_2$O$_3$. *Jpn. J. Appl. Phys.* **2016**, *55*, 1202B7. [CrossRef]
100. Carey, P.H.; Yang, J.; Ren, F.; Hays, D.; Pearton, S.; Kuramata, A.; Kravchenko, I. Improvement of Ohmic contacts on Ga$_2$O$_3$ through use of ITO-interlayers. *J. Vac. Sci. Technol. B* **2017**, *35*, 061201. [CrossRef]
101. Carey, P.H.; Yang, J.; Ren, F.; Hays, D.; Pearton, S.; Jang, S.; Kuramata, A.; Kravchenko, I. Ohmic contacts on n-type β-Ga$_2$O$_3$ using AZO/Ti/Au. *AIP Adv.* **2017**, *7*, 095313. [CrossRef]
102. Carey IVP, H.; Ren, F.; Hays, D.C.; Gila, B.; Pearton, S.; Jang, S.; Kuramata, A. Band offsets in ITO/Ga$_2$O$_3$ heterostructures. *Appl. Surf. Sci.* **2017**, *422*, 179–183. [CrossRef]
103. Carey IVP, H.; Ren, F.; Hays, D.C.; Gila, B.; Pearton, S.; Jang, S.; Kuramata, A. Valence and conduction band offsets in AZO/Ga$_2$O$_3$ heterostructures. *Vacuum* **2017**, *141*, 103–108. [CrossRef]
104. Kim, S.; Kim, S.J.; Kim, K.H.; Kim, H.; Kim, T. Improved performance of Ga$_2$O$_3$/ITO-based transparent conductive oxide films using hydrogen annealing for near-ultraviolet light-emitting diodes. *Phys. Status Solidi A* **2014**, *211*, 2569–2573. [CrossRef]
105. Sui, Y.; Liang, H.; Chen, Q.; Huo, W.; Du, X.; Mei, Z. Room-temperature ozone sensing capability of IGZO-decorated amorphous Ga$_2$O$_3$ films. *ACS Appl. Mater. Interfaces* **2020**, *12*, 8929–8934. [CrossRef]
106. Deng, Y.; Yang, Z.; Xu, T.; Jiang, H.; Ng, K.; Liao, C.; Su, D.; Pei, Y.; Chen, Z.; Wang, G. Band alignment and electrical properties of NiO/β-Ga$_2$O$_3$ heterojunctions with different β-Ga$_2$O$_3$ orientations. *Appl. Surf. Sci.* **2023**, *622*, 156917. [CrossRef]
107. Ingebrigtsen, M.E.; Vines, L.; Alfieri, G. Bulk β-Ga$_2$O$_3$ with (010) and (201) surface orientation: Schottky contacts and point defects. *Mater. Sci. Forum* **2017**, *897*, 755–758. [CrossRef]
108. Yatskiv, R.; Tiagulskyi, S.; Grym, J. Influence of crystallographic orientation on Schottky barrier formation in gallium oxide. *J. Electron. Mater.* **2020**, *49*, 5133–5137. [CrossRef]

109. Wong, M.H.; Nakata, Y.; Kuramata, A.; Yamakoshi, S.; Higashiwaki, M. Enhancement-mode Ga$_2$O$_3$ MOSFETs with Si-ion-implanted source and drain. *Appl. Phys. Express* **2017**, *10*, 041101. [CrossRef]
110. Downey, B.P.; Mohney, S.E.; Clark, T.E.; Flemish, J. Reliability of aluminum-bearing ohmic contacts to SiC under high current density. *MicroElectron. Reliab.* **2010**, *50*, 1967–1972. [CrossRef]
111. Liu, L.; Ling, M.; Yang, J.; Xiong, W.; Jia, W.; Wang, G. Efficiency degradation behaviors of current/thermal co-stressed GaN-based blue light emitting diodes with vertical-structure. *J. Appl. Phys.* **2012**, *111*, 093110. [CrossRef]
112. Moens, P.; Banerjee, A.; Constant, A.; Coppens, P.; Caesar, M.; Li, Z.; Vandeweghe, W.; Declercq, F.; Padmanabhan, B.; Jeon, W. Intrinsic reliability assessment of 650V rated AlGaN/GaN based power devices: An industry perspective. *ECS Trans.* **2016**, *72*, 65. [CrossRef]
113. Li, Y.; Ng, G.I.; Arulkumaran, S.; Kumar, C.; Ang, K.; Anand, M.; Wang, H.; Hofstetter, R.; Ye, G. Low-contact-resistance non-gold Ta/Si/Ti/Al/Ni/Ta Ohmic contacts on undoped AlGaN/GaN high-electron-mobility transistors grown on silicon. *Appl. Phys. Express* **2013**, *6*, 116501. [CrossRef]
114. Piazza, M.; Dua, C.; Oualli, M.; Morvan, E.; Carisetti, D.; Wyczisk, F. Degradation of TiAlNiAu as ohmic contact metal for GaN HEMTs. *MicroElectron. Reliab.* **2009**, *49*, 1222–1225. [CrossRef]
115. Jeong, Y.J.; Yang, J.Y.; Lee, C.H.; Park, R.; Lee, G.; Chung, R.; Yoo, G. Fluorine-based plasma treatment for hetero-epitaxial β-Ga$_2$O$_3$ MOSFETs. *Appl. Surf. Sci.* **2021**, *558*, 149936. [CrossRef]
116. Zhang, L.Q.; Shi, J.S.; Huang, H.F.; Liu, X.; Zhao, S.; Wang, P.; Zhang, W. Low-temperature Ohmic contact formation in GaN high electron mobility transistors using microwave annealing. *IEEE Electron. Device Lett.* **2015**, *36*, 896–898. [CrossRef]
117. Hou, M.; Xie, G.; Sheng, K. Improved device performance in AlGaN/GaN HEMT by forming ohmic contact with laser annealing. *IEEE Electron. Device Lett.* **2018**, *39*, 1137–1140. [CrossRef]

Disclaimer/Publisher's Note: The statements, opinions and data contained in all publications are solely those of the individual author(s) and contributor(s) and not of MDPI and/or the editor(s). MDPI and/or the editor(s) disclaim responsibility for any injury to people or property resulting from any ideas, methods, instructions or products referred to in the content.

MDPI
St. Alban-Anlage 66
4052 Basel
Switzerland
www.mdpi.com

Inorganics Editorial Office
E-mail: inorganics@mdpi.com
www.mdpi.com/journal/inorganics

Disclaimer/Publisher's Note: The statements, opinions and data contained in all publications are solely those of the individual author(s) and contributor(s) and not of MDPI and/or the editor(s). MDPI and/or the editor(s) disclaim responsibility for any injury to people or property resulting from any ideas, methods, instructions or products referred to in the content.

www.ingramcontent.com/pod-product-compliance
Lightning Source LLC
LaVergne TN
LVHW070625100526
838202LV00012B/728